PLANT TAXONOMY

Theory, Principles & Practices

Dr. D. A. Patil

M.Sc., Ph.D.
Former Professor & Principal,
Post-Graduate Department of Botany,
S.S.V.P.Sanstha's L.K.Dr.P.R.Ghogrey Science College,
Dhule-424005 (Maharashtra), India.
e-mail: dapatil_10aug@yahoo.com

SCIENTIFIC
PUBLISHERS

Published by:

SCIENTIFIC PUBLISHERS (INDIA)
5A, New Pali Road, P.O. Box 91
Jodhpur 342 001, India
E-mail: info@scientificpub.com
Website: www.scientificpubonline.com

ISBN: 978-93-90495-41-2 (HB)
eISBN: 978-93-90495-51-1 (EBook)

Printed in India

Respectfully Dedicated

To

Professor R.M. Pai

My Most Venerated Guru (Teacher)
Who Always Worked With The Highest Morals
&
A Botanist Who Nurtured
Plant Science Ceaselessly

Preface

Agriculture and human settlements were initiated hand in hand probably by the Mesolithic age. Agriculture became well developed during this age, nearly 3000 B.C. When man started tilling the earth, he also began the study of plants by grouping similar kinds of plants and selecting better ones. Taxonomy is thus as old as human civilization. It is in this early start and instinctive activity of man, elements of folk taxonomy are met with. This was period of acknowledging the plant diversity. By about 16th century, students of medicine started description and classification of the medicinal herbs. Herbal taxonomy was thereby started by the literate people. Botanists realized the need to bring order into the diversity of nature/ plants by about 17th century. This was a real beginning of classical taxonomy based on exomorphology. Data from other disciplines of botany was integrated with time in classical taxonomy, which then eventually transformed it into modern taxonomy.

Taxonomy is appreciated as mother science, old classical science, as basic science, dynamic and all-embracing, as also queen and servant. This science is also abused (erroneously) and called the oldest science, dead science, monotonous and repugnant, a science without data of its own and questioned as lame duck. This sorry state of affair appears due to the faulty curricula, methods of teaching and questionable expertise and interest of the exponents/ teachers assigned for teaching taxonomy. Taxonomy has remained controversial, misunderstood and maligned in the past. Inspite of the vicissitudes, it remained prominent in biological realm. Its everchanging nature always depended upon the needs, interest and technology employed. The Earth Summit (Anonymous, 1992) introduced and coined the term 'biodiversity' and thereby awakened the richness of biotic diversity of each nation. This helped redeem the situation and once again taxonomy regained its lost status and glory.

'Taxonomy is often thought as a purely academic or even superficial subject of little relevance to the urgent needs of modern man. Such a view is, however, far from the truth, for plant taxonomists have an important role to play in man's betterment, a role emerging as a key one now that many of our traditional natural resources are in short supply or running out (Stace, 1989)'. We as taxonomists celebrate diversity. We celebrate the wildness of the planet. We celebrate the numerous human attempts to understand this wildness, and mourn its loss through human miscalculation. We sense the aesthetic of life and much of our efforts are aimed at reflecting this composition (Stuessy, 1990)'. Both of these epilogues explicitly convey the role to be played by taxonomy and the taxomomists.

Taxonomy, now being synthetic science, draws information from diverse fields, like anatomy, embryology, palynology, cytology, cytogenetics, phytochemistry, phytogeography, ecology, etc., apart from exomorphology. It behaves amoeba-like and engulfs data from all possible sides. The newer aspects e.g. micromorpholgy, ultrastructure, numerical taxonomy, molecular taxonomy, cladistics, etc. are providing more information and are being sought to bear on understanding relationships. Although some recent books on this science stepped in proper direction, they do not cover all aspects and necessary details required for students and even teachers of higher education. In my endeavor, an attempt has been made to accommodate and balance these all. This book is largely restricted to a consideration of angiospermic plants, although few evidences/ examples have been drawn from other plant groups wherever found relevant and necessary. This has been obvious for angiosperms being the most numerous, familiar and well researched group of plants. Emphasis has been laid on the use of diverse data for taxonomic purposes and to instill their further use by students and researchers in this subject. Relevant specific statements, expressions or opinions made by stalworts in the subject have been purposefully provided at length with a view to increase impression and intensity of thoughts on the minds of learners. Needless to say that while presenting the modern concepts and scope of the subject every care has been taken not to ignore the fundamentals which are so important as a basis in any modern discipline of biology.

Hopefully the book will be useful for those interested in knowing general principles and as a source-book for taxonomic methodology. The coverage for this book has gradually evolved over three decades in my engagements from undergraduate to post-graduate courses, and to doctoral assignments under the umbrella of University of Pune and then North Maharashtra University (India). Moreover, the indelible academic impression on me of my much revered teachers Professor R.M.Pai and Dr.V.N.Naik of Dr.Babasaheb Ambedkar Marathwada University (Maharashtra, India), which cannot be ignored in such an academic pursuit. My role has been to stay only as a living link between the students, books and nature.

During development of curricula for undergraduate and post-graduate classes through organized seminars, workshops and University academic meetings, it was possible to imbibe much from colleagues in the subject. This has also been reflected in this endeavour. Nevertheless, the errors and deficiencies, if experienced, the responsibility obviously lies on me. Comments on this line are most welcome.

Book written in hurry, particularly in one's academically younger stage, inadvertently intercalate some 'academic sin'. With this idea in mind, I wrote no book for beginners/ learners in my early career although I was persuaded sometimes by students, friends and colleagues. Encouragements received from my teachers has always proved fruitful. I am deeply indebted to them and remain to be so forever. Most generous help by way of literature is provided by my institution (Shri Shivaji Vidya Prasarak Sanstha, Dhule, Maharashtra, India). I acknowledge it with deep sense of gratitude. Students of undergraduate and post-graduate classes, besides research students interacted with me on many botanical aspects in class, laboratory and field. These live queries and discussions always stimulated me. I remember them all with love. I cannot conclude this preface without expressing the co-operative silence observed by my family members during the course of the writing of this book.

Prof.D.A.Patil
S.S.V.P.Sanstha's L.K.Dr.P.R.Ghogrey
Science College, Dhule-424005
(M.S.) India

Contents

1

Introduction To Taxonomy

(I) Taxonomy *vis-à-vis* Systematics:

The term '*taxonomy*' was coined first by the famous French botanist Augustin Pyrame de Candolle. He introduced the term to the plant science in his 'Theorie Elementaire de la Botanique' (1813) and meant 'the theory of plant classification'. This term stems from the two Greek words, 'taxis' meaning arrangement and 'nomos' or 'nomous' meaning law/rule. Literally, it is a lawful arrangement of plants or organisms. Later, by constant usage, the term became more inclusive and presently it encompasses *identification*, *nomenclature* and *classification* of plants. It is also used interchangeably with other terms '*Systematics*' or '*Systematic Botany*'. Thus, '*taxonomy*' may be comprehensively defined as the study and description of the variation of plants or organisms, the investigation of the causes and consequences of this variation, and the manipulation of the data obtained to produce a system of classification (*cf.* Lawrence, 1951). This wider definition of 'taxonomy' coincides with meaning of the term 'Systematics'. In fact, nowadays, these two terms are used synonymously.

Different views have been expressed as to the applicability of the two terms *viz.*, Taxonomy and *Systematics* : (i) Simpson (1961), Heywood (1967), Mayr (1969), and Ross (1974) conceived the term '*Systematics*' as a more inclusive field covering the scientific study of the diversity and differentiation of organisms and the relationships that exist between them. Taxonomy, in their opinion, is a part of systemactics. (ii) Radford (1986) opined that systematics is the study of phenetic, genetic and phylogenetic relationship among taxa. (iii) Solbrig (1966) considered the two terms with different concerns. In his opinion, taxonomy includes *classification* and *nomenclature* but, relies on systematics for its concepts. (iv) Thus the science of systematics is basic to the act of classification and probably that is why the two terms are often equated and used synonymously.

Human beings, since their appearance on the earth, have been dependent on the plant wealth for their necessities of life. This dependency forced them to identify the plants. They were also able to classify plants into various groups such as food plants, toxic plants, medicinal plants, etc. Thus illiterate human beings were necessity taxonomists during prehistoric times. This is the earliest start in plant taxonomy. The methods of recognition of useful plants by the earliest men were simple. Today, it has become all-embracing and more complex. The horizon of plant taxonomy has considerably widened over the centuries.

Taxonomy is at the same time the most basic and the ultimate of all the biological sciences. It is basic because no start on understanding the wealth of variation can be made until some sort of classification is adopted, and ultimate because taxonomy is not complete until the data from all other fields of investigation have been integrated. *Taxonomy* has thus to depend for its improvement and existence completely on information from other domains such as morphology, anatomy, embryology, palynology, cytology and phytochemistry, besides data accrued from genetics, breeding systems, geography and ecology. It has no data of its own. In a nutshell, taxonomy is both the 'queen' and 'servant' of biology. It is queen because it is the ultimate as stated earlier. At the same time, it is servant because one of the primary functions of taxonomy is to provide basic service by way of information on the identity, probable close relatives and things like that to the needy in other fields of research, Radford *et al.* (1974) rightly remarked that taxonomy is the pedestal upon which biology is built.

(II) Functions of *Taxonomy*

Taxonomy primarily deals with the identification, nomenclature and classification. These are dilated in the following.

(i) ***Identification***: Determination of a *taxon* involves comparison with a previously known plant. The comparison reveals overall resemblances and differences. A representative specimen of a given taxon is compared directly or indirectly. Direct method includes comparison with the authoritatively predetermined plant specimens, whereas indirect method involves use of keys, descriptions, illustrations, etc. The specimen under examination may not show similarity to any known specimen. In such circumstances, it may be considered to be completely new to science. Generally, the gross morphological characters suffice to provisional determination of a taxon. Further conformation of such a determination may be also carried out by consideration of other characters. Correct determination of a taxon is a prerequisite for any study based on it.

(ii) ***Nomenclature***: The name of a plant is the key to its literature. Therefore, after correct identification of a *taxon*, it is necessary to give it a valid scientific name. The rules for scientific naming of a taxon are framed by the International Botanical Congress (which meets at regular intervals) and contained in an International Code of Botanical Nomenclature (ICN). The ICN is aimed at providing an internationally acceptable code of practice on nomenclature. A new name is given to a new taxon and then added to an already existing list of taxa known so far.Thus,the term 'nomenclature' means the study of the system and methods of naming taxa, and the construction, interpretation and application of the regulations governing this system.

(iii) ***Classification***: The mantle of the earth globe is covered by the enormously high number of plant species. It is a humanly impossible task to study and understand them individually. This is possible only when they are considered in discret groups. It is a natural human instinct to classify organisms in their ambience. Any act of grouping them into categories is an act of classification. Placement of plants or organisms into small or large groups based on their resemblances and differences is, therefore, necessary. These groups are then arranged according to their levels into categories in a sequential manner. Each category has a name according to the rules of nomenclature. Classification is thus a production of a logical system of categories each containing any number of plants or organisms, which permits easier reference to its kinds of plants / organisms.

The three aspects *viz., identification, nomenclature* and *classification* sometimes cloud the minds of the pioneers or beginners in plant taxonomy. These terms, *sensu stricto*, refer to their different things, although they appear(erroneously) overlapping, for example, when one says that a certain plant specimen is *Solanum nigrum* L. It means it is identified, named and classified as well. Nevertheless, identification means recognition of a specimen as the same or different from another known entity. Nomenclature involves attributing a name to it. Likewise, classification involves its proper placement in the taxonomic system.

(III) Principles of Taxonomy

Taxonomy was in practice since the oldest times and used by our ancestors in various parts of the earth, particularly in Rome, Greece, Germany, U.K., China and India. The main purpose of taxonomy was to provide a convenient method of classification and communication. Medicinal, toxic or non – toxic, edible or non-edible plants, ease of classification and set of readily observable characters, etc. were simple principles employed while devising systems of plant classification. Classification has been the natural instinct since the appearance of man on the earth and still continues to be so even in modern periods. The various systems of plant classification proposed to date generally fall in three main kinds of classification *viz., artificial, natural* and *phylogenetic*. The *artificial classifications* were developed prior to 19th century. Later, there was a shift from artificial over to the natural or phylogenetic systems. Concomitantly, there has been also a change in principles. The principles of taxonomy have always centered around the criteria employed in classification. Now, certain fundamental principles have been formulated (*cf.* Cronquist, 1968) :

(i) *Taxa* are properly established on the basis of multiple correlation of characters.

(ii) Taxonomic importance of a character is determined by how well it correlates with other characters. This means that the taxonomic importance of a character is determined 'a posteriori' rather than 'a priori'.

(iii) An important feature of taxonomy is its predictive value.

The first principle lays emphasis on correlation of characters. *Characters* and *character-states* show a tendency to adhere and are transmitted together. Modern taxonomy depends heavily on multiple correlation rather than on single character differences. For example, the number of cotyledons in angiosperms is correlated with the type of leaves, orientation of vascular bundles, nature of flowers, etc. In dicotyledonous angiosperms, the two cotyledons are correlated with *dorsiventral*, articulately veined leaves, 4 to 5-merous flowers and *collateral open vascular bundles* arranged in a ring. Likewise, the single cotyledon in monocotyledonous angiosperms are correlated with *isobilateral, parallel veined* leaves, 3- *merous* flowers, and closed, scattered vascular bundles of stems. A character bearing no correlation with other character/s is / are thought anomalous and taxonomically insignificant e.g. regular flower in Scrophulariaceae, irregular flower is Ranunculaceae, simple leaves in Bignoniaceae, alternate leaves in Oleaceae and such others are conceived to be anomalous as they do not exhibit correlation with general characters of the taxa.

The second principle states the importance of *character weighting*. Innumerable characters are expected in plants / organisms, However, all of them are not employed in taxonomy. We generally select characters while classifying taxa and do not prefer them randomly. Usually selection of characters is an intuitive process. Characters of greater taxonomic significance, more stable and conservative ones thought to be evolutionary markers and of greater diagnostic value are selected. This type of selection of characters for their supposed significance is called 'a *priori weighting*' of characters. However, the empirical taxonomists reject 'a priori' considerations of characters. Instead, all characters are considered to be of equal weight to begin with and classification is based on similarity-dissimilarity ratio (i.e. similarity co-efficient).The comparative importance of each character would be known subsequently after detection and estimation of the degree of correlation with other characters. This type of weighting of characters is called 'a *posteriori weighting*'. It is employed especially in numerical taxonomy.

The third principle informs about productivity of criteria. For instance, if one discovers that a plant belongs to the Gramineae (Poaceae), one can immediately infer a great deal of information about it. Its peculiar floral morphology, anatomy, unique seed structure and such other general anatomical, cytological and chemical characteristics. Virtually, we know all grasses possess a basic uniformity which applies equally to undiscovered and to fully studied species. Thus the very statement that 'a plant belongs to the Gramineae', automatically predicts many characters that will be applicable to that plant. Predictability is one of the most obvious criteria that can be used to assess how good a classification is. A *classification* which enables one to carry out this operation successfully is a predictive classification. Obviously, an artificial system of classification has an extremely low predictability. *Natural classifications* have high level of predictability.

(IV) Aims of Taxonomy

The two approaches *viz.*, the empirical and the interpretative, are combined in modern taxonomy, and are usually aimed at : (i) providing a convenient method of identification and thereby communication, (ii) providing classification established on possible natural affinities of organisms, (iii) providing an inventory of plant taxa by means of flora, (iv) detecting evolution at work, discovering its processes and interpreting its results, (v) providing an integrating *vis-a-vis* unifying role in the training of biology students.Thus it appears that the jurisdiction of modern taxonomy is not confined to classification of organisms as they exist now. It includes an understanding of the origin and evolution of variations, isolation and speciation among them.

(V) Approaches To Taxonomy

General purpose of taxonomy is to arrange plants in such a way as to provide an information retrieval system. It informs the name and rank or position in an ordered hierarchy. Traditionally taxonomy was based on gross morphological characters. However, in later periods, it virtually started engulfing information from other disciplines. It was artificial in the beginning and became later natural and phylogenetic. Now, there appear to be two main approaches to taxonomy *viz.,* (i) The Empirical Approach, (ii) The Interpretative Approach.

(i) **The Empirical Approach :**One of the aims of taxonomy is to provide a classification system based on natural relationship. The great majority of systems of classification are hierarchical. They combine species into an ascending series of successively larger and wider categories, ultimately arriving at single all-embracing group covering all plants. Most plants (or animals) have been thus classified. Observations of a large number of characters will facilitate obviously a more acceptable classification. This is an 'empirical approach.'

(ii) **The Interpretative Approach :**There is another school of thought which subscribe to an evolutionary interpretation, and hence called an 'interpretative approach'. A classification is so constructed that it can interpret the evolution of a taxon. While working out a vivid picture of the existing organic diversity, it reveals numerous evolutionary phenomena, provides information that enable the reconstruction of phylogeny and makes important conceptual contributions to biology. This thus leads to a phylogentic classification which also demands data from past history of a taxon.

(VI) Need and Importance of Taxonomy

Taxonomy is the most basic and the most all-embracing of the biological sciences. It has no data of its own but use those from every other field of biological investigation. It was a common understanding during the last few decades that the prominence of taxonomy in biology has changed much over the past century, with changing technology, tastes and requirements. It was also further commented that *taxonomy* has declined in popularity as new fields of investigations emerged. At present, we are witnessing interspecific, biochemical or genetical differences. The study of such differences can only be properly documented, and their significance fully grasped, by the employment of taxonomic expertise.Later, these are expressed as systematic information within taxonomic classifications. It is apparent, at present, to claim that the decline of taxonomy has been slowed or even halted. There is an increasing acceptance of the fundamental importance of taxonomy, however, it needs a firmly based and comprehensible methodology associated with it.

One of the primary functions of taxonomy is to provide basic service by way of information on the identity, probable close relatives and things like that to the needy researcher in various areas of biology. Research workers in genetics, plant breeding and biochemistry get their first information reports from the available classificatory systems. The system also informs them on the probable wild relatives which are essential in all our crop improvement programmes. The wild relatives enable them to breed for desirable qualities. Indiscriminate ecological destruction cause considerable loss to the wild relatives. This force coupled with the importance of the wild relatives has alarmed the biologists considerably, who are busy establishing '*germplasm banks*' in different parts of the world,

Taxonomy depends on other branches of biology and takes data to construct classificatory systems. In turn, it is expected to provide them with the correct identifications along with information on their probable relatives and the degree of relationships, apart from accurate name and natural relationships.

In the last few decades, most of the disciplines of biology emphasized unity within diversity of life. However, taxonomy alone have been exploring and revealing the diversity of organisms.It will have a continued role to play and significant contributions to make with regard to our understanding of the different kinds of organisms on this planet and their variability. Introduction of the term '*biodiversity*' instead of simply saying '*flora*' or '*fauna*' has accelerated this role of taxonomy especially after 'Earth Summit' held at Rio de Janero, Brazil (Anonymous,1992).

Studies on the different aspects of structure, organization and life processes have yielded voluminous information. It would also continue to accrue and that too rapidly because of development in the different biological research fields. Classificatory systems are also so developed. They can function as an effective mechanism for the storage of all types of information on the various taxonomic groups. The needy can easily retrieve the necessary bits of information. The names of taxa inform much about the structure and properties of the groups denoted.

Linnaeus (1753) incorporated about 6000 species under 1000 genera in his 'Species Plantarum'. Jussieu (1789) estimated the total number of species of organisms not more than 20,000. This has proved to be a gross underestimate. The number has greatly exceeded and still a great number of species yet to be known. Species are now thought dynamic entities. As a consequence, variations within the species have become a major concern. Moreover,modern taxonomy demands information on the origin and evolution of variations and all other types of information in guiding the evolution of classificatory systems.

Herbaria, museums, botanic gardens, taxonomic literature and literature from related fields are the sources of information content in biology. The former three represent collections from different areas at different times. They function as treasure houses of information *in situ* for posterity on the evolutionary levels. They also act as indicators of the level of environmental pollution of different kinds of the time when gathered. We should be, however; cautious for evolution is a long drawn process (Anonymous, 1974). Any collection represents a few centuries and might not take us for back. Likewise, our knowledge of phylogeny would be on a horizontal plane, because the collections would reveal the range of variations as existed at a particular time. Such collections would represent a very minor segment of the whole spectrum.

We have Floras (Manuals) at hand of different geographical areas which include minimal description, keys and observations. They aid us in identification. The emphasis is always on the diagnostic exomorphic features. Complete observations are hardly recorded and usually remain uninterpreted in view of their relationships. Moreover comparative data is wanting in these sources. It is very hard to speculate on the presence / absence of a particular character. One can not be certain whether they were really absent or the author has failed to comment on it (Watson, 1971).

Knowledge about plants (and animals) increased to such an extent in recent times that taxonomy, sometimes called the digestive system of biology (*cf.* Heywood, 1973), is unable to digest the whole lot. In such circumstances, the use of modern 'Electronic Data Processing Equipment' to store and retrieve information is advisable and meaningful. Establishment of data bands are of special importance in the tropical areas, where 'the fragile ecosystems' along with the biota are being indiscriminately destroyed.

We live in an age of environmental crisis caused by man. The crisis is on account of his scant respect for other organisms inhabiting the earth. The man, *Homo sapiens,* appeared later on the earth and so younger in relation to other organisms. He conquered nature in pursuit of the quick and short-term material prosperity. Several plant / animal species have gone extinct, and several others are threatened. Public attention on the endangered plant species is rather little. But the plants in preserving the ecosystem, do count much more than the animals.This is so because a disappearance of a plant species can take away with many dependent animal species. Taxonomists have alerted against this force of depletion and called for concerted efforts to discover, describe and document them before they are lost to posterity. Till date, very few species are said to be noted for reasonably detailed knowledge. A study of the existing tropical flora is more important in this context as we can gain more from them about phylogeny and evolution of flowering plants than we can gain from paleobotany. It is also a matter of concern that concomitant to the destruction / depletion of the forests is the '*genetic erosion*' of crop plants of potentially valuable wild and weedy relatives. Moreover, in the name of 'Green Evolution' the farmers are induced to cultivated standard genetically uniform strains. This leads to narrowing of the genetic base of our crop plants. Thus man-engineered destructive activities coupled with the technological and scientific advances have put more responsibility on taxonomists.

There is also lack of understanding and appreciation of taxonomy. This is particularly shocking to one who has made great efforts to acquire sufficient knowledge of plants.

This knowledge of taxonomy served his colleagues and to anyone who needed help or information about plants. The help and time are freely and willingly given.

Taxonomists are also criticized for practicing esoteric art (Kubitzki, 1975) but for no fault of taxonomy or taxonomists. This remark implies that there is no scientific merit. But this does not hold water. Survival of taxonomy against all odds is itself a proof. Biologists have been running away from taxonomy in pursuit of more fashionable (seemingly) prestigious disciplines. *Taxonomy* is thus left at the hand of a few.

The dynamic nature of populations and the process of speciation and evolution have guaranteed a continuing role for them for all times to come. Taxonomy is gradually gaining strength and its importance cannot be ignored in any of the future biological research. The researches in other disciplines discussing the systematics of the group are in themselves testimony for the recognition of taxonomy as a fundamental endeavour.

2

Classifications

Classification is one of the functions of taxonomy. Any act of grouping objects into categories is an act of classification. It is an indispensable act in human life to deal with the innumerable objects in the universe. Nay, it is an inborn occupation of mankind. The science of taxonomy of which classification is one of the components, itself means 'arrangement by rules'. Thus to cope with the disorderly objects of the universe orderly thinking is a prerequisite.

(I) Need for Classification

One can separate out and conceive the vast array of objects only by naming them first and then grouping in recognizable categories. It is an absolute need to have some system of classification. This requirement is not only meant for taxonomists or biologists. It is part of the everyday life of all kind of people. It is a human act carried out naturally and by an instinct. Biologists must know practically what kind of organisms they are working with before they pass on information about them to other people.

Obviously, this is a function of taxonomy. Some disciplines of biology e.g. cytology, genetics, comparative anatomy, palynology, embryology phytosociology, etc. are so intimately associated with taxonomy that they cannot be practiced without the use of a basic classification. If the determination of an organism is ambiguous or wrong, the value or significance of the work done in these disciplines is tremendously affected or even rendered worthless. Experiments can be repeated only when the identification of an organism is absolutely correct. Findings of a biologist are certainly influenced by the choice of a particular species. Apart from identification, the link of the various disciplines is also concerned with relationships. In applied domains such as agriculture, horticulture, forestry, pharmacology, etc., this relationship is especially close and usually of great economic importance Classification, therefore, should provide a sound basis for the biological sciences in providing a correct name and an expression of *natural relationships* as well. It is the serious responsibility of a taxonomist to classify the organisms in the most generally useful way.

(II) Process of Classification

The process of classification can be viewed as a series of operations as stated by Stuessy (1979). It usually concerns two distinct operations *viz.,* (i) Grouping and (ii) Ranking, as in the following:

Grouping

Individual Organisms (All Species) —(Selection)→ Characters —(Description and measurement)→ Character States —(Evaluation)→ Taxa

Ranking

Taxa (All Character States) —(Selection)→ Characters —(Evaluation)→ Categories

Grouping involves three specific operations:(I) One must select characteristics of the organisms to employ in evaluating the similarities and differences. (II) The second operation is describing or / and measuring these characters. Instead of character, character states are actually used for purposes of taxonomic comparison and evaluation. (III) The third operation is concerned with comparing the chosen character states to obtain the groups. These comparisons can be made in different ways. A formal method e.g. as in phenetics and cladistics or the comparison can be made more intuitively in traditional approaches.

The next step in classification is ranking of the recognized groups. It involves two specific operations: (I) All the character states of the groups are examined and some are selected for use. The character states might be the same and also include all of those employed for grouping, however, generally not all them are used for ranking. Other characteristics also might be selected for consideration which were not used for the grouping. (II) These selected character states of the groups are evaluated in terms of the categories available for use in the taxonomic hierarchy. The presence / absence of certain kinds of features generally will indicate a proper rank in the hierarchy of classification.

(III) Classification and Aesthetics

Various opinions and views have been expressed in past by the naturalists, philosophers and biologists concerning contacts between classification and aesthetics. Friedman (1966) stated that classification is strongly aesthetic particularly in phyletic approaches and therein lies the artistic connection. In the opinion of Stearn (1964), "aesthetic appreciation rests chiefly on a sense of form. The co-ordination of varied elements into a whole is mentally satisfying and harmonious, giving an impression of completeness, is nevertheless achieved by emphasis and omission. The perception of overall resemblance implies selection by the mind from the multitude of unevaluated details presented to it by the eyes". The 'sense of form' what Stearn (*lo.cit.*) thought although fundamental to artistic endeavours, is very much a part of classification. Kubler (1962) defined 'art as system of formal relations'. This, in his opinion, could apply equally well to the resultant hierarchy of classes constructed in the process of classification. Simpson (1961) construed classification as 'a useful art'. He further understood classification as a combination of art and science in taxonomic art. Mayr (1974) agreed that classifying is an art and a superior classification provides a genuine aesthetic pleasure. Stearn (1964) finds similarity in the work of a taxonomist to that of an artist in that both seek patterns within diversity. Taxonomists find patterns in nature, while the artist do so in his own head. We must note that there are many examples of excellent taxonomists who are also accomplished artists. This is evidenced in the illustrations of Hutchinson (1926, 1934 and 1969) and Burger (1967). These indicate that these taxonomists have substantial talent in both areas.

(IV) *Hierarchical Classifications*

The great majority of systems of classification combine species into an ascending series of successively larger and wider categories ultimately arriving at a single all - embracing group covering all plants. Such systems are termed hierarchical classifications. Non-hierarchical classifications are done by ordination in which variable (such as species or other taxonomic groups) are arranged in two-dimensional space in relation to major axes of variation in the data set used (e.g. DuPraw, 1964, 1965). *Hierarchical classifications* are of more interest to us since they have higher information content and are clearly more useful for human needs than non-hierarchical ones. There are many definitions for hierarchical classifications. Few appropriate ones are considered here: (I) any set of objects grouped into classes (reflecting coordinate relationships) (II) a nested hierarchy of classes (reflecting coordinate and subordinate relationships) (III) a particular scheme of grouping and ranking of a particular set of objects (IV) the relationships (expressed as some measure of similarity and/or difference) between and among taxa in a particular classification scheme. The first two are 'general' definitions, whereas the second two are 'restricted' definitions.

Different roles are played by classifications: (i) An important role of classifications is as devices for information storage and retrieval (Davis and Heywood, 1963; Heslop-Harrison, 1962). Bock (1973) opined that the most fundamental attribute of biological classification is that it must be useful. It must order and summarize biological information, must be heuristic and must provide the basis for future studies. A tremendous amount of information which can be drawn and used in different ways by different kinds of people is contained in the hierarchical structure of a system of classification. New data are added to the existing classification for a continual updating of information when new organisms are discovered or new aspects of already known organisms are brought into light. (ii) Identification services to be performed are served by classifications (Warburton, 1967). Identification is necessary step in both storing and retrieving information and hence a classification which helps identification is essential to all organismic evolutionary biologists (*cf.* Darlington, 1971). (iii) Classification serves as a guide for the labeling of unnamed objects by comparison with those items already named (iv) Another important contribution of classification lies in the reflection of evolutionary relationships (*cf.* Warburton, 1967; Nelson, 1973). Some workers believe that a classification should / can reflect phylogeny in some fashion. This represents an attempt to capture the 'reality' of the world. However, classification cannot express full relationship than others. (V) Use of classification as summarizing and predictive devices is another significant feature (*cf.* Rollings, 1965; Warburton, 1967). Warburton (1967) emphasized the importance of the predictive value of classification. In his opinion, all other biologists must trust taxonomists to provide them with classifications which maximize this probability as the validity of all observational and experimental biology depends upon such classifications.

(V) General and Special Purpose Classifications

A classification which enables one to predict many features that will be applicable to that organism or plant is called predictive classification. Predictivity is one of the most obvious criteria that can be used to assess how good a classification is. It is also clear that the best classifications are those which utilize the most characters, and these also are most predictive. Such classifications because they are best for most purposes are termed general purpose classifications. The form a classification takes depends on the purpose for which it is intended, so that a particular classification is only better than another within the purpose for which it was constructed. However, a *natural classification*, because it contains the most information will serve the greatest number of purposes. It will, therefore, be the most generally useful and be equivalent to what is often called a 'general purpose classification'. Such classifications are reasonably represented by the familiar sequences of families, genera and species, etc. which appear in Floras. These arose by a process of gradual improvement as more taxonomically valuable characters are revealed and incorporated.

Many botanists wish to know something of the evolutionary history and relationships of a group of taxa i.e. they want the classification primarily reflecting the phylogeny of the plants. A classification which comply with this is known as a '*phylogenetic* or *phyletic classification*' to distinguish it from a '*phenetic classification*' which is based upon the overall present-day resemblance and differences of the plants. Phylogenetic classifications theoretically can be thought less natural than phenetic ones because they are generally constructed from fewer characters or they are constructed from features which have phylogenetic significance.

It is stated above that a particular classification is better than another only in relation to its purpose and its form is determined by the purpose for which it is constructed. There are many sorts of classification best for some special purpose. Such classifications are associated with particular field or discipline of study. They may be far from taxonomic in outlook and have low or even no predictivity. For example, the timber trade uses systems based simply on timber characteristic. The distinction between softwoods and hardwoods is solely that between gymonosperms and angiosperms. Gardeners classify plants based on their life - duration and frost susceptibility (hardy annuals, etc.). Ecologists do so according to their growth-forms (hemicryptophytic, etc.) and phytosociologists according to their associations and alliances. Similarly, a classification based on flower colour might be the best one to use in a book used by laymen to identify wild flowers or one based on an alphabetical sequence in a seed catalogue. Thus for special purpose classifications, predictivity is not a criterion to judge as a better classification. There are various criteria useful for general purpose classifications. Even, they can be described as artificial. Linnaeus' sexual system was the best available as it was easy to understand.

Among the most well-known special purpose classifications are those based upon the degree of interfertility between plants. They are of particular (special) interest to biosystematics, geneticists and plant breeders, etc. Separate scheme of biosystematic categories have been devised by Turesson (1922), Danser (1929) and few others. These categories are quite different from the family, genus and species, etc. which are based soley on the extent of interbreeding possible. The terms used in these classifications are discussed at length in another chapter. (Taxonomic Structure)

(VI) *Horizontal* and *Vertical Classifications*

Two kinds of relationship between organisms is recognized - first among successive taxa in ancestral - descendant lineage i.e. in time; second, among taxa as they exist at any one time. The former is vertical relationship, whereas the latter is horizontal relationship. The classification based on the former is called vertical classification, while the one based on the latter is known as *horizontal classification. Vertical classification* relies on the presumed ancestry. It is largely phylogenetic or cladistic in perspective. Horizontal classification is dependent on relationships among taxa in the same time zone and determined by similarities and differences. It is largely natural or phenetic in perspective. Vertical relationships, being three-dimensional, are very difficult to express on the printed page. Many ingenious methods have been invented for depicting them (*cf.* Lam, 1936; Jaepson, 1944; Rodriguer, 1950; Voss, 1952; Sporne, 1956). It is mainly discontinuites in the fossil record or long periods of morphological stability, which permits one to delimit taxa vertically. One cannot classify a continuum. However, there is no difficulty in classifying horizontally at any one period of time. Davis (1978) opines that more phenetic overlap occurs among taxa when going up the hierarchy to the families and higher levels. He also uses the concepts of vertical and horizontal but in a different context to mean changes rather than approaches in classification. He also emphasizes that study of a group will often result in a vertical change or elevation in rank.

(VII) *Polythetic* and *Monothetic Classifications*

A classification is said to be polythetic when organisms are placed together that have the greater number of shared features and no single feature is either essential to group membership or is sufficient to make an organism a member of the group (Sokal and Sneath, 1963). A classification is known as monothetic in which groups are formed by rigid and successive logical divisions so that the possession of a unique set of features is both sufficient and necessary for membership in the group thus defined (Sokal and Sneath, 1963). Monothetic quality is in contrast to polythetic one. It is characteristic of *artificial systems* of classifications e.g. Linnaeus' Sexual System. Artificial system of classification lacks the information content which leads to higher levels of predictivity. This search for increased information content is the second need in *biological classification*, and it resulted in the natural approach i.e. *polythetic classification*. The development of such approaches was pre-Darwinian and therefore evolutionary considerations were not included.

(VIII) *Folk Classifications*

Ancient man needed to communicate to others the names and properties of plants which were of significance to him. He, therefore, must have had some sorts of system of plant classification. Classifications which grow up in communities, whether primitive or civilized, through need and without the influence of science are called folk classifications. However, anthropologists recognize two kinds of categories while classifying artifacts of culture *viz*., (i) *emic* and (ii) etic.The *emic* approach investigates how local people think; how they perceive and categorize the world, their rules for behavior, what has meaning for them, and how they imagine and explain things. The *etic* approach (scientist -oriented) shifts the focus from local observation, categories, explanation and interpretations to those of the anthropologist. This approach realizes that members of a culture often are too involved in what they are doing to interpret their cultures impartially. Folk classifications are basically based on emic categories which are cultural and community based. The degree of detail in these is largely marked by the existence or non- existence of local names for taxa which can be also used to measure the coincidence of the folk-taxa with those of modern taxonomy. For example, in English the terms 'grass' and 'sedge' are largely equivalent to the families Gramineae (Poaceae) and Cyperaceae respectively. Some groups of plants have much greater human significance in folk classification (taxonomy) to distinguish far lower even than the species, for instance, (i) cabbage, cauliflower, sprouts, broccoli, etc.; for *Brassica oleracea* (ii) beat, chard, mangold, etc.; for *Beta vulgaris.* Similar diversity can be found in other languages for plants of significance to man.

Folk classifications are being studied indepth in the ethnobiological or ethnobotanical domain in recent times. Alternatively, folk classifications are now called ethnobiological or ethnobotanical classifications. The studies, although meagre, have yielded some concrete facts or principles. Some glimpses of such classifications appear relevant in view of the subject matter. Newmaster *et al.* (2006) reviewed different categories in ethnobiological classification e.g. linguistic, hierarchical, utilitarian, morphological, ecological, spiritual, etc. For example, (i) Linguistic category : It is dialogue – based. Most communities use the same name to call commonly occurring tree categories, like oak is a common tree, then its second logical types include jack oak, post oak, blue oak, live oak, red oak and white oak, (ii) Hierarchical category : There are five taxonomic ethnobiological categories e.g. (a) unique–beginner corresponding to plant or animal kingdom, (b) Life-form categories–cacti, palms, agave, (c) aspeng oak, redwood, and monkey puzzle are typical folk genera, (d) *folk-specific* taxa are usually labeled binomially e.g. White oak, (e) Varietals are folk - specifics that are subdivided e.g. spotted white oak versus swamp white oak, (iii) Utilitarian categories : People from various cultures differently classify plants according to their use e.g. bamboo (in China) is classified according to use and taste. The aqqyul group has sweet bamboo shoots constrasting to the almaq group, which has bitter shoots, (iv) Morphological categories : Folk classification systems also use morphology as a primary criterion. For example, Malayali tribal farmers in Southern India classify the landraces of millet into folk species and varieties based on the characteristics like plant height, hairiness, ear head shape and structure, seed colour, luster, stem colour, etc. Katanaikka in Southern India identify various *Dioscorea* species by characteristics like shape, size and number of tubers and so on,

Folk or *ethnobiological* classification systems are part of indigenous local knowledge and their mechanisms of classification or identification are based on the perceived morphological, utilitarian, ecological and experiential affinities between organisms. Such classification system can be used in rapid assessment of biodiversity.

(IX) Phases of Classification

Classifications, in modern time, are based on a very wide range of characters. The incorporation of characters is a gradual and continuous process. The naturalness or predictivity of a classification depends on the extent to which the plant have been investigated. Our knowledge of the world's flora is uneven and for this reason its classification can be considered in four overlapping phases *viz.*, (I) the pioneer (or exploratory) phase, (II) The consolidation (or systematic) phase, (III) the biosystematic phase and (IV) the encyclopaedic phase. Valentine and Love (1958) recognized the first three phases. Davis and Heywood (1963) rightly added the last fourth phase.

(i) **The pioneer (or exploratory) phase :** This phase is primarily concerned with identification. It involved collection and subsequent classification from a limited range of herbarium material. Field data is, however, mearge. Morphology and distribution provide the data on which the taxonomists have to rely. Limited herbarium material means that taxonomic decisions are often extremely arbitrary and can only be tested when additional material becomes available.

(ii) **The consolidation (or systematic) phase :** In this phase, the plant material or the species are studied both in the herbarium and field for a considerable period. Herbarium representation is rich. The local and regional variation of species is rather well known. Many plants described as species in the pioneer phase on limited material are found to be variants of other species in this phase. Further exploration leads to the discovery of more new species. A compilation of a flora or a monograph is possible. Thus extensive herbarium and field studies of a wide selection of material of each taxon are carried out.

(iii) **The biosystematic phase :** Apart from the thorough knowledge of a taxon based on the earlier two phases, this phase takes into account the information from geographical variation and biosystematic investigation. The latter mainly concerns chromosome number and morphology, breeding behaviour and barriers, information obtained from transplant experiments, about the constancy and genotypic control of characters. Microevloution and variation is emphasized in this phase.

(iv) **The encylopaedic phase :** This phase co-ordinates the earlier three phases. Data obtained from a very wide range of disciplines are assembled to form a good predictive and natural classification. All available evidence is considered in order to express the taxonomic and evolutionary relationships of plants at all levels of the hierarchy.

The first two of these phases correspond to the '*alpha classification*' of Turill (1938) and is based solely upon more or less obvious exomorphic characters. The last two phases correspond to his '*omega classification*' which would be the equivalent of an ultimate perfected system based upon all available characters. It may exploit data accumulated from researches in the fields, laboratory, garden, herbarium and library. The data may be analysed with the help of computers. Davis and Heywood (1963) and Cain (1956) opined that distinction between alpha and omega classifications is one of knowledge and not of aims. In the former, little is known about the groups of organisms being studied, while in the latter the groups are more thoroughly studied.

Turill (1938) stated that some of us please ourselves by thinking we are now groping in a '*beta*' taxonomy. Davis and Heywood (1963) opined that today classification extends beyond '*alpha*' taxonomy. It would be reasonable to claim that in a few well-studied groups an omega stage is within reach, but at the same time it should be realised that for the great majority of taxa especially lower plants and tropical plants the alpha stage has not been passed. Stace (1989) expressed his view that taxonomy is both the most basic and the ultimate field of biology- 'alpha taxonomy' forms the basis of biology, while the final accumulation of all data is ultimately incorporated into '*omega taxonomy*'.

(X) Splitting and Lumping of Taxa

Man's ability to discriminate the objects in his ambience is inherent or inborn, and not from any foreign impetus. He does so often by using certain characters or attributes which generally strike him at sight, Davis and Heywood (1963) rightly remarked that classification is a natural occupation of man, recognition one of his most necessary past time. Naturally he observes similarities as well as differences amongst the objects. Every man is thus taxonomist by birth. Any data which show difference and resemblance as from species to species, whether plants or animals, are of taxonomic significance and thus constitute part of the information / evidence which may be used by especially taxonomists. Some taxonomists simply prefer to emphasise the difference (the so called splitters) and others (the lumpers) the similarities. Thus what is one taxon in the view of one taxonomist are two or more taxa to another. This is obviously due to lack of any standard or absolute yardstick. Stace (1989) attracted attention of taxonomists toward seed-plants enlisted in Index Kewensis upto 18th supplement covering the years 1753 to 1985. The number of names of species he noted is about 970000 and showed a yearly increase of nearly 4000. Of these 4000, about 2300 are calculated by their authors to be new species; the rest are new names or new combinations for already known species. He further remarked that the number of distinct species of seed – plants estimated by various authorities has dropped over the past 20 or 30 years from about 330000 to about 240000. He pointed out a rise from, on average, fewer than three synonyms per species to more than four.

3

Taxonomic Structure

(I) Taxonomic Categories

An ancestral form diversifies through time and separates into subgroups. This is the result of evolutionary divergence. This diversification leads to the modified descendants which resemble each other in descending degrees. The individuals of a species resemble more each other being closer genealogically. However, the degree of resemblance decreases progressively as one passes on to various species of a genus, genera of a family and so on. Thus natural subordination of organisms in groups under groups is a fact of nature and not artifact (*cf.* Grant, 1963).

Every individual organism is treated under a series of progressively higher categories and the lower categories are subordinate to and included under the higher categories. This is called '*taxonomic hierarchy*' and this series has been termed as a '*taxonomic structure*' (*cf.* Mason, 1950). A series of hierarchical categories are recognized. Categories are the levels (ranks) in a hierarchy to which groups are assigned. Categories are purely artificial. They are defined by taxonomists and chosen arbitrarily before groups are assigned to them. The groups placed in the categories represent discrete sets of organisms in nature sharing from common characters and are non-arbitrary. In other words, 'ranks' and 'categories' refer to certain concepts. They do not exist in nature. However, 'taxonomic group' exist in nature. Taxonomic group is an assemblage of individual organisms defined on the basis of their morphological or phylogenetic relationships.

Swiss botanists, Bauhin brothers (J.Bauhin and G.Bauhin) worked in the 16th century. Gaspard (Casper) Bauhin 1673 published the most important book 'Pinax Theatri Botanici' (1623). Its title (Pinax=register) indicates one of its most significant features-the listing of about 6000 species known to him and also all their synonyms. Thus chaotic state of nomenclature was brought to order to a considerable degree. Apart from this, G. Bauhin is noted for his recognization of genera as well as species as taxonomic levels and for using *binary nomenclature* composed of the generic name followed by a single specific epithet to designate many of the species. Thus a distinction, which was lacking in the earlier works, was clearly recognized. Actually, this marked the beginning of recognition of larger (genus) and smaller (species) groups of plant classification. However Carolus Linnaeus (1753), a Swedish botanist, is credited for the establishment of the '*binomial system*' of nomenclature as he used it consistently.

A large number of enthusiastic botanists, workers and students set out to explore the world in later centuries. As a result, a tremendously large number of plant species were gathered. A need for grouping of plants into still larger groups was felt. Taxonomists, in such circumstance, recognized several groups at various levels. The recent studies of the living populations of plants resulted into smaller groups of plants.

The categories and the endings of their names as accepted by all worldover and stated in the International Code of Nomenclature are as follows: The principal ranks are shown in capitals.

Rosa canina var. lutetiana *f.* lasiostylis shown as example.

Rank	Ending	Example
KINGDOM		Plantae
Subkingdom	-bionta	Embryobionta
DIVISION (Phylum)	-phyta	Tracheophyta
Subdivision	-phytina	Spermatophytina
CLASS	-opsida(-phyceaeinalgae)	Angiospermopsida
Subclass	-idae (-phycidae in algae)	Dicotyledonidae
(Superorder)		(Rosanae)
ORDER	-ales	Rosales
Suborder	-ineae	Rosineae
FAMILY	-aceae	Rosaceae
Subfamily	-oideae	Rosoideae
Tribe	-eae	Roseae
Subtribe	-inae	
GENUS		*Rosa*
Subgenus		*Rosa*
Section		*Caninae*
Subsection		*Caninae*
Series		
Subseries		
SPECIES		*canina*
Subspecies(subsp. or.ssp.)		
Variety (var.)		*lutetiana*
Subvariety (subvar.)		
Form (f.)		*lasiostylis*
Subform (subf.)		

The Plant Kingdom (Regnum Vegetable) is divided into the subordinate ranks. They are arranged in a descending manner from higher to lower categories. Article 5 of IBN states that this relative order of ranks must not be changed. The 'International Code of Botanical Nomenclature', the internationally recognized rule-book of green plant (and fungal) nomenclature, recognizes 12 main ranks in the hierarchy *viz.*, Kingdom, Division, Class, Order, Family, Tribe, Genus, Section, Series, Species, Variety and form. This number of ranks can be increased by adding the prefix 'sub' to the terms denoting the ranks. This doubles the number of ranks. These are designated below each rank e.g. Subdivision, Subfamily, Subgenus, Subspecies, etc. Some authors have added super categories e.g. Superoder above the principal ranks. (*cf.* Takhtajan 1980, Thorne, 1976, Dahlgren, 1980). The hierarchy in botanical, zoological and bacteriological nomenclature is substantially the same, except that in zoology the division is known as the phylum. This has also gained favour with some botanists (*cf.* Tippo, 1942; Hutchinson, 1959, 1973).

The names of ranks should not be used in their loose everyday contacts. They should be used in their strict senses. Instead, taxonomically 'neutral' words e.g. taxon (taxa), group/s, variant/s should be used where a certain rank is not implied. The names of ranks are usually provided with characteristic endings / suffixes as in above Table. These endings are not the same in all organisms, but within green plants the exception is found in the algae. Differences in this context exist in fungal and zoological nomenclature. In bacteriology, these endings are based on those of green plants. The names of genera are nouns in the singular number. The names of taxa at lower ranks may be nouns or adjectives in the singular or plural and none is provided with characteristic endings devoting its precise rank.

Among the major categories in classification, 'family' is the smallest category. However, it is encountered more commonly in the taxonomic studies.

Article 18 of the ICN states that the name of a family is a plural adjective used as a substantive. It is based on a legitimate name of an included genus with a suffix –aceae, e.g. *Magnolia* – for the family Magnoliaceae, *Solanum* - for the family Solanaceae, etc. However, there are certain families with irregular terminations. These are exceptionally sanctioned by ICN. These are allowed because their names were in use long before the present accepted rules of nomenclature. They have been retained or conserved by the rules. Those who adhere to the rules of ICN, have been provided with alternative names as follows :

Earlier Family Names (Prior To ICN)	Alternative Family Name
(i) Cruciferae	Brassicaceae (Based on type *Brassica* L.)
(ii) Guttiferae	Clusiaceae (Based on type *Clusia* L.)
(iii) Leguminosae	Fabaceae (Based on type *Faba* Mill.)
(iv) Umbelliferae	Apiaceae (Based on type *Apium* L.)
(v) Compositae	Asteraceae (Based on type *Aster* L.)
(vi) Labiatae	Lamiaceae (Based on type *Lamium* L.)
(vii) Palmae	Arecaceae (Based on type *Areca* L.)
(viii) Gramineae	Poaceae (Based on type *Poa* L.)

(II) Major Categories

The rules of ICN prescribe the term 'Division' (Diviso) to represent the category of the highest magnitude within the plant kingdom. Nearly equivalent term 'Phylum' has been prescribed for animal kingdom. The term 'Phylum' has been also preferred by some taxonomists (*cf.* Hutchinson, 1959. Tippo, 1942). Traditionally, there are two kingdoms *viz.,* Plantae and Animalia, but these have been challenged frequently and many other classifications have been suggested in its place in the recent past. Copeland (1938) recognized four kingdoms *viz.,* (i) Monera for prokaryotes, (ii) Protoctista for those with no or little tissue differentiation, (iii) Metaphyta and (iv) Metazoa for higher plants and animals respectively. Wittaker (1969) proposed a five kingdom classification of organisms in which plants are divided into the four groups *viz.,* Monera, Protista, Fungi and Plantae;and animals being placed under Animalia. Marguilis (1971, 1974) accepted and modified this classification slightly. Treub (1975) suggested two superkingdoms *viz., Cellulare* and *Acellulare* (includes viruses.) The former is divided further into two kingdoms *viz.,* (i) Procaryotae (ii) Eucaryotae. Plantae and Animalia have been proposed to the rank of subkingdoms. Edwards (1976) has recognized seven kingdoms under two superkingdoms *viz., Prokaryota* and *Eucaryota.* It is proposed that the plants be assigned to seven kingdoms, subordinate to the super kingdoms Procaryota and Eucaryota which are based on the prokaryotic and eukaryotic types of cellular organization. The Cyanochlorobionta for photosynthetic plants containing chlorophyll a only placed within the super kingdom procaryota. Photosynthetic plants within the super kingdom Eucaryota have been divided into three groups, the kingdom Erythrobionta for organisms with no accessory chlorophyll. Or occasionally chlorophyll d (and lacking flagella), the kingdom Chlorobianta for organisms with accessory chlorophyll b, and the Ochrobionta for those organisms with accessory chlorophyll c. Vascular plants have been assigned to a single division, the Tracheophyta, since in addition to having a similar biochemistry (chlorophylls a and b, similar carotenoids, starch as a reserve food product) they also process a basic morphological ground plan-a vascular system containing xylem and phloem, and a life history consisting of an alternation of generations. The fungi have been assigned to three kingdoms as opposed to most extant schemes of classification in which they are allocated to a single polyphyletic assemblage. The true fungi have been divided into kingdoms, Fungi 1 and Fungi 2 since it is thought that they have evolved from the Chlorobionta and Ochrobionta, respectively; the myxomycetes have been placed in a third kingdom, the Myxobionta.

The number of divisions within the plant kingdom is rather variable. Eichler (1883) divided it into five divisions, while Engler and Prantl (1887-1915) suggested 14 major divisions in their system. Generally, all seed plants are grouped under one division *viz.,* Spermatophyta. This division is characterized by the dominant sporophytic generation and presence of ovules which develop into seeds. A division may be subdivided into two or more subdivisions. Some even prefer to drop the category of subdivision and propose the division directly into two or more classes. The division Spermatophyta is usually divided into two subdivisions *viz.,* Gymnospermae and Angiospermae. But division

Pteridophyta is considered to be composed of directly classes and no subdivisions are proposed.

Class is a subordinate in rank to division. The names of classes are latin. The subdivision Angiospermae comprises two classes *viz.*, Dicotyledons and Monocotyledons. The class may be further divided into subclasses e.g. Class Dicotyledons are divided into Polypetalae, Gamopetalae and Monochlamydeae.

The class or subclass may include one or more orders which are subordinate to them. Some authors e.g. Bentham and Hooker (1862-1883) used this term instead of the term family. However, it is considered to be illegitimate by the rules of ICN. Although the names of orders here generally ending-ales and are usually based on one of the constituent families, some of the earlier names of order e.g. Centrospermae, Tubiflorae, Polygalineae, Caryophyllineae, etc. do not have such an ending. Instead, they have ending –ae. These names have long usage and hence have been allowed by ICN. Alternatives are also provided for such orders like Polygalales, Caryophyllales, etc. An order may be further divided into suborders. Some authors, as previously stated, have also proposed the category 'superorder' to which the category 'order' is subordinate. An order usually consists of one or more families.

The concept of family is generally encountered in the systematic studies. It is generally encountered in the systematic studies. It is the smallest of the major categories and more inclusive. It may contain one or more genera and has to be more natural and monophyletic. Family names are established on any of the constituent genera with a suffix-aceae, e.g. *Solanum*-Solanaceae, *Lilium*-Liliaceae, etc. Names of certain families, as mentioned earlier, have irregular terminations. These have been conserved by the rules of ICN. Their alternative names are also allowed. Families are recognized as 'definable' and 'indefinable' (Walters, 1961). The former ones are very natural and can be readily recognized from other families, for example, Umbelliferae, Leguminosae Compositae, Gramineae, etc. Their demarcation form other families is very distinctive by type of inflorescence, type of fruit and some such characteristics. Such families are thought 'successful families as they reached 'adaptive peaks' during evolution (Stebbins, 1974). The indefinable families are not as distinctive as the definable ones since they include considerable diversity of structure, for example, Ranunculaceae. Within the limits of Ranunculaceae, plants are herbs or climbers, leaves simple or compound, flowers regular or irregular, stamens one to many, fruits achenal or capsular, etc. These families are rather heterogenous. It is to be noted that the constituent genera are, however, easily definable on account of obvious morphological discontinuities. Reverse is the case of definable families. The constituent genera are less natural and ill-defined. In certain cases, the question of placement of a genus in a particular family becomes largely a matter of taste. The taxonomic position of such genus leads to controversy, for example, inclusion of *Barringtonia* in Lecythidaceae or Barringtoniaceae, *Nyctanthes* in Verbenaceae or Oleaceae, etc. Larger families are divided further into subfamilies. Names of subfamilies are based on one of the constituent genera and end with the suffix – oideae. Likewise, large subfamilies are subdivided into tribes and are based on any one constituted genus with an ending – eae.

(III) Minor Categories

Conventionally, categories at and below generic levels, are referred as minor categories. The name of a plant is formed by the name of minor category. It may be a genus, a species or any other category of subordinate rank. The category of the highest magnitude among the minor categories is obviously a 'genus' which is subordinate to the family. One or more genera constitute a family. It is usually agreed that different genera within the family are closely related and have often a common ancestral stock. The genera are delimited by a few well-defined characters. The first of the two words comprising a binomial is the generic name of a plant. The names of genera are nouns in the singular number, whereas the names of taxa at lower ranks may be nouns or adjectives in the singular or plural, and none is provided with distinctive endings denoting its precise rank. The rank should, therefore, be stated whenever it is not clear from the context.

(i) Genus: The concept of genus satisfied the practical need of plant identification and nomenclature. The concept is as old as folk science and has brought species together under a single entity. Several authors endeavoured to define genus. The definition based on morphological discontinuity is widely accepted. Tournefort (1700) opined that the genus is a fundamental category of classification. Plant having two or three characters in common were conceived as belonging to the same genus. Generic description and form limits of the groups were given for the first time by Carl Linnaeus (1737). The genera described by him and others were delimited especially on morphological ground. It was also thought that their constituents (species) have more characters in common with each other than those of the constituent species of other genera of the same family. This concept is still followed by many taxonomists in present time.

Rollins (1953) defined genus as a group of closely related species with emphasis on the 'naturalness' of the group shown by its homogeneity and distinctiveness. A genus should be bound by bonds of 'natural affinity'. It should be delimited naturally from other genera of the same family by well-defined 'gaps' or 'breaks'. Davis and Heywood (1963) considered three primary criteria important to determine generic status. These are: (i) naturalness, (ii) delimitation of closely related genera, and (iii) practicability of keeping them distinct or including them in other genera. Legendre and Vaillancourt (1969) defined genus as a monophyletic group of species that occupies a given adaptive zone. These authors also introduced the concept of 'phyletic closure' by which all the species descending from a common ancestor have to be included in the same genus. The genus is not a mere taxonomic category. The constituents of the genus warrant phylogenetic relationships amongst themselves as well as those of the other genera of the same family. Today, the genus is thought as a biological category. The taxonomic studies of the genus should include, apart from morphological similarities, origin, migration, geological history and ecological, physiological, cytological and genetical behaviours. Re-assessments, whenever necessary, should be based on these trends. A large genus may be divided into further subordinate categories like subgenus, section, subsection and series. However, these are not fundamental to the hierarchy and are not always used in classification within a particular group. These are helpful especially in providing infrageneric structure within large genera.

(ii) Species: The species is the lowest category in the taxonomic hierarchy. It is the fundamental category and consistently used and recognized by all worldover. It is one of the primary levels of integration in many branches of biology, as in systematics, genetics, ecology, physiology and in the study of behaviour (*cf.* Mayr, 1957). Moreover, species are also viewed as the basic units of the evolutionary process (*cf.* Stebbins, 1977). Because of the importance and consistent usage, the species category has been defined more successfully and explicitly than any other unit. Darwin (1859) also devoted to decipher the nature of species in his classic work. Davis (1978) called the species '*building bricks*' in *biological classification.*

(IV) Historical Development of Concept of Species

It is necessary to trace back to the time of Greek civilization, and to Plato in particular, to understand a fuller progression of the species concepts. Plato developed a concept of 'eidos' or 'species' to refer to any different kind of thing. He considered all objects as shadows of the 'eidos'. As a consequence, he overlooked variation in favour of '*typological species*' concept.

The principle of logical division was laid down by Aristotle. It was the basis of taxonomy for several years. It was based partly upon the idea of Plato. It served as a 'schema' upon which his concept of species was developed. A species, according to the principle of logical division, was thought any unit possessing a common essence ('Essense' means an abstract idea or concept which makes the unit what it is). The logical species was considered a relative term being applicable to various levels in a classification scheme. This logical relationship also existed between the genus and species. Consequently a species was defined on a 'priori' basis. It was regarded as fixed and unchanging. However, Lennox (1980) argued that Aristotle had two concepts of species and genera – the one just described above and another one based on relative differences. The principle of logical division hold good for material objects (e.g. tables, houses, books, etc.), however, it was inadequate to deal effectively with the variability of living organisms. This principle of logical division was a contribution in the further development of the 'typological species concept' as it maintained that each class of organisms possessed an essence.

Ray (1686-1704) believed that species bred true, and any variations that occurred were to be treated as accidents resulting from either environmental factors or factors inherent within the plant itself. Ray did not change the rigidity of the species concept. He added new dimensions by trying to objectively define the species in terms of morphology and reproductive relationship instead of continuing to utilize the 'essence' of Aristotle. He believed in use of all parts of the plant in a classification. He further stressed reproductive isolation as the basic characteristic distinguishing the species. Thus his species concept was non-arbitrary and completely non-variable.

Species concept of Linnaeus was based upon the idea that originally all species had been created by God and each one possessed an 'essence' (*sensu* Aristotle and even in religious sense). He believed that the elements of order consist of the fixed, discrete, 'natural' kinds created by God. But his view underwent a change during his lifetime. He became less dogmatic with statements on the constancy of species. Due to his experiences in gardening the tulips, he later became cognizant of the differences which could result in a species due to the efforts of man. He also became aware of variations and aberrant forms in the specimens he gathered. It was clear to Linnaeus that these deviant forms were not deserving of the rank of species but rather of variety.

Cuvier (1835) conceived species as a specific form 'conditioned by function' and governed by laws of metaphysics and mathematics thus acquiring the coherence and necessity of a geometrical definition from which we can deduce all the characters of the object of the definition. The consideration of all parts of organism for indicating the whole and eventually for indicating the species itself was an important new idea. This gave more momentum to the *typological species* concept.

Darwin is most intimately connected with problem of species concept in the entire history of biology. He believed in variation and overlapping or ever-changing pattern of species. His view also underwent a change in his life time. He experienced difficulty in delimiting species with his idea of evolution. He ultimately regarded species as something purely arbitrary and subjective. Darwin believed that species were extremely plastic and mutable. He stressed the evolutionary integrity of species through descent from a common ancestor, and this is the beginning of the evolutionary species concept still in vogue.

The concept of 'fixity' of species was substituted by the one that they are represented by variable populations which are evolving. This led to the development of the 'biological species concept' by Mayr (1963) and its general acceptance. He stressed that 'the non-interbreeding of natural populations rather the sterility of individuals be taken as the decisive species criterion. Unlike the earlier concepts of species based on phenetic relationships, biological species is defined by genetic kinship as evidenced by breeding behaviour (Mayr, 1969). *Biological species* is defined as 'a community of cross-fertilizing individuals linked together by bonds of mating and isolated reproductively from other species' (Grant, 1957). This concept emphasized, first, interbreeding between members of the same species and second, reproductive isolation between the members of different species.

(V) Concepts of Species

Various attempts have been made to define a species. Following are current species concepts:

1. **Morphological Species Concept:** Morphological species (morphospecies) is the one most frequently employed especially by revisionary workers or museum taxonomists. This has been also called the 'classical phenetic species concept' (Sokal, 1973) or 'Linnean or classical species concept' (Burger, 1975). Although this title of concept suggests no variation in the current usage of concept, variation is regarded as the expected result of dealing with populations. Workers have emphasized the importance of recognizing species on morphological ground alone. This is because we generally do not have sufficient information on reproductive behaviour to permit the biological species concept to be applied successfully. The morphological species concept has been expressed in various forms as in the following :

   (a) **Cronquist (1978):** Species are the smallest groups that are consistentlyand persistently distinct, and distinguishable by ordinary means.

   (b) **Shull (1923):** Species may be defined as the easily recognized kinds of organisms, and in the case of macroscopic plants and animals their recognition should rest on simple gross observation such as any intelligent person can make with the aid only, let us say, of a good hand-lens.

   (c) **Regan (1826):** A species is a community, or a number of related communities, whose distinctive morphological characters are, in the opinion of a competent systematist, sufficiently definite to entitle it, or them, to a specific name.

   (d) **Du Rietz (1930):** Species are distinct discontinuity in the series of biotypes

   (e) **Sneath (1976):** The species level is that at which distinct phenetic clusters can be observed.

   Various opinions have been expressed by the experts in the field. Heywood (1958) made it clear that the species recognized in the flora must be definable on morphology. Stuessy (1972) expressed a view that the morphological discontinuities recognized formally do not reflect biological limits of isolation, commonality of interbreeding, and genetic divergence.

   Heywood (1967) also pointed out, "we delimit species in practice on the basis of the differences shown by populations irrespective of the evolutionary factors that contribute to the development of these differences." Burger (1975) stated that the Linnean or classical species concept of readily recognized and morphologically defined species has served as a practical and efficient system for information retrieval in most flowering plants.

2. ***Biological Species*** **Concept:** This is the one concept held conceptually by many systematists. As stated earlier, species is a group of interbreeding populations which are isolated reproductively from other groups. This concept takes into consideration particularly reproductive isolation, which is important in evolutionary theory. This concept and its operation especially in botany has been a subject of very critical evaluation in the recent past. Mayr

(1982) has suggested a modification of the definition of biological species to stress ecological aspects along with reproductive isolation. He viewed a species as a reproductive community of populations (i.e. isolated from others reproductively) that occupies a specific niche in nature. It is rationale to emphasize adaptations to new environments in which the new populational system arise, besides importance of isolating mechanisms, Hengeveld (1988) criticised the use of niche in this definition on the grounds that it is too difficult to define accurately. Many criticisms have been leveled at the original biological species concept. The problems are obvious especially about**(a)** the accurate determination of interbreeding among population **(b)** the real extent of gene flow among populations **(c)** the common occurance of interspecific hybridization between species of flowering plants **(d)** and the applicability of the concept to asexual species. Sokal and Crovello (1970) opined that the biological species definition is extremely difficult and often impossible to apply to natural groups. In taxonomic practice, the most useful species concept is phenetic one. Many of the phenetic data are morphological in nature, Ehrlich and Raven (1969) suggested that although a species may have some form of common gene pool, it is selection rather than gene flow that maintains the general phenetic nature of the species, Davis and Heywood (1963) opined that taxonomic and biological species, although based on differently conceived abstractions, may coincide in practice. Morphology provides the evidence for putting the genetical definition into practice (Simpson, 1961).

It also appears pertinent to attract attention to '*sibling species*'. Genotypic and phenotypic evolution need not correspond among organisms. They might evolve at different speed and rate. There are certain cases where morphological distinctiveness has not gone hand in hand by genetic isolation. On the other hand, there are instances where 'morphologically similar populations are isolated reproductively '. This latter type of species are known as '*sibling species*' by Mayr (1948, 1963). Steyskal (1972) suggested an alternative term, 'aphanic species' to designate such groups, meaning 'hidden', '*cryptic*' or 'inscrutable'.

3. ***Genetic Species* Concept:** This concept is closely related to the biological species concept, According to this concept, the biological factors of gene flow and reproductive isolation are operative, but that the way to define species is by a measure of the genetic differences or distance among populations or groups of populations. In effect, this is the numerical phenetic species concept using a quantitative measure of genetic distance as the yardstick rather than morphological one.

   Newer techniques of measuring at least part of the genome via allozyme electrophoresis are helpful here. The genetic divergence based on allelic frequencies can be measured by various statistics such as Nei's (1972) genetic distance. At the higher levels, this approach may be severely limited due to parallel point mutations. However, at the specific and infraspecific levels, they should be more helpful. The direct measure of genetic distance from DNA sequences is gaining momentum in plants especially from chloroplast DNA studies (*cf.* Palmer and Zamir, 1982).

4. ***Paleontological Species* Concept:** Fossil materialis often fragmentary and rarely shows populational variation even at the morphological level. Moreover, they deal with the time dimension in which species appear and later disappear in the fossil record. Therefore, paleontologists can not directly deal with species concept based on gene flow and reproductive isolation. They often speak of '*paleospecies*' (Simpson, 1961) or '*chronospecies*' (George, 1956) in which arbitrary time and/or morphological limits are employed to define '*paleontological species*' (Cook, 1899). In practice, the paleospecies is generally a time-oriented *morphospecies*. Vaughanc (1905) termed a collection of paleospecies in a monophyletic succession as 'genes'.
5. ***Evolutionary Species* Concept:** Simpson (1951) defined an evolutionary species as a lineage (an ancestral-descendent sequence of populations) evolving separately from others and with its own unitary evolutionary role and tendencies. Simpson (1961) later emphasized that this concept avoids the difficulties with determining actual or potential levels of interbreeding and gene flow. It also permits some degree of interspecific hybridization. In his opinion, it is the fact of evolution that has made genetical species separate and that keeps them from always being sharply separate. The genetical definition of species has evolutionary significance but the definition does not actually involve any evolutionary criterion or anything about evolution.
6. ***Cladistic Species* Concept:** Wiley (1978) adopted and altered Simpson's (1951) concept of species on evolutionary base. In his opinion, 'A species is a single lineage of ancestral descendant populations of organisms which maintains its identity from other such lineages and which has its own evolutionary tendencies and historical fate'. Wiley (*loc.cit.*) made the definition of species more compatible with the cladistic viewpoint. The evolutionary species concept might be especially appealing to the cladists, who would search for a concept to relate to dichotomous branches on a cladogram. The emphasis on 'single' for the lineage more nearly equates this to a single branch on a *cladogram*. The use of 'maintains its identity from other such lineages' rather than 'evolves separately from others' opens the possibility of use of synapomorphies in detecting such lineages rather that the more general phrasing. The stress 'historical fate' instead of 'evolutionary role' is a significant shift from an ecological viewpoint to a historical context resulting from apomorphic changes within single branches of a *cladogram*. This concept of species is supported by some cladists e.g. Bremer and Wanntorp (1979). Donoghue (1985) and Mishler (1985) although supported it but called it the 'phylogenetic concept'. Willis (1981) opined that each species is an internally similar part of a phylogenetic tree.

7. ***Biosystematic Species* Concept:** Numerous and special categories have been proposed to deal with the units resulting from biosystematic investigations. In these, much effort is placed on interpreting reproductive limits of taxa. These species concepts reflect a desire to have units that more nearly reflect the diversity of reproductive relationships beyond the limitations by the *Linnaean hierarchy*. Most of them have not received wide usage, but some have become helpful informal descriptors in specific situations.
   (a) Cain (1953) and Grant (1960) racognised biosystematic categories as (1) ecotype (2) ecospecies (3) coenospecies. The ecotype refers to closely related but ecologically distinct populations that are largely interfertile. Ecospecies are similar but hybrids between them are of reduced viability, and coenospecies are not interfertile, even artificially.
   (b) Camp and Gilly (1943) defined about twelve kinds viz., homogeneon, phenon, parageneon, dysploidion, euploidion, alloploidion, micton, rheogameon, cleistogameon, heterogameon, apogameon, and agameon. All of these concepts are based largely on morphological and interbreeding criteria. The apogameon and agameon apply to apomictic groups.
   (c) The comparium and commiscuum of Danser (1929) are similar to the coenospecies, but more stress is placed on the ability to hybridise and on geographic factors.
   (d) The coenogemodeme and syngamodeme of Gilmour and Heslop-Harrison (1954) are equivalent to coenospecies and comparium respectively.
   (e) Van Valen (1976) defined multispecies concept as a set of broadly sympatric species that exchange genes in nature.
   (f) Harlan and De Wet (1963) proposed the concept of the compilospecies, which is genetically aggressive, plundering related species of their heredities, and in some cases may completely assimilate a species, causing it to become extinct.

(VI) Infraspecific Categories

Subspecies, Variety and Form: The designation of species within a taxon in many cases is fully adequate. However, there is a real need to circumscribe infraspecific taxa especially when complex patterns of variation occur. This is important particulary when we are emphasizing the local population as the focus for evolutionary change. The ICN recommended use of no more than five infraspecific categories *viz.,* subspecies, variety (varietas), subvariety (subvarietas), form (forma) and subform (subforma) Greuter *et al.* (1988). Of these, subspecies, variety and form are more commonly used. The usage of these has changed over the years. Some workers use only subspecies to describe initial patterns of infraspecific variation, others use only varieties, and still others use both categories. Some workers believe that subspecies and variety are unnecessary as separate categories and treat them more or less as synonyms. Others advocate the use of forms to describe more minor morphological variations. Some others do not recommend using this category. These different perpectives have led to enormous confusion in the way infraspecific taxa are circumscribed.

The final recommendation for use of subspecies, variety and form is as follows:

(i) Forms should not be used in a formal nomenclatural sense except perhaps in groups with strong economic value such as wild relatives of crop plants, in which the formal highlighting of unusual variation might be a stimulus for incorporating these into an ongoing breeding programme of great potential for increased food production or horticultural improvements.

(ii) If subspecies and / or variety have been used already within a group to describe the infraspecific patterns of variation, then this precedent should be followed insofar as possible.

(iii) Changes of rank from subspecies to variety or vice - versa without strong evidence should be resisted.

(iv) If no infraspecific classification has ever been proposed within a group, and the pattern of morpho-geographic variation leave a question as to whether subspecies or varieties should be recognized, then use of subspecies as the initial category will be a choice. Several criteria need to be utilized for useful definitions of subspecies, variety, and form for reproducing angiosperm taxa.

Table-I: Characteristics Useful for Distinguishing Subspecies, Varieties, and Forms in Sexually Reproducing Flowering Plants

Category	Morphological distinctions	Geographical patterns	Genetic divergence	Likelihood of natural hybridization	Fertility of hybrids
Subspecies	Several conspicuous differences	cohesive; largely allopatric or peripatric	Usually markedly multigenic	Possible along contact zones	Markedly reduced fertility
Variety	One to few conspicuous differences	Cohesive; largely allo patric with some overlap,	Multigenic or with some simple control	Probable in overlap region	Reduced fertility
Form	Usually a single conspicuous difference	Sporadic; sympatric	Single control (usually single gene)	Always expected	Complete fertility

(VII) Deme Terminology

A new system of terminology is designed by Gilmour and Gregor (1939). This system provides an infinitely flexible series of categories which could be used to define any group of individuals on the basis of any set of criteria. It is non-hierarchical and does not recognize the formal taxonomic categories like genus and species. It therefore avoids the use of root-words like 'Species' and 'type'. It uses a neutral root, a '*deme*' which implies nothing in itself. It means simply a group of related individuals of a particular taxon. Meaning of the terms are precise and provided by the different prefixes. Originally, only three categories were proposed by these authors as the following :

(i) ***Topodeme***: A deme occurring within a specified geographical area.

(ii) ***Ecoedeme***: A deme occurring within a specified kind of habitat.

(iii) ***Gamodeme***: A deme composed of individuals which interbreed in nature.

This terminology is later expanded by Gilmour and Heslop-Harrison (1954) and provided more terms as the following:

(iv) ***Agamodeme***: A deme composed of predominantly apomictic (asexually reproducing) individuals.

(v) ***Autodeme***: A deme composed of predominantly autogamous (self-fertilising) individuals.

(vi) ***Clinodeme***: A deme which together with other such demes forms a gradual variational trend over a given area.

(vii) ***Endodeme***: A deme composed of predominantly endogenous (closely inbreeding) but dioecious individuals.

(viii) ***Genodeme***: A deme differing from others genotypically.

(ix) ***Phenodeme***: A deme differing from others phenotypically.

(x) ***Plastodeme***: A deme differing from others phenotypically but not genotypically.

(xi) ***Genoecodeme***: An ecodeme differing from others genotypically.

(xii) ***Plastoecodeme***: An ecodeme differing from others phenotypically but not genotypically.

A further term, 'cytodeme' has been also used by some workers. It is as follows.

(xiii) ***Cytodeme***: A deme composed of individuals all with the same karyotype

A comparison of all such biosystematic categories proposed by different authors are tabulated below:

Table-II: Comparison of The Biosystematic Categories of Turesson, Danser, and Gilmour, Gregor and Heslop-Harrison.

Behaviour of individuals within the group	Turesson	Danser	Gilmour, Gregor and Heslop-Harrison	Behavior of group towards other such groups
Individuals capable of hybridzing with one another		Comparium	Syngamodeme	Group incapable of hybridizing with any other groups
Individuals capable of hybridizing with one another to give hybrids showing some fertility	Coenospecies	Commiscuum	Coenogamodeme	Group capable of hybridizing with other groups but such hybrids are sterile
Individuals capable of hybridizing with one another to give fully fertile hybrids	Ecospecies		Hologamodeme	Group caable of hybridizing with other groups to give hybrids showing some fertility
Individuals occupying a particular habitat and forming an interbreeding population which differs genotypically from other such populations	Ecotype	Convivium	Genoecodeme	Group capable of hybridizing with other groups to give hybrids showing complete fertility
Individuals occupying a particular habitat and adapted to it phenotypically but not genotypically	Ecophene		Plastoecodeme	As previous category

Davis and Heywood (1963) admitted that this basic terminology has so far received only limited support. Briggs and Block (1981) also supported them. Later Briggs and Walters (1984) opined that it will never become widely adopted. Attempts to define species have been many, but none of them are completely successful. Various concepts have been proposed in past. The concept of 'fixity' of species is replaced by the one that they are represented by variable populations which are evolving. This led to the development of the '*Biological species* concept' '*Genetic species* concept', '*Evolutionary species* concept' and few such others to date. There is no magic formula to decide the 'species', Taxonomists generally use following one or the other criteria :

(i) The individuals should bear a close resemblance to one another such that they are always readily recognizable as members of that group.

(ii) There are gaps between the spectra of variation exhibited by related species; if there are no such gaps then there is a case for amalgamating the taxa as a single species.

(iii) Each species occupies a definable geographical area, whether wide or narrow, and is demonstrably suited to the environmental conditions which it encounters.

(iv) In sexual taxa, the individuals should be capable by interbreeding with little or no loss of fertility, and there should be some reduction in the level or success (measured in terms of hybrid fertility or competitiveness) of crossing with other species.

No criteria is thought absolute. It is generally left to taxonormist's acrobatics. Species have changed along different evolutionary pathways. They have reached all grades of distinctness at the present time. And hence they are truly comparable entities. Heywood (1963) opined that 'they are' in fact, equivalent only by designation, and must therefore be regarded to a considerable degree as convenient categories to which a name can be attached'. The attempts to define a species to date has led to the cynical definition of a species (*cf.* Regan, 1925). According to this definition, 'a species is a group of individuals sufficiently distinct from other groups to be considered by taxonomists to merit specific rank'.

4

Systems of Classification: Genesis and Development

The world into which we live is full of individual living and non-living objects. Their rich variety is incomprehensible, nay almost impossible for any human-being to have sufficient knowledge about each of them. Nevertheless, man is blessed by nature in having a wonderful ability to discriminate between things in his ambience. Instinctively in childhood and later more self-consciously, he categorise things together and attach general terms to them. In other words, he classifies them into convenient groups. Davis and Heywood (1963) therefore stated that 'classification is a natural occupation of man, recognition one of his most necessary pastime'. Every human-being is thus a taxonomist by birth. 'Classification is a pervasive human quality, like the predisposition to sin, it accompanies us into the world at birth and stays with us to the end' (Hopwood, 1959).

Primitive human beings were necessity taxonomists. Certainly they endeavoured to discover plants to meet their daily necessities of life. In the ancient past, the process of discovering food plants also revealed others of nutritious in themselves but useful in increasing the palatability and digestibility and still others having powerful effects on the functioning of the human body and mind, inclusive of some which could be dangerous and fatal poisons. Ancient societies must have analysed the distinction between plants useful for medicinal purposes and plants which could be eaten or could be used in making tools, utensils, dwellings, etc. This basic dichotomy in the use-value of plants has a twin stimulus and influence in the development of plant science in general and plants classification in particular.

The history and development of systems of plant classification have been viewed with different considerations by the earlier authors and their accounts are available in books (*cf.* Stace, 1989; Stuessy, 1990; Lawrence, 1951; Naik, 1984; Sivarajan, 1984, etc.). Such accounts have attempted to divide the said subject matter in different ways, into a number of periods, each marked by a characteristic common goal or principle and frequently separated from one another by notable scientific events, works or authors like Darwin, Linnaeus and Camerarius. The historical development of systems of classification has really been a continuous one. None of them signalled the end of one period and the start of a new one because development has, on the whole, been ratter gradual. However, it is doubtless that all of them are notable milestones in the history of classifications.

The bases of groupings of systems herein are obviously the monumental works by both authors *viz.*, Charles Darwin and Carolus Linnaeus in natural sciences. Their contributions, as stated earlier, have far-reaching effects on the scientific thinking on the past and will continue so ever in future. The history of development of classifications is a fascinating subject. This review will reveal the progress of the plant science and responsible contributors of such progress. It will also help understand the reasons for their structure. The entire realm of systems has been preferably divided in this chapter first, into (I) Pre-Darwinian systems and (II) Post-Darwinan systems. The former fall into convenient groups (I) Pre-Linnean systems, (II) Linnean system and (III) Post-Linnean systems, whereas the latter into (I) *Ranalian systems* and, (II) *Englerian systems*.

The works of Darwin and Linnaeus have far reaching effect on thinking men. Darwinism had an almost immediate effect on every branch of biology. It profoundly influenced the aims and strategy of research and the interpretation and integration of the results. The evolutionary view made a great advance in the study of plant form when morphological categories came to be defined by the mutual relations of organs in their position and development within the plant body. From the standpoint of evolution, the morphological categories could be seen as the expression of actual inherited regulations in the differentiation of form within a group of phyletically related organisms. Evolution also entailed changes in morphological organization with time. Consequently, notions of fixed and absolute morphological categories became isolated from development and could no longer be entertained.

Botany, according to Linnaeus, is based on the double foundation of classification and nomenclature (*dispositio et denominatia*), and the foundation of *nomenclature* is *classification*. He further opined that the classification of plants must be based on the fructification. Fructification, as Linnaeus conceived, comprises the flower (all four parts-calyx, corolla, stamen and pistil) and the fruit. Number, form, proportion and situation of the fructification, according to him, provides many characters for classification. He emphasized these essential characters on which classification is based. He was quite aware about the functions of the floral parts – especially about sexuality in plants, a contribution from Camerarius (1694).

(I) Pre-Darwinian Systems of Classification

(i) Pre-Linnean Systems: The history of the development of plant classifications has gradually evolved from a small beginning, just like the plants themselves. Theophrastus (c.370-285 BC), a student of Plato and Aristotle, is the first to write down a classification in a logical and permanent form. We are fortunate in possessing his botanical teaching, in the form of the works, (i) Enquiry Into Plants (9 books) and (ii) The Causes of Plants (6 books). He made masterly use of (a) the scientific and technical literature of his own and earlier times, (b) information about plants of several foreign regions – reports of individuals who accompanied Alexander on his military conquests, some from merchants and travellers abroad, (c) some possibly from Babylonian and Egyptian documentary sources. He placed the facts in a coherent framework and created a true science of plants. Especially, his observations on seed germination are original, first-hand and suggestive of being eye-witness accounts. His work is thus not the job of classification by observing a few plants grown in the garden of Lyceum, Athens. He classified about 480 taxa. Some of names used by him still appear as generic names e.g. daukan (Daucus), asparagos (Asparagus), Narkisos (Narcissus), etc. He used primarily the most obvious characters of gross morphology e.g. trees, shrubs, subshrubs, herbs and successively more and more cryptic features. He also recognized differences based on (i) superior and inferior ovaries, (ii) fused and separate petals, (iii) radial and bilateral symmetry of flowers, (iv) sepaloidy and petaloidy, (v) fruit tupes, (vi) centripetal (racemose) and centrifugal (cymose) types of inflorescence, (vii) plants as annuals, biennials and perennials, (viii) difference between dicotyledons and monocotyledons, (ix) gross anatomical features and (x) form of leaf. He brought plants together by grouping them on these characters and made possible more generalized discussion. The groups he established were strictly artificial. His works mentioned above are probably the most complete and orderly of all ancient biological works. Every aspect of modern botany is, at least, indicated – morphology, anatomy, systematics, physiology, ecology, pharmacognosy, agriculture, applied botany and plant pathology. The study of fungi as a part of botany and their inclusion in the plant kingdom begins with Theophrastus. He made distinction between flowering and flowerless plants, fruiting and fruitless plants and evergreen and deciduous plants. His grouping together different species of oaks, figs, etc. indicate that he had some conception of a genus. All these evidences of his keen knowledge, that too in his times, are not astounding if his recognized as the 'Father of Botany'.

A Roman admiral Caius Plinius Secundus or known as The Elder Pliny (23-79 A.D.), one of the contemporaries of Disoscorides, published an encyclopaedic account on plants in his 'Historia Naturalis' (37 volumes) in A.D.77. He mentioned indiscriminately any plant to which he found reference in any previous book. His work was chiefly a compilation. His interest in plants was essentially utilitarian. He devoted nine volumes of this work to medicinal plants. Like Theophrastus, he begins his account of plants with the trees because of their value and importance to man. Theophrastus did so because he considered organization of trees the highest and most completely expressive of plant nature.

'Erh Ya', a Chinese dictionary was compiled about the time of Theophrastus or later. It mentioned 334 plants and classed as either trees or herbs. The literary and popular name of each plant was given, together with a short descriptive definition. This dictionary built a systematic botanical terminology. Sometimes the terms show a degree of particularity more primitive than advanced, as when different words are used for the flowers of trees, herbs and cereals, or for corresponding part in different species. Chinese botanists developed a descriptive morphology and a systematic method of naming plants fundamentally on the similar close line to that of Theophrastus.

Pedanios Dioscorides (1st century A.D.) was a physician in the times of the Roman Emperor Nero and Vespasian. His native land was Asia Minor but he travelled widely. The final harvest of Greek pharmacology was gathered and preserved by him in his 'Materia Medica', completed about A.D.60. It is enriched by his personal knowledge and experience. He described nearly 600 plant taxa and their uses. He gave information on roots, stem, leaves, flowers of these taxa. His work also deals with cereals, spices, oils, perfumes and wines. The actual descriptions are very slight, and it is only those with particularly salient characteristics which can be recognized with any ease. His 'Materia Medica' showed some attempt at order, but the arrangement is seldom at all natural. Occasionally, he grouped together plants which are nearly related e.g. Labiatae, Umbllifers, but this is rare. He had some idea about natural relationship of organisms. 'Codex Anciae Julianae' published in 512 A.D., containing his text in Greek is another contribution. The plant names used by Dioscorides is also taken by Linnaeus which are in use even today. e.g. *Zingiber, Anemone, Aloe*, etc. His work served to be one of the most standard reference and for Europeans for the next one half millennia. It can be considered the first herbal. Dioscorides followed Theophrastus in the frequent mention of habitat and geographical distribution of species. This information is very useful for herb-gatherers.

The mediaeval philosopher, Albertus Magnus (1193-1280) was a follower of Aristotle and Theophrastus. The title 'Albertus Magnus', was conferred on him during his life time by the unanimous consent of the schools. He was also known as Albert of Bollstadt. His work is available in his 'De Vegetabilis' (7 books). He dealt with medicinal plants and provided descriptions besed on first-hand information. He had remarkable instinct for plant morphology. For instance, He pointed out, in the vine, a tendril occurs in place of the bunch of grapes. He interpreted the tendril as a bunch of grapes incompletely developed. His realised thorns as stem structures and the prickles merely as surface organs. He showed marked originality in his descriptions of flowers. He distinguished, in the flowers of Borage, the green calyx, the corolla with its ligular outgrowths, the five stamens and the central pistil. He showed absence of calyx in the Lily and pointed out the petals exhibiting transitions from green to white. He noticed the early fall of the calyx in the Poppy and its persistence until the ripening of the fruit in the Rose. He was aware of aestivation of floral parts. He showed the successive whorls of sepals and petals alternated with one another and also concluded that this was a device for better protection of the flower. He classified flowers under three types : (i) Bird-form (e.g. *Viola, Lamium, Aquilegia*), (ii) Pyramid – and Bell-form, (iii) Star-form. He also divided the vegetable world into trees, shrubs, undershrubs, bushes, herbs and fungi. But in his opinion, this was a arbitrary scheme since they cannot always be distinguished from one another. He evolved a system, which is presented as follows(the modern equivalents of his different groups are given in parenthesis) :

I. Leafless plants (Cryptogams in part)

II. Leafy plants (Phanerogams and certain cryptogams)

   1. Corticate plants (Monocotyledons)
   2. Tunicate plants (Dicotyledons)

      (a) Herbaceous

      (b) Woody

The words '*corticate*' and '*tunicate*' indicated that he realized an anatomical distinction between dicotyledons and monocotyledons. The botanical portion of his work is called 'De Vegetabius' in which he described vegetables e.g. radishes, carrots, cabbage, lettuce, cucumbers, etc.

The plant science, during 15th and 16th century witnessed a great renaissance in medicine and classification. Through the Middle Ages new written works were very rare. The invention of printing in Europe enabled books to be produced in large numbers. This was the period of plant description and illustration. Many world famous herbals were published during these two centuries – the Age of Herbals'. 'Materia Medica' of Dioscorides was copied and followed prior to this period. Later, it proved too inadequate for identification as plant collections from several parts of the world were being added. The various herbals provided an original text and illustrations drawn from nature, At least, some of these herbals provided the rudiments of a natural classification. Some of them were arranged artificially and alphabetically. This period saw important herbals those of O.Brunfels, (1530), J. Bock (1539), L. Fuchs (1542), P. Mattioli (1544), W. Turner (1551), M.de L'Obel (1570), J. Gerard (1597) and C. L'Ecuse (1601). The technical innovations adopted in this period materially furthered the progress of plant science. The first was the method of preserving plants by drying, pressing and mounting on stiff paper or cards. This helped for permanent reference of plants for this examination and re-examination. It became the essential and basic tool of description and systematic botany ever since. The second innovation was the use of realistic illustrations of plants drawn from living specimens. Accurate pictures of medicinal plants were provided and the services of skilled artists and craftsmen were available to publishers.

Otto Brunfels (1464-1534) was a Corthusian monk at the beginning but later he was headmaster of the grammar school at Strasburg, Germany. He ultimately turned his attention to medicine. He became practicing town physician at Bern. He published 'Herbarium Vivae Eicones' (3 volumes) in 1530. It contained 135 superb wood-engravings of living plants. Later, It reached a total of 260 in later editions. In this book with its beautiful and naturalistic illustrations, there is, as its title indicates, a real return to nature. Brunfels knowledge of botany was chiefly derived from the study of certain Italian authors. Brunfels identified plants growing around the Strasburg district and the left bank of the Rhine. His work has been rightly ascribed importance in the history of botany for two reasons. First, it established the requisites of botanical illustrations – verisimilitude in form and habit, and accuracy of significant detail. Secondly, it provided a corpus of plant species which were identifiable with a considerable degree of certainty. These plants gave a firm basis on which a general flora could be founded. He was the first to recognize the 'Perfecti' and 'Imperfecti' group of plants based on the presence and absence of flowers respectively.

Jerome Bock (1498-1554), also called Hieronymus Bock, was a contemporary of Brunfels. He himself called 'Tragus' in his Latin writings. He was a school teacher of Zweibrucken, Germany and also overseer of the Count's garden. He then moved to Hornbach where he had an extensive medical practice and devoted his spare time to botany. He was forced to leave Hornbach owing to his protestanism. His great work is the 'Neue Kreutterbuch' published in 1539 at Strasburg. Subsequent editions were titled as

'Kreuter Buch'. The first edition was without illustrations, but a second contained many woodcuts. Many figures were new. The third part of the book is particularly noticeable for the woodcuts of trees. He is chiefly remembered for his descriptions, which were a great advance on those previously published. He carefully noted the mode of occurrence and localities of the plants mentioned. On this feature, his work showed some approach to a flora in the modern sense of the word. He described plants which had come under his own personal observation. He noted flower parts as corolla, stamens, pistils and apprehended something of natural relationships. His study on *Verbena* and *Artemisia* clearly reflected the independence of his thought. He included some details of stages of growth, number of petals, time of flowering and type of habitat. He classified plants based on habit and tried to bring together related plants within these categories of habit.

Leonhart Fuchs (1501-1566) was born in Bavaria. He began his career at Munich, Germany. In 1526, he became Professor of medicine at Ingolstadt and late in 1535 at Tubingen. Later, he was invited as a physician to the king of Denmark. He found time to produce a botanical masterpiece, which appeared in 1542 from the press of Isingrin of Basle, under the title 'De Historia Stirpium'. This Latin herbal dealt with about 400 native German and 100 foreign plants. In the next year, he published a German edition, called the 'New Kreuterbuch'. He critically studied plant nomenclature of classical authors. The wood-cuts were of extraordinary beauty. Some represented first European figure of American plants, e.g. *Zea mays L, Cucurbita maxima* Duch, His figures were copied by many authors of herbals, He arranged his work alphabetically. His methods of description influenced others and were a model for the order of description of each plant.

Valerius Cordus (1515-1544) wrote an account of 500 plants 'Historia Plantarum' which was published in 1561. He studied forest species in Germany and Italy. He described plant in this order : features of stem and leaves, the flower and time of flowering, the fruit and seeds, the number of loculi in the fruit, the lines of dehiscence, the appearance and number of rows of seeds, the root, taste, smell and habitat. He has established in principle the basis of scientific plant description.

Mathias De L'Obel(1538-1616) published his work as the 'Stirpium Adversaria Nova' in 1570, with Pena as joint author. He gave system of classification in this book. The main feature of his scheme is that he distinguished different groups by the peculiarities of their leaves. He led to make a rough separation between the classes which we now call Dicotyledons and Monocotyledons. He made a more conscious effort to arrive at a natural classification.

Herbalists in their endeavour could obviously add little from classification point of view. It is, however, certain that their attempts at descriptions, illustrations, nomenclature and terminology paved way towards classification. They also tried in their own way to some groupings of plants. It is particularly from 16th and 17th centuries onwards, plants, as a science of botany, were emphasized more. It also divorced from utilitarian view and gave impetus to plants exclusively. Plants began to be the focus of attention of many scientists for their intrinsic interest rather than their importance as food or medicine.

Andraea Caesalpino (1519-1603), an Italian and student of Luca Ghini, has been called 'the first plant taxonomist'. He taught botany, medicine and philosophy at the University of Pisa for about 40 years. The most powerful contribution to botany and further development were given by him in his 'De plantis' consists of 14 chapters in which he gave the basic principles of botany. The other 15 books gave descriptive account of 1520 plants, arranged according to a principled system of classification. The whole work is of the greatest scientific and historical significance.

In his opinion, classification must be based on characters arising from the fundamental nature of plants themselves i.e. on morphological characters of plant organs and parts, not on the properties of plants in relation to man or any other accidental or superficial feature. He further emphasized that the morphological characters must be those that are stable, not those that vary with changes of habitat, soil or climate. He gave importance to use of all organs and parts of plants for classification. He used characters of fruits and parts involved in fructification (reproductive parts) in defining the higher classes, whereas characters from leaves, stems and roots in dividing the higher classes into their ultimate species. He considered the number, position and form of various parts in defining the characters of plant organs as the most important. He always intended to classify plants using the characters of the fructification as a basis, and dividing first trees and shrubs, and then undershrubs and herbs, into higher classes and species. His primary division is founded on the presence or absence of seeds as the means of reproduction. According to modern views, this is fundamentally sound. He included seedless plants in Book 16. He listed pteridophytes (18 species), fungi (19 species) and other species (17). He considered the algae and fungi as the less perfect forms. The logical relation between the method of description of species and the natural system for a general classification at which Caesalpino aimed, were very perceptively analysed by him. He emphasized the practical advantages of a natural classification. His work was not appreciated by his contemporaries for it was in advance of the times. His views influenced thinking men like Tournefort, Ray and Linnaeus. Robert Morison (1620-1683) was in charge of the Duke Orlean's Garden Blois (1650-1660) king's physician and Regius Professor of Botany of King Charles II, England, and keeper of the Physic Garden of Oxford (1669). He published 'Plantarum Historia Universities (1680) at Oxford. His schemes resembles that of Caesalpino. He followed division of plants into woody and herbaceous. He subdivided the latter into 15 sections on the basis of habit, fruit and seed characters. He also published 'Plantarum Umbelliferarum Distribution Nova (1672) at Oxford. It was the first systematic monograph of a limited plant group.

Jean Bauhin (1541-1631) and Gaspard (Caspar) Bauhin (1560-1624) were brothers and Swiss botanists. They worked independently but rather on similar lines. J. Bauhin studied at the University of Basle and learned botany from Leonard Fuchs at Tubingen. He then was associated with Gesner at Zurich in some journeys in the Alps. He did not live to see published his a most ambitious undertaking 'Histoire universelle des plantes'. His son-in-law Cherler, a physician of Basle, brought out a preliminary sketch of it in 1619, 1650 and 1651. It was published under the name of Historia plantarum universalies' it is a compilation from all sources and includes descriptions of 5000 plants.

Gaspard Bauhin studied at Balse, Padua, Monpelier, Paris and Tubingen. He was Professor of Botany, Anatomy and Medicine at Basle. In 1623, G.Bauhin published his most important botanical work, the 'Pinax theatri botanici'. The 'Pinax' occupied an important place in the development of systematic botany. It was specially significant as it included every species with comprehensive references to previous descriptions and figures by other authors. He established a name for each species and indicated with which names in the literature he considered it synonymous. For most species, he employed binomial nomenclature, consisting of generic name followed by specific name. There are some similarities of arrangement at plants in the works by both brothers, but they are identical. They followed division into trees, shrubs and herbs but not consistently. They based whole groups on single morphological character e.g. leaf shape, form of fruit, type of inflorescence. They mixed up with groups based on habit, habitat, pharmacology and utility to man. They recognized a few more or less natural assemblages e.g. legumes, umbelifers, labiates, bulbous monocotyledons, grasses, fungi, etc.

G. Bauhin's 'Pinax' was an important contribution to floristic progress since it contained (i) a reliable list of known plants (ii) a convenient standard method of naming of plants and (iii) a more complete synonymy. It marked the transition to new stage in plant systematics and coincides with the beginning of a several extension of the field of botanical enquiry. Thus, in two major ways, G. Bauhin's 'Pinax' foreshadowed the great works of Linnaeus- first, the chaotic state of nomenclature was brought to order to a considerable degree, and second, use of a *binary nomenclature* composed of generic and specific names to designate the species.

John Ray (1628-1705) was an English philosopher and naturalist. He travelled in different European countries and produced many books on plant classification such as (i) Methodus Plantarum Nova (1682) (ii) Historia Plantarum (1704) and (iii) Synopsis Methodica Stirpium Britannicorum (1690). The last one represents the first British Flora. The second edition of Ray's 'Methodus' dealt with about 18000 species arranged using his system of classification based on floral and vegetative characters. His contributions are briefed in the following.

Ray's work influenced both the theory and the practice of botany more decisively. He combined systematic collecting from all types of habitat and many localities with careful and detailed comparison between each species found and similar species already recorded in the literature. He described habit, morphological characteristics, time of flowering, whether annual or perennial and medicinal properties if any, He tried to know plant in the field, and avoided exclusive reliance on the dried specimens Like Caesalpino, Ray rejected the use of accidental characters (as habitat, size, time of flowering, medicinal powers, utility to men or features like taste, smell or color, etc.). Instead, he employed the essential characters. These characters are taken for the principal parts of plants (as flowers, calyx, seed-vessel, etc.). He used these features as they are, in his opinion, constant in form, readily observed and have many conspicuous features of position, form and number; These, he thought, associated with other characters common to a particular group (genus). Ray showed the greatest originality and insight in the discrimination of the highest, most fundamental and natural divisions of the plant kingdom. He showed separation of dicotyledons and monocotyledons, as also in between Angioperms (plants with enclosed seeds) and Gymnosperms (plants with naked seeds). He also recognized ferns as a group. Fungi are defined as a new class and classified them based on the morphology of the sporophore. It is thus evident that Ray broke new ground in many ways, and made very real steps towards the first natural system of plant classification.

J.P. de Tournefort (1656-1708), a French botanist, was Professor of Botany at the Jardin des Plantes from 1688. He was in-charge of a scientific expedition to Asia Minor and parts of Europe. He gave descriptions of 698 genera, 9000 species and grouped them into 22 classes in his 'Institutiones Rei Herbariae' (1700). It was largerly artificial but practical system. He was the first to give names and descriptions to his genera, many of which are continued to be recognized by Linnaeus, although re-defined and re-named. He divided plants into two categories *viz.,* trees and herbs. These were further subdivided into groups on the basis of characters such as flowers petalous or apetalous, regular or irregular, and polypetalous or gamopetalous. He introduced the term 'pistil' to designate the central organ of the flower comprising ovary, style and stigma.

The method of classification of plants has actually early start in ancient Indian civilization. As the world at large was ignorant about the pristine language – Sanskrit, these Indian classifications were not exhumed in right period. They remained unknown to scientific communities worldover. As in other parts of the world, the medicinal, dietic and other miscellaneous significance was apparent in ancient times as shown by some Sanskrit texts and treatises. These contained reckonable elements of classification, nomenclature and terminology for plant description. Some of them are viewed from this point in the following.

Ancient Indian scholars classified substances or organisms based on their then prevalent utilitarian importance. They classed the plants or animals based on observations made by the local folks. They primarily also emphsised plant organs in their classification. For example, pushpa-varga (group based on flowers), etc. First of all, they started with classification of plants used in food. Then they switched over plants used in medicine. This was the general trend in their scientific thinking.

The 'Vrksayurveda' (Science of Plant Life) of Parasara (250-120 BC.), an Indian scholar, named different parts of a plant body and plant products : mula (root), kanda (stem), patra (leaf), puspa (flower), phala (fruit), tvak (skin,bark), sara (heartwood), svarasa (sap), niryasa (exudation), sneha (fats and oils), kantaka (prickle), bija (seed), paroha (seedling), etc. Parasara attempted to classify the plants according to their habitat into three broad groups : (i) Jangala (dry place), (ii) anupa(marshy, swampy or watery region), (iii) Misra or Sadharana (mixed or normal region) (Sircar and Sarkar, 1996).

Parasara's Vriksayurveda is a singular Sanskrit work which is exclusively devoted to plant sciences. He also classified plants broadly into four groups : (i) Vanaspati (plants with flowers hidden) (ii) Vanaspatya (plants bearing visible flowers and fruits) (iii) Virudha-valli(creeper with tendrils), and (iv) Culma or Ksupa (plants with short branches and roots but no tendril). He gave elaborate system of classification based on morphological characteristics of flowers and fruits. The texts actually contains descriptions of nearly 30 flowering families (called gana or ganiyam). For example, (i) Samiganiyam : flowers *hypogynous, gamosepalous*, petals five, free and of different sizes, stamens ten, fruits legume with seeds on side. This is further divided into (a) Vakrapuspam (curved flower), (b) Vikarnikapusparm (c) Sukapuspam on the basis of forms of flowers. It is to be noted that this description refers to modern Fabaceae (Leguminosae). (ii)Swastikaganyam: Calyx resembling a 'Swastika', ovary superior, 4 free sepals, 4 free petals, six stamens of which two shorter, and carpels two forming bilocular fruit. This refers to modern Brassicaceae (Cruciferae). Linnaeus (1753) is credited as the first scientist to develop a system of classification based primarily on floral characters. But Parasara outlined his system of classification on the basis of characters of flowers and fruits many centuries earlier. It deserves critical estimation of the plant taxonomists. Linnaeus would have probably adopted the Hindu method had he known the Sanskrit language (Sircar and Sarkar, 1996).

In the Rigveda (5000 B.C.), various terms have been used to indicate the different habits of plants. In the Atharva Veda (4500-2500 B.C.), sasa (i.e. herb), one of the category of habit, are again divided into four types *viz.,* (i) prastrnati (expanding), (ii) kandini (jointed), (iii) ansumati (having many branches, and (iv) ekasunga (one sheathed) (Sharma, 1993). Charaka in his 'Charak Samhita' (1000 B.C.), grouped 'dravyas' or plants into four categories based on inductive method. He emphasized functional importance in nomenclature of these groups or classes.. These are then divided into two groups based on medicinal properties *viz.,* (i) virecana (purgative) and (ii) anupana (astringent). The latter are further grouped into ten 'Vargas' (major groups) and 50 classes depending upon their actions on organs of man or particular symptoms of disease. Although the total number of 'dravyas' is 500, they belong to total 370 plant species included in his treatise (Sharma, 2005). In the Manu – Samhita (200 B.C.), the first ethical law book, the plants are divided into mainly eight groups on the basis of the mode of growth and the external forms. This encyclopedic compendium is enriched with proposals of the basic principles of *binomial nomenclature* and a scientific basis of biological classification (Singh, 2008). Susrut (Nagarjuna)in his 'Susrut Samhita' (1000 B.C.) categorized ' dravyas'or plants first into two groups *viz.*, (i) Drava-dravya (ii) Anna-dravya. He recognised total 37 groups or classes and named them on the basis of main or important 'dravya' (Sharma, 2001).

The 'Kautiliya Arthasastra' (c.4th century B.C.) contained utilitarian classification of plants. Kautilya classified forest flora into nine groups : (i) soradaru varga (timber yielding plants), (ii) venu varga (bamboo), (iii) valli varga (twiners), (iv) valka varga (bark or fibre yielding plants, (v) rajjibhanda varga (perhaps plants utilized as utensils, (vi) patra varga (plants yielding materials for writing upon), (vii) puspa varga (plants yielding dye from flowers, (viii) ausadhe varga (medicinal plants, and (ix) visa varga (poison yielding plants) (Shamashastry, 1915; Ganapatishastri, 1924).

The facts referred above about ancient Indian scholars indicate that the ancient Indians were well versed with plant classification. They used habit, habitat, utility to man and other miscellaneous characteristics of plants.

(ii) Linnean System: Certain developments and circumstances prevailing at the beginning of this period of classification are worth notable. Because of the efforts of the earlier botanists, the number of described plant species increased greatly. Vast literature was developed. Different systems of classification were put forth by various taxonomists, Bibliographic documentation was not systematically arranged. Rules of nomenclature were underdeveloped. Thus the 18th century was marked by the bewildering array of literature, systems of classification and accumulation of plant material. Moreover, sexuality in plants was discovered by Camerarius (1694). During this chaotic period emerged a Swedish naturalist Carolus Linnaeus (1707-1778) who confronted these problems. In recognition of his contribution to the science of taxonomy, he was knisghted (1753) and later (1761) granted a patent of nobility. He was known as Carl Von Linne. His importance in the history of taxonomy rests not only upon his contributions, but also upon the enthusiasm he instilled in his many students. To record the works of the Creator was his mission of life.

The two well known works *viz.,* (i) Genera Plantarum (1737) and (ii) Species Plantarum (1753), inclusive of their later editions, are the most important to plant taxonomists. He classified plants according to his artificial '*Sexual System*'. As stated earlier, sexuality in plants was proved in 1694 by Camerarius. The characters of stamens and pistils in the flowers of different plant groups appeared to him relatively constant and of potential importance in classifying plants. He could notice constancy in their number and constructed basically on arithmetical or numerical ground. Actually, his system was first published in 1735 in his 'Systema Naturae' in which he classified plants, animals and minerals as well. The plant genera recognized by him are listed and briefly described in the Genera Plantarum. He gave prominence to the rank of genus. He but carried forward the work of Bauhin and Tournefort. Many of his generic names mentioned are those of the works of these two authors. Other generic names are, however, were taken from the literature or commemorated the names of distinguished botanists. His two works, *viz.,* Genera Plantarum and Species Plantarum included about 7700 species belonging to 1105 genera. Generic descriptions are not included in the latter work. It provided, under each genus, the numbers, names and brief description of species. It also gave synonyms, habitats, country of origin and references to important earlier literature. The description of each species is in the form of a phrase – name of up to twelve word, commencing with the generic name. For new species, the entry is concluded by a longer description, while genera containing only a single species were given no description, for that species. Each species was provided with a one word specific epithet or trivial name. These names 'originated as indexer's paper – saving device. But their use proved convenient and quickly became the standard name of each species. Such a binary name, formed of the generic name and the specific epithet, is known as a binomial. It is distinct from the phrase-name which is a polynomial. Binomials, as mentioned earlier, were quite often used by Gaspard Bauhin in his 'Pinax' (1623).

An outline of the classes of the Linnaeus' '*sexual system*' of plant classification is given as an illustration of the mechanics by which it operated, together with a few examples showing the disposition of some typical plants :

Class 1. Monandria : Stamen one – *Canna; Salicornia*
Class 2. Diandria : Stamens two; *Olea, Veronica*
Class 3. Triandria : Stamens three; *many grasses*
Class 4. Tetrandria : Stamens four; *Protea, Galium*
Class 5. Pentandria : Stamens five; *Ipomoea, Campanula*
Class 6. Heptandria : Stamens six; *Narcissus, Lilium*
Class 7. Heptandria : Stamens seven; *Trientalis, Aesculus*
Class 8. Octandria : Stamens eight; *Vaccinium, Dirca*
Class 9. Enneandria : Stamens nine; *Laurus, Butomus*
Class 10. Decandria : Stamens ten; *Rhododendron, Oxalis*
Class 11. Dodecandria : Stamens 11-19; *Asarum, Euphorbia*
Class 12. Icosandria : Stamens 20 or more on the calyx; *Opuntia*
Class 13. Polyandria : Stamens 20 or more on the receptacle; *Ranunculus*
Class 14. Didynamia : Stamens didynamous; many Labiatae
Class 15. Tetradynamia : Stamens tetradynamous; *Brassica*
Class 16. Monadelphia : Stamens monadelphous; Malvaceae members
Class 17. Diadelphia : Stamens monadelphous; Legumes
Class 18. Polyadelphia : Stamens polyadelphous; *Theobroma, Hypericum*
Class 19. Syngenesia : Stamens syngenesious; many composites
Class 20. Gynandria : Stamens united to the gynoecium; Orchids
Class 21. Monoecia : Plants monoecious; *Carex, Morus*
Class 22. Dioecia : Plants dioecious; *Salix, Juniperus*
Class 23. Polygamia : Plants polygamous; *Acer, Nyssa*
Class 24. Cryptogamia : Flowerless plants; *Pteris, Agaricus,* Lichen

In his system, plants are grouped into total 24 classes which are based on the number, arrangement and fusion of reproductive parts. 'Didynamia' and 'Tetradynamia' are based on relative length of the stamens. The classes' *viz.,* 'Monadelphia', 'Diadelphia' and 'Polyadelphia' are characterized by fusion of anthers only and 'Gynandria' on fusion of androecium and pistil. Sex distribution *viz.,* monoecious, dioecious and polygamous characterize the classes 'monoecia', 'Dioecia' and 'Polygamia' respectively. All the non-flowering plants are classed under the 'Cryptogamia'.

Species Plantarum (1753) was the first work in which all plants were given a latin binomial. ICN considered it as the starting point for the nomenclature of all plants, except mosses and algae. For the first time, the botanical world was provided with a work which catalogued all known plants. *Linnean classification* continued to dominate taxonomic works until well into the 19th century. The great majority of floras employed his system. A fourth edition was prepared mainly by C. L. Willdenow and published between 1797 and 1830. Its six volumes included 16,000 species, more than thrice as many as in the first edition. These were the last successfully completed world monograph of all plants including the Poaceae. He died in 1978. His collections and books were purchased by J. E. Smith, an English botanist, after the death of son of Linnaeus. In the next year, these were sent to England. Smith founded the Linnean Society of London in 1788 and after his death in 1828; the collections were sold to the Society. It still houses the Linnean collections.

Linnaeus chose one or two characters and hence the system was simple and convenient. Linnaeus himself admitted that these characters cannot be considered to show natural affinities of plants. He classified plants by using two principles : (i) logical division (the rules of dichotomy-the character is present or absent), (ii) *a priori* reasoning employing functional organs of the plants. He conceived the stamens to be the most important. Stearn (1959), however, stated that there was a conflict in the mind of Linnaeus in his choice of *a priori* reasoning and modern empirical approach. His system, although artificial, provided a very easy means of identification and filing up.

Linnaeus adhered to the concept of divine creation. In his opinion, species were static entities, *monotypic*, genetically unitary (true breeding) and equivalent. In later period of his life he modified his view. He was dissatisfied with the static, biblical concept. He recognized variation within species but considered it to be phenotypic. He further maintained that many genera and species had arisen by hybridization. He was also aware that his sexual system was artificial. He visualized a natural system that could reveal it best. He substantially made his first attempt at natural classification in his 'Classes Plantarum.' In the 6th edition of his 'Genera Plantarum' (1764) he proposed 58 natural orders with their included genera. Many of these correspond to modern families such as Umbelliferae, Compositae, Palmae, Gramineae, etc. But this concept of Linnaeus has been largely ignored by the botanists. He retained the sexual system only for its practical value of identification. He had his own ideas of evolution by hybridization which he thought was mainly responsible for the origin of new forms. This was evidenced by his work 'Plantae Hybridae' (1751). It is said that his ideas about the science of evolution was read by Darwin.

(iii) Post – Linnean Systems: Prior to switching over to the post-Linnean period, it appears pertinent to know the circumstances prevailing particularly in the second half of the 18th century. The sexual system of Linnaeus was very convenient tool available to a large number of his enthusiastic students and other botanists to explore the plant wealth of the unexplored regions of the world. They went out collecting plants in different regions of the world. They extended their researches to the parts such as parts of America, tropical Indonesia, Island of Amboina, East Indies, North America, Egypt, Arabia, Cape of Good Hope, South America, Tropical Africa, Australia, South-Eastern Asia, China and India. The wealth of plant material gathered by the botanists could not be identified justifiably with the help of Linnaeus' system A need of another better system was immediately felt. To suffice that need, better systems were developed in the following period.

A large proportion of the plant materials collected were of species new to science, each to be given name, to be described, and placed in a classificatory system. The situation was made more severe due to man's increased knowledge and understanding of the organography and functioning of plants, and awareness of the biological significance of reproductive parts. The systems put forward by the different taxonomists in this era are natural. They served the desire of the human mind for the order, besides as an aid to identification.

Michel Adanson (1727-1806) was a French botanist who explained tropical flora in Africa. He spent six arduous but fruitful years in Africa and returned with a remarkable store of information on the geography, geology, natural history, languages, and resources of the country, in addition to thousands of botanical and zoological specimens. On his return from Africa in 1754, Adanson was invited by Bernard de Jussieu to live in his house. He remained there ten years. It was during this household intimacy with Bernard that Adanson wrote 'Families des Plantes' (1763) in which he made Bernard's taxonomic procedure explicit and consistent and gave it theoretical justification. He championed the idea that in classification one should use a great range of characters covering all aspects of the plant. This is called an empirical approach. He rejected all artificial classifications and for the first time a natural system of classification was published. He described taxa more or less equivalent to modern orders and families. He divided plants into 58 families, of which 52 contain flowering plant genera and other six contain gymnosperms and cryptogams. He gave circumscription of each families. This was epoch-making advance and became a model which profoundly influenced the progress of taxonomy.

He made a thorough investigation of all earlier systems of classification in preparation for establishing a natural system. As a result, he challenged the validity of some opinions expressed by Linnaeus. Adanson's work differed in two fundamental ways. First, he considered all methods, whether they were founded on the fructification or on other parts of the plant as well, secondly, he presented in another light some methods, especially those of Caesalpino, Morison, Ray, Knaut. Hermann and Boerhaave, which have been misunderstood or not sufficiently faithfully treated. For instance, according to Linnaeus all these authors have based their system on the fructification alone. Adanson pointed out, especially about Caesalpino, that apart from fructification, characters of roots, leaves, flowers, latex, cotyledons, radical, etc. were also employed. He thought over critically on another as points of consideration to be used in classification. For example, (i) He abandoned completely the idealist notion of essential parts (like seed or fructification) which play an important role in the life or being of a plant that they constitute its 'essence'. Thus he accepted materialist view without reservation. He opined that 'there can be only one natural method in botany which considers the totality of all the parts of plants'. (ii) He stressed that microscopic characters must be used if appropriate. (iii) He never held that all characters have equal importance or equal 'weight' in classification. Only he insisted that all characters should be given equal consideration as first step. Thus there must be no *a priori* selection or rejection.

Jean de Lamarck (1744-1829), a French biologist, is best known to taxonomists for his theory of evolution, called Lamarckism, whereby characters acquired during life become inherited. This theory was finalised after his main taxonomic contribution appeared. These comprised his 'Flore Francoise' (1778) and the botanical part of Panckoucke's 'Encyclopedie Methodique' (1783-1798, completed by J.L.M. Poiret, 1804-1817). His taxonomic fame rests on his realization that a natural system of classification was not the best for rapid identification. His 'Flore Francoise' contained an analytical method,floras, were meant to be simply a method of identification.

Antoine-Laurent de Jussieu (1748-1836) was a student of his uncle Bernard de Jussieu (1699-1777), a contemporary of Linnaeus. Bernard de Jussieu was unsatisfied with Linnaeus's artificial system. He tried to improve it but could not publish it. Antoine-Laurent published his uncle's system, with some improvements in 'Genera Plantarum' as 'Genera Plantarum Secundum Ordines Naturals Disposita' (1789). This work marked the beginning of the natural systems of classification. He divided plants into three main groups *viz.*, (i) Acotyledons (cryptogams and a few misunderstood mono cotyledons), (ii) Monocotyledons, and (iii) Dicotyledons (Dicotyledons and gymnosperms). He employed many of the familiar modern characters within the last two groups. These characters were e.g. ovary superior or inferior, stamens free or attached to corolls, petals free or fused, etc. All known plants were divided into 15 classes and 100 natural orders (families). Many of these are delimited similarly in modern classification. His system is outlined below :

1. Acotyledons – Fungi, Algae, Musci

   Monocotyledons
2. Stamina hypogyna – Cyperoideae, Gramineae
3. Stamina perigyna – *Lilia, Narcissus*, Iridales
4. Stamina epigyna – Musae, Orchideae, Dicotyledons

   Apetalae
5. Stamina epigyna – Aristolochiae
6. Stamina perigyna – Proteae, Lauri, Polygoneae
7. Stamina hypogyna – Amaranthi, Nyctagines

   Monopetalae
8. Corolla hypogyna – Labiatae, Solaneae, Borragineae
9. Corolla perigyna – Rhododendra, Ericae

   Corolla epigyna
10. Antheris connatis – Cichoraceae
11. Antheris distinctis – Rubiaceae

    Polypetalae
12. Stamina epigyna – Araliae, Umbelliferae
13. Stamina hypogyna – Ranunculaceae, Cruciferae

14. Stamina perigyna – Rosaceae, Leguminosae
15. Diclines irregularis – Amentaceae, Coniferae

He was associated with his uncle and Adanson as well. Adanson was also a student of his uncle. He was fortunate inheritor of their ideas and many-sided knowledge of plants. He added hitherto unknown families which poured into the herbaria during his lifetime from the many voyages of exploration. He also adopted the Linnean nomenclature. Fundamentally he accepted the principle of Adanson that natural relationship is determined by multiple affinity and that affinity is found by considering many (all possible) stable characters. He also gave a full diagnostic description of all family groupings. He accepted the division into monocotyledons and dicotyledons, and tried to find other natural, supra-family groupings of the same kind.

Prior to the 19th century, knowledge of plant structure was limited. It was restricted to exomorphic aspects and to a very elementary endomorphic features established by the 17th century anatomists such as Nemeiah Grew and M. Malpighi. The male and female nature of the stamens and carpels was proved experimentally by Camerarius in 1694. Fairchild (1717) produced probably the first artificial interspecific hybrid. The true nature of cryptogams particularly about their sex was largely unknown. It is only in the 19th century in different European countries, more details about sex in plants was revealed by Brown, Hofmeister, and Strasburger. Brown, during the early part of the 15th century, studied floral morphology of several groups in detail. He recognized the naked ovule of the gymnosperms and discovered the nature of the cell nucleus. Hofmeister, during the mid part of the 19th century revealed the life cycle pattern of many cryptogamic groups such as bryophytes and pteridophytes. Hofmeister's brilliant work threw more light on the relationships amongst cryptogams and to the phanerogams. He indicated the position of the gymnosperms between the higher cryptogams and the angiosperms. Strasburger, in the later part of the 19th century, carried further Hofmeister's work. He was the first to record sexual fusion in higher plants. It is also notable that explorations in America, Africa and Asia brought out an enormous number of new species. Many classifications appeared during the 19th century which were devised in a lively and dynamic environment.

In the first decade of the 19th century, significant advances in natural classification were made by Auguste Pyrame de Candolle and Robert Brown. They were close contemporaries, knowing each other's work. Their paths, however, were quite independent. De candolle outlined his views in 'Theorie Elementire de Botanique' (1813). It is an exposition of the principles of natural classification and of the art of describing and studying plants'. He coined the term '*Taxonomy*' to designate the theory of plant classification. His most important work, 'Prodromus Systematic naturalis Regni' Vegetabilis (1823-1873), a 17-volume book written by 35 authors under the general editorship of A.P. de Candolle (vols. 1-7) and after his death, by his son, Alphonse de Candolle (1806-1893) was published. He divided plants into two main groups *viz.*, (i) Cellulares and (ii) Vasculares. The former represented non-vascular and the latter vascular plants respectively. Cellulares were again grouped into two classes – class

Foliaceae and class Aphyllae, whereas Vasculares were categorized into classes *viz.*, Exogenae (Dicotyledons) and Endogenae (monocotyledons). Endogenae also included cryptogams.This included all species of Dicotyledons in the world. It accounted for over 15000 species in 161 families. He included gymnosperms and placed close to the catkin – bearing Dicotyledons. The task was not completed and could not cover the monocotyledons and cryptogams. He patterned his system in 1819 as below:

I. Vasculares (Vascular plants with cotyledons)

Class 1: Exogenae (vascular bundles in a ring, dicotyledons)

A. Diplochlamydeae (both calyx and corolla present)

(a) Thalamiflorae (polypetalous, hypogynous)

Orders 1-46 including Ranunculaceae, Cruciferae, Malvaceae, etc.

(b) Calyciflorae (perigynous or epigynous, polypetalous or sympetalous)

Orders 47-84 including Rosaceae, Leguminosae, etc.

(c) Corolliflorae (gamopetalous, hypogynous)

Orders 85-108 including Primulaceae, Labiatae, etc.

B. Monochlamydeae (calyx only present)

Orders 109-128 including Chenopodiaceae, Coniferae, etc.

Class 2: Endogenae (vascular bundles scattered, monocotyledons)

A. Phanerogamae (flowers present)

Orders 129-150 including Liliaceae, Iridaceae, Orchidaceae, etc.

B. Cryptogamae (flowers absent, hidden or unknown)

Orders 151-155 including Ferns, etc.

II. Cellulares (Plants without vascular bundles or cotyledons)

Class 1: Foliaceae (leafy, sexuality known)

Orders 156, 157 as mosses, Liverworts

Class 2: Aphyllae (not leafy, sexuality unknown)

Orders 158-161 as Algae, Fungi, Lichens, etc.

Few more salient features of his scheme may be noted. He arranged families in a linear sequence from Polypetalae, Gamopetalae and Apetalae to be succeeded by Coniferae. In linear sequence, the Ranalian group was the starting point. Although he recognized physiological characters useful in classification, he also accepted that morphological characters are the essential ones in practice. He opined that the specific positional relations of organs remain basically unaltered, forming as it were a ground-plan common to a group of related species. He used the term 'symmetry' for this concept of morphological relationships and applied it with particular success to the analysis of floral structure. De Candolle's doctrine of symmetry had a profound effect on the development of floral morphology, and of the natural system at generic and family level. His selection of the Ranalian group as the starting of the linear sequence of his scheme was accepted by many botanists as reflecting the natural course of interrelationships.

It appears relevant to attract attention to the system by Stephen Endlicher (1836-1850) in his 'Genera Plantarum'. It was widely used on the continent of Europe for a period, but was followed by Bentham and Hooker's 'Genera Plantarum', and finally by Engler and Prantl's 'Naturlichen Pflanzenfamilien'. In 1836, he was keeper of the Herbarium in the imperial collections of Natural History in Vienna.

In his scheme, instead commencing with families in which both sepals and petals are present, as in the Jussieu de Candolle systems, he began with a group without petals. Thus commenced the great divergence of the two main system still in use. Bentham and Hooker in their Genera Plantarum followed that of Jussieude Candolle, while in Engler in the Naturalichen Pflanzenfamilien followed Endlicher, later modified by Eichler. Endlicher's system showed an advance in the treatment of the cryptogams, which are placed first and arranged in a fairly natural sequence.

Bentham and Hooker's scheme (1862-1883) was the last major *natural system* of plant classification. The publication of Darwin's theory of evolution automatically ended this era in the history of classificatory system. Although their system is post – Darwinian chronologically, it remained pre-evolutionary in concept. As we know, the pre-Darwinian systems were based on the doctrine that the species are special creations and therefore constant and immutable. Botanists in this period were still imbued with the idea of the fixity of species and on the other hand, the Darwinian theory of descent was available to them as guide. In such circumstances, Hooker was inclined to reorganize the whole system but Bentham persuaded him to maintain it.

George Bentham (1800-1884) and Sir Joseph Dalton Hooker (1817-1911) presented their system of classification as a 3-volume work 'Genera plantaram (1862-1883) in Latin. Bentham was an amateur but well-trained botanist. He was an accomplished linguist and Latinist. He published world monographs of some families and authored a 7-volume flora of Australia. Hooker, son of the botanist Sir William J. Hooker, was the plant explorer and plant geographer. He was Director of the Royal Botanic Gardens, Kew (England), like his father before him. He explored many parts of the world and published, 'The Student's flora of the British Islands (1870)' and 'The Flora of British India' (1872-1897).

Bentham and Hooker's 'Genera Plantarum' is based upon de Candolle's system but differed as well in many respects. They set new standards in descriptive botany. Every genus known to them was described afresh from the study of herbarium specimens in the British and Continental herbaria. They provided a first-hand, accurate full information in respect of all the genera. Their work did not represent a compilation made from then existing literature. All descriptions are original and based on their plant dissections.

In their 'Genera Plantarum', only seed plants were dealt with. It described total 202 families and classified into 3 classes, 3 subclasses, 21 series and 25 orders (cohorts). Families *viz.*, Vochysiaceae and Cyrilleae were incorporated in later and numbered as '20 a' and '26 a' respectively. Initially it was intended to include 200 families. Each family is given a definite number. It included a total of 7569 genera, each in meticulous detail. Some genera are subdivided into subgenera and / or sections. The Dicotyledons were grouped into three major group's *viz.*, Polypetalae (with free petals), Gamopetalae (with united petals), and Monochlamydeae (with no petals). Their first order (cohort) being the Ranales, composed of the well-known Ranunculaceae and allied families, characterized by *apocarpy* and *hypogyny* of the flower. From these they proceeded to a discifloral series and finally, in case of Polypetalae, to the calycifloral *perigynous* and *epigynous* types of flower. They then enumerated successively the Gamopetalae, Monochlamydeae. (Apetalae), Gymnosperms and finally Monocotyledons. In elaborating their system, the Monochlamydeae were (unfortunately) retained as a group apart from the Polypetalae. This treatment has been mostly criticized by many botanists. This is also the case of their placement about the Gymnosperms. Actually, authors were aware and even pointed out, quite clearly, affinities at the end of each family. The erection of a new series Disciflorae and its incorporation between the Thalamiflorae and the Calyciflorae is a refinement over that of de Candolle's system. They experienced difficulty in placement of certain families (Orders) e.g. Coriareae, Moringeae and Plantagineae, and hence kept under 'Ordines anomali', although unsatisfactorily. The beginning of monocotyledons after the dicotyledons correspond well with the present state of knowledge on phylogeny of angiosperms. Their work was primarily concerned with a practical handbook of plant genera to facilitate their determination at best.

During this phase in the classification of cryptogams rather little progress was made. In the post-Linnaean period, *Lycopodium* was placed in mosses, cycads among ferns, *Salvinia* among liverworts and *Equisetum* among the conifers. De Candolle (1813) was the first to place all the pteridophytes together as a distinct group. Brogniart (1843) advocated the division of plants into major groups *viz.*, Cryptogamae and Phanerogamae. Braun, A. (1864) recognized three groups *viz.*, Bryohpyta (algae, fungi, and bryophytes), Cormophyta (pteridophytes) and Anthophyta (spermatophytes). Real progress towards the modern system was visible after Eichler's classification (1883) which belongs to next period.

Synopsis of Bentham and Hooker's system of classification is provided in the following:

I. Class : Dicotyledons

A. Subclass : Polypetalae (Corolla of separate petals. Orders 1-82, 20a and 46a)

Series 1: Thalamiflorae (Stamens hypogynous and usually many. Orders 1-33 and 20a)

Cohort 1 : Ranales (Gynoecium usually apocarpous)

Orders :
1. Ranunculaceae
2. Dilleniaceae
3. Calycanthaceae
4. Magnoliaceae
5. Annonaceae
6. Menispermaceae
7. Berberidaceae
8. Nymphaeaceae

Cohort 2 : Parietales (Placentation parietal)

Orders :
9. Sarraceniaceae
10. Papaveraceae
11. Cruciferae
12. Capparidaceae
13. Resedaceae
14. Cistineae
15. Violaceae
16. Canellaceae
17. Bixineae

Cohort 3 : Polygalineae (Ovary 2-locular)

Orders :
18. Pittosporeae
19. Tremendreae
20. Polygaleae
20a. Vochysiaceae

Cohort 4 : Caryophyllineae (Placentation free-central or axile)

Orders :
21. Frankaniaceae
22. Caryophylleae
23. Portulaceae
24. Tamariscineae

Cohort 5 : Guttiferales (Stamens many, sepals imbricate)

Orders :
25. Elatineae
26. Hypericineae
27. Guttiferae
28. Ternstroemiaceae
29. Dipterocarpeae
30. Chlaenaceae

Cohort 6 : Malvales (Stamens many, sepals valvate)

Orders :
31. Malvaceae
32. Sterculiaceae
33. Tiliaceae

Series 2: Disciflorae (Hypogynous disc often present, stamens definite, as many as or twice the number of petals. Orders 34-53 and 46a)

Cohort 7 : Geraniales (Androecium obdiplostemonous, ovules pendulous, raphe ventral)

Orders :
34. Lineae
35. Humariaceae
36. Malpighiaceae

37. Zygophyllaceae
38. Geraniaceae
39. Rutaceae
40. Simarubeae
41. Ochnaceae
42. Burseraceae
43. Meliaceae
44. Chailletiaceae

Cohort 8 : Olacales (Androecium diplostemonous, ovules pendulous, raphe dorsal)

Orders : 45. Olacineae
46. Ilicineae
46a. Cyrillaceae

Cohort 9 : Celastrales (Androecium haplostemonous, ovules erect, raphe ventral)

Orders : 47. Celastrineae
48. Stackhousieae
49. Rhamneae
50. Ampelideae

Cohort 10 : Sapindales (Androecium diplostemonous, ovules ascending, raphe ventral or inverted)

Orders : 51. Sapindaceae
52. Sabiaceae
53. Anacardiaceae

Ordines

Anomaly: 54. Coriareae
55. Moringeae

Series 3 : Calyciflorae (Sepals united, stamens peri- or epigynous. Orders 56-82)

Cohort 11 : Rosales (Stamens indefinite, often twice or more the number of petals, styles distinct)

Orders : 56. Connaraceae
57. Leguminosae
58. Rosaceae
59. Saxifrageae
60. Crassulaceae
61. Droseraceae
62. Hamamelideae
63. Bruniaceae

Cohort 12 : Myrtales (Stamens definite, rarely indefinite, flowers perigynous or epigynous)

Orders : 64. Halorageae
65. Rhizophoreae
66. Combretaceae
67. Myrtaceae
68. Melastomaceae
69. Lythrarieae
70. Onagrarieae

Cohort 13 : Passiflorales (Placentation parietal)

Orders : 71. Samydaceae
72. Loaseae
73. Turneraceae
74. Passifloreae
75. Cucurbitaceae
76. Begoniaceae
77. Datisceae

Cohort 14 : Ficoidales (Perianth undifferentiated, ovules on axile or basal placentae)

Orders : 78. Cacteae
79. Ficoideae

Cohort 15 : Umbellales (Inflorescence umbellate)

Orders : 80. Umbelliferae
81. Araliaceae
82. Cornaceae

B. Subclass : Gamopetalae (Corolla of united petals. Orders 83-127)

Series 4 : Inferae (Stamens as many as petals and alternating with them, ovary inferior. Orders 83-91)

Cohort 16 : Rubiales (Stamens epipetalous, anthers distinct, ovary 2-many locular, locules 1-many ovuled)

Orders : 83. Caprifoliaceae
84. Rubiaceae

Cohort 17 : Asterales (Stamens epipetalous, anthers distinct or united, ovary 1-locular, 1-ovuled)

Orders : 85. Valerianeae
86. Dipsaceae
87. Calycereae
88. Compositae

Cohort 18 : Campanales (Stamens not adnate to corolla, ovary 2 to 6 locular, locules many ovuled)

Orders : 89. Stylideae
90. Goodenoveae
91. Campanulaceae

Series 5 : Heteromerae (Stamens as many as corolla lobes and opposite them or many, ovary superior or inferior, carpels more than 2, generally isomerous with the corolla lobes. Orders 92-103)

Cohort 19 : Ericales (Stamens twice as many as corolla lobes or isomerous and alternating with them)

Orders : 92. Vacciniaceae
93. Ericaceae
94. Monotropeae
95. Epacrideae
96. Diapensiaceae
97. Lennoaceae

Cohort 20 : Primulales (Stamens as many as corolla lobes and opposite them, ovary 1-locular)

Orders : 98. Plumbagineae
99. Primulaceae
100. Myrsineae

Cohort 21 : Ebenales (Stamens as many as corolla lobes and opposite them or many, ovary 2-many locular)

Orders : 101. Sapotaceae
102. Ebenaceae
103. Styraceae

Series 6 : Bicarpellatae (Stamens as many as corolla lobes and alternating with them or fewer, ovary mostly bicarpellary and superior. Orders 104-126)

Cohort 22 : Gentianales (Corolla actinomorphic, leaves mostly opposite)

Orders : 104. Oleaceae
105. Salvadoraceae
106. Apocynaceae
107. Asclepiadaceae
108. Loganiaceae
109. Gentianaceae

Cohort 23 : Polemoniales (Corolla actinomorphic, leaves mostly alternate)

Orders : 110. Polemoniaceae
111. Hydrophyllaceae

112. Boraginaceae
113. Convolvulaceae
114. Solanaceae

Cohort 24 : Personales (Corolla zygomorphic, posterior stamen often reduced to staminode, ovules often more than 4)

Orders : 115. Scrophulariaceae
116. Orobanchaceae
117. Lentibulariaceae
118. Columelliaceae
119. Gesneriaceae
120. Bignoniaceae
121. Pedaliaceae
122. Acanthaceae

Cohort 25 : Lamiales (Corolla zygomorphic, posterior stamen reduced, ovules 4)

Orders : 123. Myoporineae
124. Selagineae
125. Verbenaceae
126. Labiatae

Ordo anomalous : 127.Plantagineae

C. Subclass : Monochlamydeae (Perianth 1 or two seriate, mostly sepaloid, minute or absent. Orders 128-163)

Series 7 : Curvembryeae (Seeds mostly with mealy endosperm, embryo curved, lateral or peripheral; ovary mostly one ovuled)

Orders : 128. Nyctagineae
129. Illecebraceae
130. Amaranthaceae
131. Chenopodiaceae
132. Phytolaccaceae
133. Batideae
134. Polygonaceae

Series 8 : Multiovulatae aquaticae (Immersed aquatic herbs, ovary syncarpous, many ovuled)

Orders : 135. Podostemaceae.

Series 9 : Multiovulatae terrestris (Terrestrial herbs, ovary syncarpous, many ovuled)

Orders : 136. Nepenthaceae
137. Cytineae
138. Aristolochiaceae

Series 10 : Micrembryeae (Carpels 1-2 ovuled, ovules with copious endosperm and minute embryo)

Orders : 139. Piperaceae
140. Chloranthaceae
141. Myristiceae
142. Monimiaceae

Series 11 : Daphnales (Ovary monocarpellary, rarely syncarpous and 2-4 ovuled, plants woody or herbaceous, perianth mostly sepaloid, stamens perigynous)

Orders : 143. Laurineae
144. Proteaceae
145. Thymelaeaceae
146. Penaeaceae
147. Elaeagnaceae

Series 12 : Achlamydosporeae (Ovary 1-locular, 1-3 ovuled, seeds endospermic, without testa)

Orders : 148. Loranthaceae
149. Santalaceae
150. Balanophoreae

Series 13 : Unisexuals (Flowers strictly unisexual or polygamous)

Orders : 151. Euphorbiaceae
152. Balanopseae
153. Urticaceae
154. Platanaceae
155. Leitnerieae
156. Juglandaceae
157. Myricaceae
158. Casuarinaceae
159. Cupuliferae

Series 14 : Ordines Anomali :
160. Salicineae
161. Lacistemaceae
162. Empetraceae
163. Ceratophyllaceae

Gymnospermae :

Orders : 164. Gnetaceae
165. Coniferae
166. Cycadaceae

Class : Monocotyledons

Series 15 : Microspermae (Inner perianth petaloid, ovary inferior, seeds minute, numerous)

Orders : 167. Hydrocharitaceae
168. Burmanniaceae
169. Orchideae

Series 16 : Epigynae (Inner perianth petaloid, ovary inferior, ovules large and few to many)

Orders : 170. Scitamineae
171. Bromeliaceae
172. Haemodoraceae
173. Irideae
174. Amaryllideae
175. Taccaceae
176. Dioscoreaceae

Series 17 : Coronarieae (Inner perianth petaloid, ovary free)

Orders : 177. Roxburghiaceae
178. Liliaceae
179. Pontederiaceae
180. Philydraceae
181. Xyridaceae
182. Mayaceae
183. Commelinaceae
184. Rapataceae

Series 18 : Calycinae (Inner perianth sepaloid, rigid or herbaceous)

Orders : 185. Flagellariaceae
186. Juncaceae
187. Palmae

Series 19 : Nudiflorae (Perianth absent or reduced)

Orders : 188. Pandaneae
189. Cyclanthaceae
190. Typhaceae
191. Araceae
192. Lemnaceae

Series 20 : Apocarpae (Perianth 1-2 seriate or absent, carpel solitary or gynoecium apocarpous, seeds non-endospermic)

Orders : 193. Triuridaceae
194. Alismaceae
195. Naiadaceae

Series 21 : Glumaceae (Flowers in dense inflorescences subtended by bracts or glumes, perianth reduced, glumaceous or absent, ovary or locules 1-ovuled)

Orders : 196. Eriocauleae
197. Centrolepideae
198. Restiaceae
199. Cyperaceae
200. Gramineae

II. Post-Darwinian Systems of Classification

Organismal classification revolved round the various theories and concepts before Darwinism (1859). The theory of evolution marked the beginning of a revolution in systematics. All the earlier theories lost their validity. The theory of evolution brought about drastic changes in the conceptual basis of taxonomy. It set a platform for the science of taxonomy to lead to a new era. The place of 'typological thinking' is taken away by 'population thinking' which highlights the changes that a single individual might undergo during different conditions and stages of life. Natural selection of better adapted variants, as proposed by Darwin, also offered a rational mechanism by which the evolution of new forms was brought about.

As far as systems of plant classification concerned, the era of arranging plant groups either randomly, in a liner sequence or a stepwise progressive manner have been replaced gradually by the phylogenetic systems.The most striking feature of post - Darwinian systems is that they attempted not only to be phylogenetic but also to be evolutionary.

Post-Darwinian systems mostly fall into two groups depending on the supposed nature of the primitive angiosperm flower. Two widely different concepts of primitive flower have been developed by the two schools more or less simultaneously over a considerable long period. These two schools of thought are *viz.*, **(i)** Ranalian school **(ii)** Englerian school. Their views are expressed below:

(i) Ranalian School: The ranalian concept of primitive flower was proposed by Charles E. Bessey (1915). It was then modified greatly by John Hutchinson (1926). This school of thought maintained a view that the angiospermic flower arose from a gymnosperm with a bisexual strobilus bearing microsprophylls below and megasporophylls above. The upper sporophylls developed into stamens and carpels, while the lower ones developed into sepals and petals by their progressive sterilization. The receptacle in flower is the result of compaction and shortening of the axis of a strobilus. The primitive flower bore numerous distinct perianth segments with many separate stamens and carpels. All these were spirally arranged on the receptacle. Flower in Magnolian – Ranalian complex, living today is the nearest approach to this primitive type of flower. This concept of primitive flower was supported theoretically by Arber and Parkin (1907). They suggested that the ancestor of angiosperms was an undivulged gymnosperm distantly related to that rare fossil *Cycadeoidea dacotensis* in bennettitalean group, a mesozoic plant that had produced anthostrobili in the axils of its cycad – like leaves. Bessey, Hallier,Hutchinson, Takhtajan Cronquist and others belong to this school and patterned their systems based on this concept of primitive flower.

(ii) Englerian School: Another concept of primitive flower was proposed first by Alexander Braun, a German taxonomist, in his 'Flora der Provinz Brandenburgs' in 1859. A.W. Eichler, was influenced by this concept but it was fully developed subsequently by Adolf Engler. According to this concept, unisexual and naked flowers generally borne in *ament* (*catkin*) are the most primitive within angiosperms. Probably, they have been derived from gymnospermous ancestors with unisexual storbilus bearing mega – or microsporophylls. Bisexual flowers are thought derived from a cluster of male and female flowers held together in the same inflorescence. They resembled a flower (*pseudanthium*) and perianth developed *de novo* The unisexual flowers were wind pollinated. Various groups of angiosperms, according to this school, originated from different gymnosperm groups and even pteridphytes. Braun, Eichler, Engler and his associates, Rendle, Wettstein and Mechior belong to this school and have based their system on this concept.

The developmental history of systems of classification of these two schools is presented in the following :

(I) Systems of Ranalian School: Charrles E. Bessey (1845 - 1915) was the first American to make a major contribution to plant classification and rélationships. He worked at the University of Nebraska as Professor. He was trained student of Asa Gray. He put forward his system of angiosperm classification which he believed definitely to be phylogenetic. He derived angiosperms from his Cycadophyta and the latter from implied bennettitalian ancestry. Earlier versions of his system appeared in 1883 and 1897 and the final treatment appeared as 'The Phylogenetic Taxonomy of Flowering Plants' in 1915. He considered the Angiosperms to be monophyletic. This claim is supported by the allegedly unique double fertilization, characteristic 8-nucleate embryo sac and the formation of secondary endosperm in angiosperms. He produced a Cactus Diagramin which every order is shown as being directly derived from another (Fig.1).

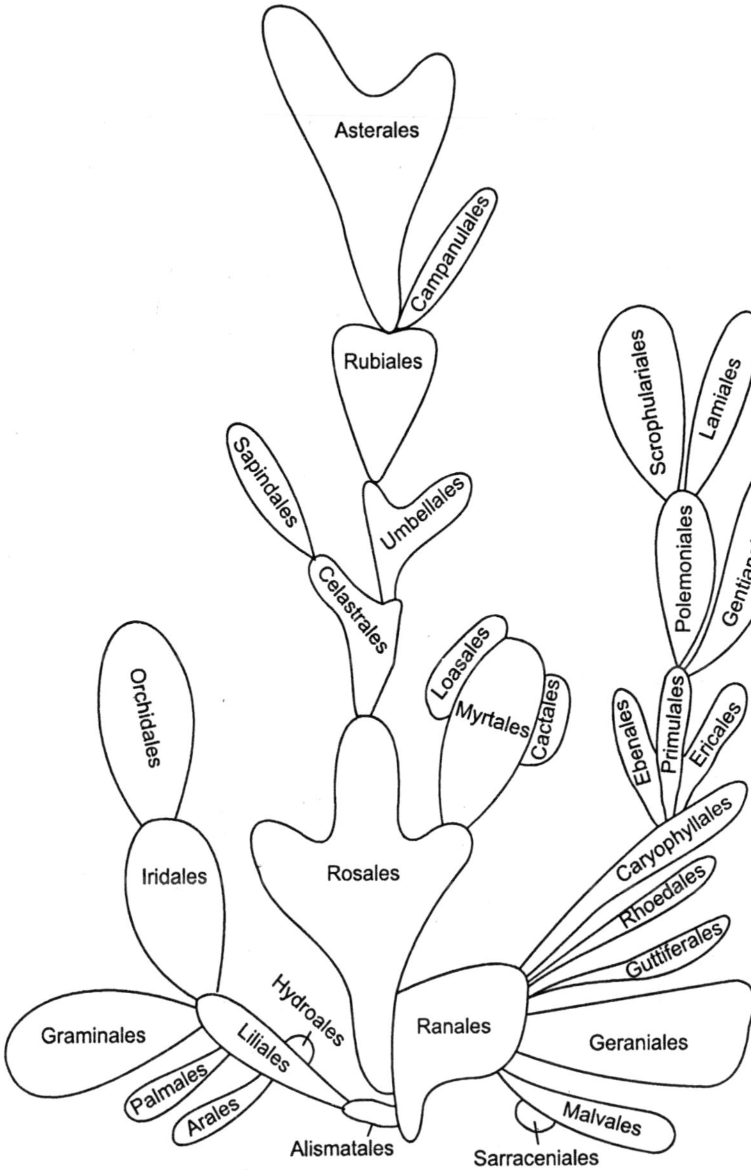

Fig. 1. Showing relationships of orders recognized by Bessey. The areas are approximately to the number of species in the order. (Modified after Bessey, 1915)

He holds a view that the Ranales (*sensu lato*) were primitive and that the rest of the monocotyledons and dicotyledons were derived from them. Angiosperms, in his opinion, constituted two major groups *viz.*, Oppositifoliae (Dicotyledons) and Alternifoliae (Monocotyledons). He employed data accrued from historical, ontogenetical and morphological studies of homologies. He based his system on his own dicta, guiding principles of primitive and advanced character states (*cf.* Bessey's Dicta) Three major series are derived from a single stock – the Ranales. The left line represents monocotyledons, whereas the remaining two suggest two major groups within the dicotyledons *viz.*, Strobiliodeae (the Ranalian line) and Cotyloideae (the Rosalean line). He abolished the Polypetalae, Gamopetalae and Monochlemydae as natural groups. He gave emphasis on *perigyny* and *epipgyny*. This has been critcised severely as these have been evolved independently in several phyletic lines within angiosperms. However, his system gained more support for choice of the Ranales as a primitive group. He derived the monocotyledons from the herbaceous Ranales by way of the Alismatales, a group of aquatic monocotyledons resembling the Ranales in free carpels. His system and also of Hallier were the first allegedly phylogenetic systems to start with the Ranunculaceae and / or Magnoliaceae and therefore to implicate them as the most primitive Dicotyledons. The following provides the synopsis of his classification (Figures in parentheses indicate number of families included)

Division	:	Anthophyta (300)
Class	:	Oppositifoliae (Dicotyledoneae - 255)
Subclass	:	Strobilioideae (146)
Superorder	:	Apopetalae – Polycarpellateae
Order	:	Ranales (24)
		Malvales (12)
		Serraceniales(2)
		Geraniales (21)
		Guttiferales (21)
		Rhoeadales (7)
		Caryophyllales (17)
Superorder	:	Sympetalae – Polycarpellatae
Order	:	Ebenales (5)
		Ericales (6)
		Primulales (5)
Superorder	:	Sympetalae – Dicarpellatae
Order	:	Gentianales (6)
		Polemoniales (6)
		Scrophulariales (10)
		Lamiales (4)
Subclass	:	Cotyloideae(109)
Superorder	:	Apopetalae

Order	:	Rosales (23)
		Myrtales (15)
		Loasales (5)
		Cactales (1)
		Celastrales (24)
		Sapindales (15)
		Umbellales (3)
Superorder	:	Sympetalae
Order	:	Rubiales (5)
		Campanulales (4)
		Asterales (14)
Class	:	Alternifoliae (Moncocotyledoneae - 45)
Subclass	:	Strobiloideae (31)
Order	:	Alismatales (9)
		Liliales (13)
		Palmales (1)
		Graminales (5)
Subclass	:	Cotyloideae (14)
Order	:	Hydrales (1)
		Iridales (11)
		Orchidales (2)

Bessey's Dicta:

(I) General Dicta :

1. Evolution is not always upward, but often it involves degradation and degeneration.
2. In general, homogeneous structures (with many and similar parts) are lower, and heterogeneous structures (with fewer and dissimilar parts) are higher.
3. Evolution does not necessarily involve all organs of the plant equally in any particular period, and one organ may be advancing, while another may be retrograding.
4. Upward development is sometimes through an increase in complexity, and sometimes by a simplification of an organ or a set of organs.
5. Evolution has generally been consistent, and when a particular progression or retrogression has set in, it is persisted in to the end of the phylum.
6. In any phylum, the *holophytic* (chlorophyll-green)plants precede the colorless (*hysterophytic*) plants, and the latter are derived from the former.

7. Plant relationships are up and down the genetic lines, and must constitute the framework of phylogenetic taxonomy.

(II) Dicta Having Special Reference To The General Structure Of The Flowering Plants :

1. The stem structure with collateral vascular bundles arranged in a cylinder is more primitive than that with scattered bundles, and the latter are to be regarded as derived from the former.
2. Woody stems (as of trees)are more primitive than herbaceous stems, and herbs are considered to have been derived from trees.
3. The simple, unbranched stem is an earlier type, from which branching stems have been derived.
4. Historically the arrangement of leaves in pairs on the stem is held to have preceded the spiral arrangement in which the leaves are solitary at the nodes.
5. Historically simple leaves preceded branched ("compound") leaves.
6. Historically leaves were first persistent ("evergreen") and later deciduous.
7. The reticulated venation of leaves is the normal structure, and the parallel venation of some leaves is derived from it.

(III) Dicta Having Reference on the Flowers of Flowering Plants:

1. The *polymerous flower* structure precedes, and the oligomerous structure follows from it, and this is accompanied by a progressive sterilization of sporophylls.
2. Petaly is the normal perianth structure, and apetaly is the result of perianth reduction (aphanisis)
3. The *apochlamydeous* perianth is earlier and the *gamochlamydeous* perianth is derived from it by symphysis of the members of perianth whorls.
4. *Actinomorphy* is an earlier structure than *zygomorphy* and the latter results from a change from similar to dissimilar growth of the members of the perianth whorls.
5. *Hypogyny* is the more primitive structure, and from it epigyny was derived later.
6. *Apocarpy* is the primitive structure, and from it syncarpy was derived later.
7. *Polycarpy* is the earlier condition, and oligocarpy was derived from it later.
8. The endospermous seed is primitive and lower, while the seed without endosperm is derived and higher.
9. Consequently, the seed with a small embryo (in endosperm) is more primitive than the seed with a large embryo (in scanty or no endosperm).
10. In earlier (primitive) flowers, there are many stamens (*polystemonous*) while in later flowers there are fewer stamens (*oligostemonous*).

11. The stamens of primitive flowers are separate (*apostemonous*), while those of derived flowers are often united (*synstemonous*).
12. The condition of powdery pollen is more primitive than that with coherent or massed pollen.
13. Flowers with both stamens and carpels (*monoclinous*) precede those in which these occur in separate flowers (*diclinous*).
14. In diclinous plants the *monoecious* condition is the earlier, and the *dioecious* later.

Hans Hallier (1868-1932), a German botanist of Hamburg, independently proposed in 1905 and 1912 a system for the flowering plants which resembled Bessey's scheme in many respects. It did so particularly in the commencement of the dicotyledons with the Ranunculaceae – Magnoliaceae alliance. However, it differed in that the dicotyledons preceded the monocotyledons. He based his system primarily on the study of herbarium material and literature on paleobotany, anatomy, serology and ontogeny. It is thus a synthesis of researches then available. He placed more significance on ovular morphology. He thought the strobiloid type of flower to be more primitive in both the dicotyledons as well as the monocotyledons. He also considered polycarpy and spiral arrangement of floral appendages more primitive than the synarpy and cyclic arrangement. Seed plants are considered under two phyla. Each phylum is thought monophyletic and derived from an unknown extinct cycads related to the Bennettitales and exhibiting affinities with the Marattiales. He derived monocotyledons from extinct stocks ancestral to Lardizabalaceae. The dicotyledons are thought originated through the order of his Proterogenes (which included Magnoliaceae) and included 29 orders under four primary subdivisions. His scheme especially regarding the monocotyledons is rather unsatisfactory except for the treatment of Amaryllidaceae as an unnatural taxon and consideration of Gramineae and Cyperaceae in separate orders. He followed Bessey's dicta.

John Hutchinson (1884-1972), formerly of the Royal Botanic Gardens, Kew, England, was a leading contemporary exponent of phylogenetic system of classification. He himself concerned primarily with the angiosperms. He considered angiosperms monophyletic and sought their origin in hypothetical angiosperms of Arber and Parkin. His two volume work was first published in 1926 and 1934 as 'The Families of Flowering Plants'. The last revised scheme appeared in 1973 after his death in 1972. His classification is based on Bessey' ideas but superimposed upon them his own ideas. In his earlier versions, the dicotyledons were divided into two subclasses *viz.*, Archichlamydeae and Metachlamydeae. The former was based on petal free or absent and the latter on petals united. The former contained 213 families belonging to 59 orders, while the latter included 51 families placed under 23 orders. The monocotyledons were categorized into three major groups *viz.*, (i) Calyciferae, (ii) Corolliferae and (iii) Glumiflorae on the basis of nature of perianth. In his final version of 1973, he transferred the family Cythraceae from his Herbaceae to the Lignosae under the order Myrtales. He accepted smaller families proposed by other authors or even expressed his own views.

In the last revision, the dicotyledons are presented by 83 orders and 349 families and monocotyledons belonged 29 orders and 69 families. Fig.2 gives the linear sequence of orders in Hutchinsons scheme as revised in 1973. The number of families included under each order is given in parenthesis.

Fig. 2. Linear sequence of orders in Hutchinson's System (1973)

In the second edition of 'The Families of Flowering Plants' in 1959 he stated that 'all that has been done in this new version is to apply them with added conviction, the result being that the fundamentally woody subphylum the Lignosae and the fundamentally herbaceous subphylum, the Herbaceae are now completely separated in the linear sequence. This indicated that he was firm on his subdivision of the dicotyledons into Lignosae and Herbaceae. He further started that "I am well aware that I may be regarded as a crank". In his opinion, this made possible to establish more natural groups of closely allied families. For instance, the Flacourtiaceae woody and the Violaceae (mostly herbaceous), which have been closely associated on account of a single character parietal placentation are now more widely separated. Many more examples are now on record such as the herbaceous Umbelliferae and woody Cornaceae and Araliaceae, the herbaceous Labiatae and woody Verbenaceae, etc. He considered that the resemblance of families such as mentioned above is the result of evolutionary convergence.

It is worth to note that in his phylogenetic tree, Hutchinson derived one order not directly from the order but from its ancestral stock. Direct derivation of one from another order is certainly not safe and acceptable. In his final version, the number of families increased to 418 from 342 in earlier version. He critically delimited them and their relationships seem to be more natural. His sound judgments with flowering plants are the result of his critical observation in the field and herbarium as well his decisions although based primarily on exomorphic features, later investigators in other related fields supported him greatly with his deep knowledge of tropical floras. He was obviously in a strong position to assess such relationships. He agreed with Bessey in that angiosperms are monophyletic and that monocotyledons evolved from a primitive dicotyledonous stock. It may be further noted that Hutchinson put less emphasis on the position of ovary in the monocotyledons than he did in the dicotyleedons. He contributed a real service by his careful and critical appraisal of family and ordinal limitations, and in his excellent artifical key to dicotyledonous families he adopted certain general principle (dicta) for the classification of flowering plants as the following:

Oswald Tippo(born in 1991), University of Illionois, projected his classification in 1942, It is based on classification of the *non-vascular* plants by Smith (1938) and the basic arrangement of *vascular* plants proposed by Eames (1936). Actually, it is a synthesis of several proposals. He emphasised vegetative characters such as plant form, branching and vascular anatomy in arriving at the natural groupings of the larger categories. The system emphasizes that Pteridophyta is a heterogeneous group. It further maintains that there is no clear demarcation between pteridophytes and spermatophytes. A sound study of paleopbotanical information and its comparison with the data from study of living plants is used to indicate their relationships. Tippo thought the Amentifereae to be recent and advanced. He also considered Magnoliales to be very primitive, Synopsis of his system is provided in the following :

Kingdom Plantae

Subkingdom-I : Thallophyta (Plants forming no embryos)

Phylum :
1. Cyanophyta
2. Euglenophyta
3. Chlorophyta
4. Chrysophyta
5. Phaeohpyta
6. Rhodophyta
7. Schizomycophyta
8. Myxomycophyta
9. Eumycophyta

Subkingdom-II : Embryohpyta (Plants forming embryos)

Phylum: Bryophytaor Atracheata (3 classes-Musci, Hepatica, Anthocerotae)

Subphylum : Tracheophyta or Tracheata
1. Psilopsida
2. Lycopsida
3. Sphenopsida
4. Pteropsida

Class :
1. Filicineae
2. Gymnospermae
3. Angiospermae

Subclass :
1. Monocotyledonae
2. Dicotyledonae

Karl Christian Mez (1866-1944) was Professor of Botany, at the University of Koenigsberg, Germany. An outline of his classification based on *serodiagnostic* data was presented in 1926. It was revised and amplified in 1936. His classification was based on his theory that relationships between the larger groups of genetically related plants could be determined by study and analysis of their protein reactions. The information obtained can be employed for groupings. This physiological approach is known as serodiagnosis. These prove fruitful especially in case of doubtful affinities. It is interesting to know that the relationships indicated by this technique concurred to a degree with those proposed by Bessey and Hutchinson, This data contributes important corroborating evidence to the formation of an ultimate system of classification.

Alfred Gunderson (1950) of the United States of America published a phylogenetic system especially of the dicotyledons in his 'Families of Dicotyledons' and the sources of information he utilized was cytology and anatomy, apart from morphology. He divided them into ten groups like Magnoliflorae, Cistiflorae, Theaflorae, Rosiflorae and such others. The first one, in his opinion, is primitive. He further divided these ten groups into 42 orders. In his treatment, he overlooked polypetalous and gamopetalous features of flowers.

Lyman Benson (1957), Professor of Botany, Pomona College, California, presented a 'Chart of Relationships of The Orders of Flowering Plants' in his book 'Plant Classification.' He did not claim it a new system of classification. He rearranged the angiospermic families of Bessey, Engler and Prantl in an essentially horizontal two-dimensional chart as in Fig.3 instead of a speculative phylogenetic tree–like of other authors. The chart was prepared considering the chart distance of orders from the Ranales or the Alismatales, number of features of flower structure, and the size of figure representing each order in proportion to species number composing it.

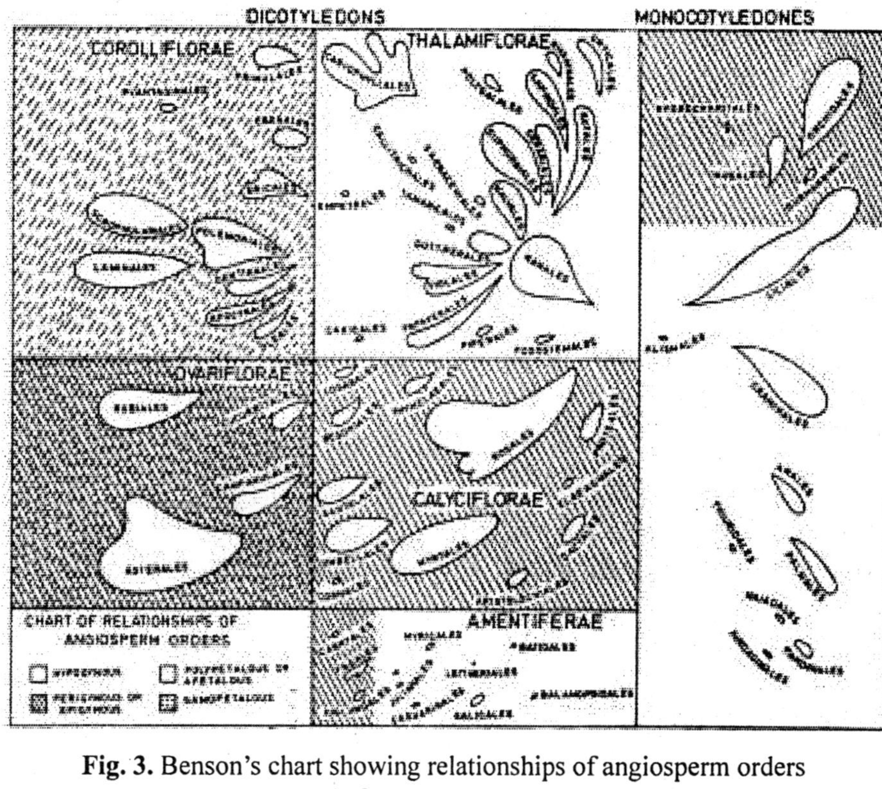

Fig. 3. Benson's chart showing relationships of angiosperm orders (After Benson, 1957)

He divided the angiosperms first into Dicotyledoneae and Monocotyedoneae. The former is subdivided into five group *viz.*, (1) Thalamiflorae (2) Corolliflorae (3) Calyciflorae, (4) Ovariflorae and (5) Amentiferae. He also gave dichotomous indented key to these groups mostly based on floral characters (given inside back cover of the book). He thought no living flowering plants derived from any other living order. Instead, he considered related orders evolved from pre- existing groups with some members similar to the species now composing the living taxa. In his opinion, the arboreal Ranales have retained more primitive features than any other group. He further opined that the ancestors of flowering plants had these characters, apart from other primitive characters some of which have disappeared and some are still persisting in other orders. The origin of angiosperms, according to him, is obscure. In general, his presentation appears appreciable, felicitious and suitable for teaching purposes, but still can not escape from criticism especially in respect of uniformity, relationships and grouping of angiospermic plants.

Armen Takhtajan, Komarov Botanical Institute of The Academy of Sciences, Leningrad (Russia), is a leading taxonomist. He proposed a preliminary phylogenetic classification in 1942 in a research paper. His system based on his 67 phylogenetic principles (1953) appeared in 1954 in Russian. It was translated into English and published in 1969 as 'Flowering Plants : Origin and Dispersal' and attracted attention worldwide. A new version of his system appeared in 1980 in a journal 'Botanical Review'. It is Hallier's contribution (1905, 1908 and 1912) that stimulated him to prepare a synthetic evolutionary classification of Angiosperms. The version appeared in 1980 is more satisfactory. The treatment particularly for *petaloid* monocotyledonous taxa is detailed and justifiable in their placement and interrelationships. In general, he derives the monocotyledons and the Nymphaeales from a hypothetical common dicotyledonous ancestor. with wood devoid of vessels and monocolpate pollen.

He considerd Magnoliophyta (Angiosperms) to be *monophyletic*. He divided it into two classes *viz.,* Magnoliopsida (Dicotyledons) and Liliopsida (Monocotyledons). The latter are thought to have derived from fomer primitive one. The former are further categorized in the new version of 1980 into seven subclasses *viz.,* (i) Magnoliidae, (ii) Ranunculadae, (iii) Hamamelidae, (iv) Caryophyllidae, (v) Dilleniidae, (vi) Rosidae and (vii) Asteridae Likewise, Liliopsida are divided into three subclasses *viz.,* (i) Alismatidae, (ii) Lilidae and (iii) Arecidae. A subclass Commelinidae recognised in the year 1969, is abolished in 1980 version.

He thought the subclass Magnoliidae as the basic group from which all other subclasses have evolved. Hamemelidae contains most of the families of the traditional amentiferae, a group mostly *anemophilous*. Floral reduction is associated with *wind pollination* in this alliance. It is suggestive of their early adaptation to temperate conditions. A strong tendency for basal or free central placentation and mostly associated with the occurrence of betalains in their flowers earmark the subclass Caryophyllidae. The subclass Rosidae and Dillenidae are parallel groups. *Centripetal* staminal development and a strong trend toward the development of uniovulate ovarian loculii and nectariferous disc do help delimit the Rosidae. Similarly, *centrifugal* staminal development, usually ovarian loculi with ovules more than solitary and rare occurrence of nectariferous disclead one to identify the subclass Dillenidae. Asteraceae are, however, the specialized amongst all other subclasses of Magnoliopsida His subclass of Liliopsida is also remarkable by certain characteristics of their own. The subclass Alismatidae includes all aquatic plants, whereas other subclasses are generally terrestrial.The subclass Liliidae is characterized by much more common growth forms such as bulbs, tubers, corms or rhizomes, apart from their *entomophilous* flowers. The subclass Commelinidae (which is but abolished in new version of Takhtajan's scheme) exhibits floral reduction e.g. reduction of ovule number, stamen number, and size of outer floral appendages, etc, besides *anemophily*. The subclass Arecidae usually have arboreal plants bearing gigantic petiolate foliage and giant crowded inflorescences.

His subclasses are divided into superorders, orders and families. Intercalation of category of classification *viz.,* superorder is a new step in his scheme. The subclass Commelinidae once recognized by him with inner petaline perianth is merged within the subclass Liliidae, whereas the latter have generally both whorls of perianth *petaloid*. In his version in 1959, he discussed about the close affinities of these classes and in later verdict of 1980 he actually merged them into a composite subclass Liliidae. The relative position of ovary is emphasized in this alliance. In a nutshell, his system is a welcome being based on well defined evolutionary principles.

Synopsis of his system is presented in the following (numbers in the parenthesis suggest the number of families included) :

Division : Magnoliophyta or Angiospermae (419)

Class : Magnoliopsida or Dicotyledones (342)

Subclass A	:		Magnoliidae (29)
Superorder I	:		Magnolianae (23)
Order	:	1.	Magnoliales (8)
		2.	Illiciales (2)
		3.	Laurales (10)
Suborder	:		Monimineae (6)
			Chloranthineae(1)
			Lactoridineae (1)
			Lactorineae (2)
Order	:	4.	Piperales (2)
		5.	Aristolochiales (1)
Superorder II	:		Rafflesianae (2)
Order	:	6.	Rafflesiales (2)
Superorder III	:		Nymphaeanae (4)
Order	:	7.	Nymphaeales (3)
Suborder	:		Nymphaeineae (2)
			Ceratophyllineae (1)
Order	:	8.	Nelumbonales (1)
Subclass B	:		Ranunculidae (9)
Superorder IV	:		Ranunculanae (9)
Order	:	9.	Ranunculales (7)

		10. Papaverales (1)
		11. Serraceniales (1)
Subclass C	:	Hamamelididae (26)
Superorder V	:	Hamamelidanae (23)
Order	:	12. Trochodendrales (2)
		13. Cercidiphyllales (1)
		14. Eupteleales (1)
		15. Didymelales (1)
		16. Hamamelidales (6)
Suborder	:	Hamamelidineae (4)
		Buxineae (2)
Order	:	17. Eucommiales (1)
		18. Urticales (5)
		19. Barbeyales (1)
		20. Casuarinales(1)
		21. Fagales (2)
		22. Balanopales (1)
		23. Leitneriales (1)
Superorder VI	:	Juglandanae (3)
Order	:	24. Myricales (1)
		25. Juglandales (2)
Subclass D	:	Caryophyllidae (16)
Superorder VII	:	Caryophyllanae (15)
Order	:	26. Caryophyllales (14)
Suborder	:	Phytolaccineae (10)
		Caryophyllineae (2)
		Chenopodiineae (2)
Order	:	27. Polygonales (1)
Superorder VIII	:	Plumbaginanae (1)
Order	:	28. Plumbaginales (1)
Subclass E	:	Dilleniidae (78)
Superorder IX	:	Dillenianae (47)
Order	:	29. Dilleniales (2)
		30. Paeoniales (1)
		31. Theales (19)
		32. Violales (14)

		33. Begoniales (2)
		34. Capparales (5)
Suborder	:	Capparineae (3)
		Resedineae (1)
		Moringineae (1)
Order	:	35. Tamaricales (3)
Suborder	:	Tamaricineae (2)
		Fonquierineae (1)
Order	:	36. Salicales (1)
Superorder	:	Ericanae (16)
Order	:	37. Ericales (8)
		38. Ebenales (5)
		39. Primulales (3)
SuperorderX	:	Malvanae (15)
Order	:	40. Malvales (10)
		41. Euphorbiales (4)
		42. Thymelaeales (1)
Subclass F	:	Rosidae (137)
Superorder XII	:	Rosanae (41)
Order	:	43. Saxifragales (24)
Suborder	:	Cunoniineae (4)
		Pittosporineae (9)
		Saxifragineae (11)
Order	:	44. Rosales (3)
		45. Fabales (1)
		46. Cannarales (1)
		47. Podostemales (1)
		48. Nepenthales (1)
Superorder XIII	:	Myrtanae (14)
Order	:	49. Myrtales (14)
Suborder	:	Myrtineae (11)
		Haloragineae (1)
		Rhizophorineae (1)
		Lecythidineae (1)
Superorder XIV	:	Rutanae (42)
Order	:	50. Rutales (15)

Suborder	:	Rutaneae (14)
		Coriariineae (1)
Order	:	51. Sapindales (13)
		52. Geraniales (8)
Suborder	:	Linineae (3)
		Geraniineae (2)
		Balsaminineae (2)
		Limnanthineae (1)
Order	:	53. Polygalales (6)
Superorder XV	:	Aralianae (12)
Order	:	54. Cornales (10)
		55. Araliales (2)
Superorder XVI	:	Celastranae (27)
Order	:	56. Celastrales (15)
Suborder	:	Incacinineae (7)
		Celastrineae (8)
Order	:	57. Santalales (6)
Suborder	:	Santalineae (4)
		Loranthineae (2)
Order	:	58. Balanophorales (2)
		59. Rhamnales (3)
		60. Elaeagnales (1)
Superorder XVII	:	Proteanae (1)
Order	:	61. Proteales (1)
Subclass G	:	Asteridae (47)
Superorder XVIII	:	Gentiananae (15)
Order	:	62. Gentianales (8)
		63. Oleales (1)
		64. Dipsacales (5)
		65. Loasales (1)
Superorder XIX	:	Lamianae (26)
Order	:	66. Polemoniales (7)
Suborder	:	Convolvulineae (2)
		Polemoniineae (1)
		Boraginineae (4)

Order	:	67. Lamiales (3)
		68. Scrophulariales (16)
Suborder	:	Solanineae (2)
		Scrophulariineae (13)
		Hippurineae (1)
Superorder XX	:	Asteranae (6)
Order :		69. Campanulales (4)
Suborder	:	Campanulineae (3)
		Goodeniineae (1)
Order	:	70. Calycerales (1)
		71. Asterales (1)
Class	:	Liliopsida or Monocotyledones (77)
Subclass A	:	Alismatidae (14)
Superorder I	:	Alismatanae (14)
Order	:	1. Alismatales (4)
Suborder	:	Butomineae (1)
		Alismatineae (2)
		Hydrocharitineae (1)
Order	:	2. Najadales (10)
Suborder	:	Aponogetonineae (1)
		Scheuchzeriineae (1)
		Potamogetonineae (6)
		Zosterineae (1)
		Najadineae (1)
Subclass B	:	Liliidae (57)
Superorder II	:	Triuridanae (1)
Order	:	3. Triuridales (1)
Superorder III	:	Lilianae (33)
Order	:	4. Liliales (23)
Suborder	:	Lilineae (10)
		Asphodelineae (4)
		Asparagineae (2)
		Iridineae (2)
		Haemodorineae (3)
		Pontederiineae (1)
		Philydrineae (1)

Order	:	5. Smilacales (6)
		6. Burmanniales (2)
		7. Orchidales (1)
		8. Bromeliales (1)
Superorder IV	:	Juncanae (3)
Order	:	9. Juncales (2)
		10. Cyperales (1)
Superorder V	:	Commelinanae (12)
Order	:	11. Commelinales (4)
Suborder	:	Xyridineae (2)
		Commelinineae (2)
Order	:	12. Eriocaulales (1)
		13. Restionales (5)
		14. Hydatellales (1)
		15. Poales (1)
Superorder VI	:	Zingiberanae (8)
Order	:	16. Zingiberales (8)
Subclass C	:	Arecidae (6)
Superorder VII	:	Arecanae (4)
Order	:	17. Arecales (1)
		18. Cyclanthales (1)
		19. Pandanales (1)
		20. Typhales (1)
Superorder VIII	:	Aranae (2)
Order	:	21. Arales (2)

Arthur Cronquist (1919-1992), New York Botanical Gardens (USA), was greatly inspired by his mentor Professor Ray J. Davis after his association with the then University of Idaho. He was a great botanist and his contributions in plant systematics is of lasting significance to the plant science. He died on 22nd March 1992 while studying plant specimens in the herbarium of Brigham Young University in Utah. He proposed a system of classification for Dicotyldedons in 1957. Later, it was revised in collaboration with two other botanists in 1966. Its final modification appeared in 'Evolution and Classification of Flowering Plants' (1968a). It was again revised in his book 'An Integatul System of Classification of Flowering Plants' in 1981 and The Evolution and Classification of Flowering Plants in 1988.

His system of classification which appeared in 1968 deserve special attention. He defined taxonomy as 'a study aimed at producing a system of classification of organisms which best reflects the totality of their similarities and differences'. We know other definition of taxonomy (Chapter-I). Definition given by Cronquist (1968a) itself indicates synthetic nature of his system. Obviously, it is based on literature then available from sources such as anatomy (vegetative and floral), embryology palynolgy, phytochemistry, cytology, serology, ultrastructure, apart form principal source of exomorphology. His scheme is a elaboration of Bessey's system (1915) and also a refinement over Takhtajan's system (1964).

He proposed to arrange the families and orders of Magnoliophyta (Angiosperms) avoiding intermediate category 'superorder', *sensu* Takhtajan, into total ten subclasses of which six belonged to the Magnoliatae and four to Liliatae (Fig.4).This grouping of subclass reflected some idea of relationship developed over the years by many phylogenists. It is certainly a major advance in the conceptual scheme. The circumscription and nomenclature of the subclasses find nearly parallel to Takhtajan's system except Magnolidae wherein Cronquist merged his Ranunculidae in Magnolidae. Thus Cronquist's Magnolidae consists of nearly all of which has been called the Ranalian complex and few other families. These subclasses are characterized by Cronquist as in Takhajan's system (as discussed earlier). The generalities and critical tendencies are similar. Cronquist (1968a) himself agreed that this similarity reflects a community of interest and outlook, bolstered by a mutual exchange of views over the past ten years. He also claimed that their ideas were initially developed independently. In Cronquist's opinion, the similarity of their views reflects the necessities of the present state of knowledge.

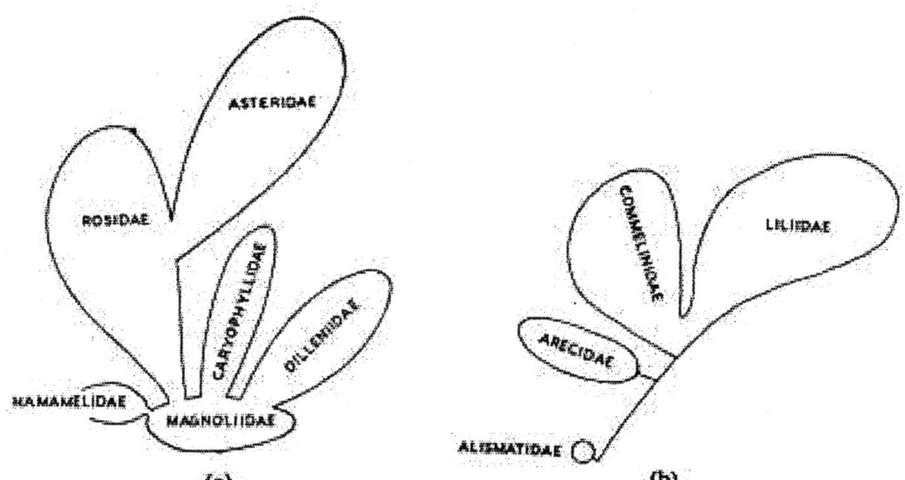

Fig. 4. Putative relationships of subclasses of flowering plants (a) Magnoliatae, (b) Liliatae by A.Cronquist (1968)

The Magnolidae are regarded as the basal complex, from which the other angiosperms have been derived. Takhtajan's four subclasses are also adopted by Cronquist but with a major change in the contents of the Liliidae and Commelinidae. The order such as

Bromeliales, Zingiberales, Juncales and Cyperales are refereed to the Cronquist's Commelinidae which are included under the Liliidae by Takhajan. He also gave clear synoptical keys for the families, orders and subclasses. They are conceptual aids and not meant for identification. This has not been presented in any system published earlier. The keys summarise what sorts of things are included in the various groups. Vast literature has been utilized to render the system synthetic and more natural. It has been widely accepted and found satisfactory. Unnecessary reliance on single character e.g. *centrifugal* stamens, *free central* placentation and submergence of some group into another has at some levels, hampered the otherwise natural classification. Synopsis of his system is presented in the following (Numbers in the parenthesis suggest the number of families included) :

Division: Magnoliophyta

A. Class: Magnoliatae

I.	Subclass	:		Magnoliidae (34)
	Order	:	1.	Magnoliales (19)
			2.	Piperales (3)
			3.	Aristolochiales (1)
			4.	Nymphaeales (3)
			5.	Ranunculales (8)
II.	Subclass	:		Hamamelidae (23)
	Order	:	6.	Trochodendrales (2)
			7.	Hamamelidales (6).
			8.	Eucomiales (1)
			9.	Urticales (5)
			10.	Leitneriales (1)
			11.	Juglandales (3)
			12.	Myricales (1)
			13.	Fagales (3)
			14.	Casuarinales (1)
III.	Subclass	:		Caryophyllidae (14)
	Order	:	15.	Caryophyllales (11)
			16.	Batales (1)
			17.	Polygonales (1)
			18.	Plumbaginales (1)
IV.	Subclass	:		Dilleniidae (69)
	Order	:	19.	Dilleniales (3)

20. Theales (13)
21. Malvales (6)
22. Lecythidales (1)
23. Serraceniales (3)
24. Violales (21)
25. Salicales (1)
26. Capparales (5)
27. Ericales (7)
28. Diapensiales (1)
29. Ebenales (5)
30. Primulales (3)

V. Subclass : Rosidae (108)

Order : 31. Rosales (17)

32. Podostemales (1)
33. Haloragales (4)
34. Myrtales (13)
35. Proteales (2)
36. Cornales (6)
37. Santalales (10)
38. Rafflesiales (3)
39. Celastrales (10)
40. Euphorbiales (5)
41. Rhamnales (3)
42. Sapindales (17)
43. Geraniales (5)
44. Linales (3)
45. Polygalales (7)
46. Umbellales (2)

VI. Subclass : Asteridae (43)

Order : 47. Gentianales (4)

48. Polemoniales (8)
49. Lamiales (5)
50. Plantaginales (1)
51. Scrophulariales (12)
52. Campanulales (6)
53. Rubiales (1)

			54. Dipsacales (5)
			55. Asterales (1)
B.	Class	:	Liliatae
I.	Subclass	:	Alismatidae (14)
	Order	:	1. Alismatales (3)
			2. Hydrocharitales (1)
			3. Najadales (8)
			4. Triuridales (2)
II.	Subclass	:	Commelinidae (25)
	Order	:	5. Commelinales (4)
			6. Eriocaulales (1)
			7. Restionales (5)
			8. Juncales (2)
			9. Cyperales (2)
			10. Typhales (2)
			11. Bromeliales (1)
			12. Zingiberales (8)
III.	Subclass	:	Arecidae (5)
	Order	:	13. Arecales (1)
			14. Cyclanthales (1)
			15. Pandanales (1)
			16. Arales (2)
IV.	Subclass	:	Liliidae (17)
	Order	:	17. Liliales (13)
			18. Orchidales (4)

Rolf M.T.Dahlgren (1932-1987), Professor, University of Copenhagen, Denmarks, presented a method of illustrating angiosperm system as an imaginary phylogenetic shrub in transaction in his 'A Textbook of Angiosperm Taxonomy' in 1974. He thereby proposed a phylogenetic system of classification and was revised in 1975 and again completely revised in 1980. In his last verdict, still he claimed his system as provisional.

He considered the group of angiosperms as class Magnoliopsida which is further categorised into two subclasses *viz.,* Magnoliidae (Dicotyledons) and Liliidae (Monocotyledons). The former is divided into 24 superorders, 80 orders and 346 families, whereas in the latter he recognised seven superorders, 26 orders and 92 families.

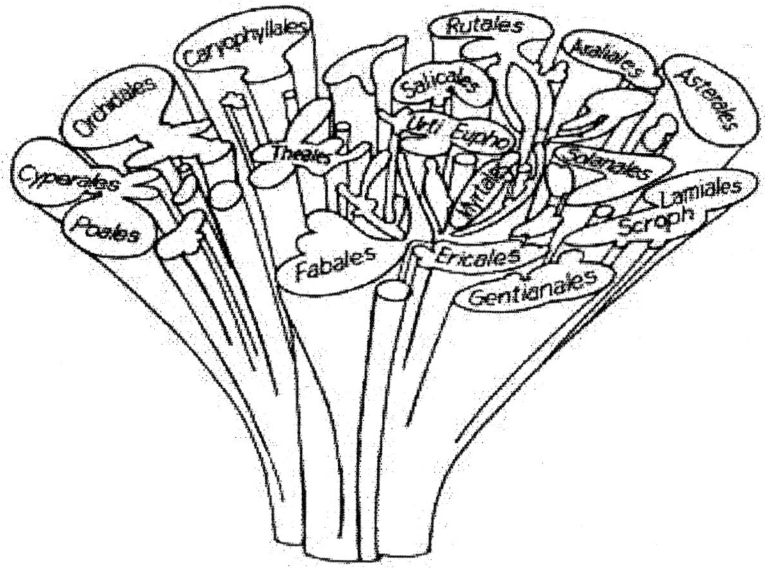

Fig. 5. Cross-section of Dahlgren's phylogenetic shrub. (Reproduced from Plant Systematics and Evolution ed. by. K.Kubitzki 1977)

His phyogenetic shrub (Fig.5) depicts the placement of variously shaped figures which stand for orders close to one another. The size of the each figure depends upon the number of species in the respective order. The figures exhibit extensions in different directions. These are suggestive of characters within them. This shrub reflects their presumed phylogenetic relationships. In its construction, an attempt has been made to accommodate information from different sources such as anatomy, embryology, phytochemistry, etc. The system is thus synthetic but agrees with most of the major features of those of Takhtajan and Cronquist. He considers angiosperms to be monophyletic in origin. Some characteristic features of angiosperms such as 8-nuleate embryosac, secondary endosperm, occurrence of *companion* cells in the *phloem*, etc., in his opinion, evolved together. From cladistic point of view, the division of class Magnoliopsida into two subclasses *viz.,* Magnoliidae and Lillidae appear haphazard. Nevertheless, the Liliidae may stand out as a possibly *monophyletic* group by virtue of features like characteristic protein bodies in the sieve tube plastids and solitary cotyledon.

Synopsis of Dahlgren's system of classification (1980) is provided below (figures in parentheis stand for the number of families included):

Class : Magnoliopsida
Subclass : Magnoliidae (346)
Superorder : Magnoliiflorae (24)
Order : Annonales (4)
Illiciales (2)

		Laurales (8)
		Nelumbonales (1)
		Aristolochiales (1)
		Refflesiales (2)
		Magnoliales (6)
Superorder	:	Nymphaeaflorae (5)
Order	:	Nymphaeales (3)
		Piperales (2)
Superorder	:	Ranunculiflorae (10)
Order	:	Papaverales (2)
		Ranunculales (8)
Superorder	:	Caryophylliflorae (9)
Order	:	Caryophyllales (9)
Superorder	:	Polygoniflorae (1)
Order	:	Polygonales (1)
Superorder	:	Malviflorae (32)
Order	:	Paeoniales (1)
		Elaeagnales (1)
		Dilleniales (1)
		Malvales (13)
		Plumbaginales (2)
		Thymelaeales (1)
		Urticales (6)
		Euphorbiales (6)
		Rhamnales (1)
Superorder	:	Violiflorae (14)
Order	:	Violales (13)
		Salicales (1)
		Tamaricales (2)
		Capparales (7)
		Salvadorales (1)
Superorder	:	Theiflorae (2)
Order	:	Theales (17)
		Droserales (3)
Superorder	:	Primuliflorae (9)
Order	:	Ebenales (4)

	Primulales (5)
Superorder :	Rosiflorae (39)
Order :	Trochodendrales (4)
	Hamamelidales (4)
	Fagales (3)
	Balanopales (1)
	Juglandales (2)
	Myricales (1)
	Casuarinales (1)
	Buxales (2)
	Cunoniales (8)
	Saxifragales (7)
	Gunnerales (1)
	Rosales (3)
	Malales (2)
Superorder :	Podostemiflorae (1)
Order :	Podostemales (1)
Superorder :	Fabiflorae (3)
Order :	Fabales (3)
Superorder :	Proteiflorae (1)
Order :	Proteales (1)
Superorder :	Myrtiflorae (14)
Order :	Myrtales (12)
	Haloragales (1)
	Rhizophorales (1)
Superorder :	Rutiflorae (45)
Order :	Rutales (6)
	Sapindales (15)
	Balsaminales (1)
	Polygalales (5)
	Geraniales (16)
	Tropaeolales (2)
Superorder :	Santaliflorae (13)
Order :	Celastrales (5)
	Vitidales (1)
	Santalales (7)

Superorder : Balanoflorae (2)
Order : Balanophorales (2)
Superorder : Araliiflorae (6)
Order : Pittosporales (3)
Araliales (3)
Superorder : Asteiflorae (4)
Order : Camanulales (3)
Asterales (1)
Superorder : Solaniflorae (13)
Order : Solanales (7)
Boraginales (6)
Superorder : Corniflorae (49)
Order : Fouquieriales (1)
Serraceniales (1)
Ericales (10)
Eucomiales (1)
Cornales (29)
Dipsacales (7)
Superorder : Loasiflorae (1)
Order : Loasales (1)
Superorder : Gentianiflorae (8)
Order : Goodeniales (1)
Gentianales (6)
Oleales (1)
Superorder : Lamiflorae (23)
Order : Scrophulariales (18)
Hydrostachyales (1)
Hipuridales (1)
Lamiales (3)
Subclass : Liliidae (92)
Superorder : Alismatiflorae (12)
Order : Hydrocharitales (3)
Zosterales (8)
Alismatales (1)
Superorder : Triuridiflorae (1)
Order : Triuridales (1)

Superorder	:	Ariflorae (2)
Order	:	Arales (2)
Superorder	:	Liliiflorae (53)
Order	:	Dioscoreales (5)
		Asparagales (31)
		Hydatellales (1)
		Liliales (8)
		Burmanniales (3)
		Philydrales (1)
		Orchidales (3)
		Velloziales (1)
		Bromeliales (1)
		Haemodorales (1)
		Pontederiales (1)
		Typhales (2)
Superorder	:	Zingiberiflorae (8)
Order	:	Zingiberales (8)
Suerorder	:	Commeliniflorae (13)
Order	:	Commelinales (2)
		Eriocaulales (2)
		Juncales (1)
		Cyperales (1)
		Poales (5)
Superorder	:	Areciflorae (3)
Order	:	Arecales (1)
		Cyclanthales (1)
		Pandanales (1)

Robert F. Thorne propounded certain guiding principles of angiosperm phylogeny in 1958 and 1963. He then presented a synopsis of his classification in 1968. Comprehensive classification appeared in 1976. In this attempt, he included 21 superorders, 50 orders, 74 suborders, 32 families and 432 subfamilies. It was realigned especially in respect of families in 1977. The nomenclature of the class Angiosperms was adopted as Annonopida. He emphasized relationships more than differences. He claimed his system different from those of Cronquist (1968), Takhtajan (1969) and Hutchinson, (1973). However, the main theme of his system appears the same as proposed by Cronquist (*loc.cit.*) and Takhtajan (*loc.cit.*). He further reiterated his claim in that his taxa are more inclusive than those of others with more disintegrative tendencies. He took under consideration all available information as the other contemporary authors. It is pertinent to know the comments made by Merxmuller (1977) on Thorne's scheme. He stated : 'At all events this inflated classification is a perfect means for confusing non- taxonomists by a fashionable application of non- existing nomenclatural rules to these higher categories'. Thorne again made refinements in his system first in 1983 and then in 1992. The diagram (Fig.6) depicts evolutionary interrelationships of angiosperms (1983). He made change in his system in respect of nomenclature of Theanae as Dillenianae, Cornanae as Aralianae, Gentiannae as Lamianae, etc. He made certain removal, reduction, expansion, elevation and conservation of spelling of some taxa included.

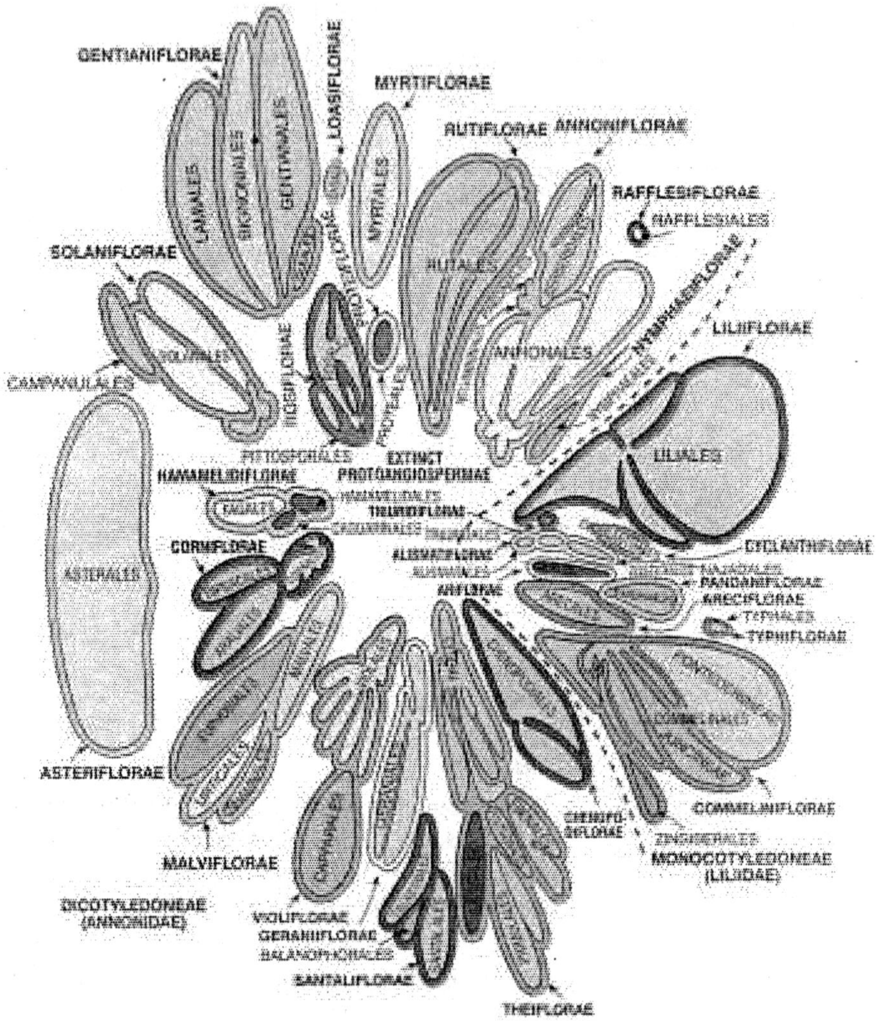

Fig. 6. A diagrammatic representation of the evolutionary interrelationships of the extinct angiosperms, as seen by Thorne (1983).

Following is the synopsis of Thorne's system as appeared in 1983 up to orders (Number in parenthesis indicate number of families):

Class	:		Angiospermae (Annonopsida)
Subclass 1	:		Dicotyledoneae (Annonidae)
Superorder 1	:		Annoniflorae
Orders	:	1.	Annonales (23)
		2.	Nelumbonales (2)
		3.	Paeoniales (2)
		4.	Berberidales (8)
Superorder 2	:		Nymphaeiflorae
Order	:	5.	Nymphaeales (2)
Superorder 3	:		Rafflesiflorae
Order	:	6.	Rafflesiales (2)
Superorder 4	:		Theiflorae
Orders	:	7.	Theales (30)
		8.	Ericales (3)
		9.	Fouquieriales (2)
		10.	Ebenales (4)
		11.	Primulales (3)
		12.	Polygonales (1)
Superorder 5	:		Chenopodiflorae
Orders	:	13.	Chenopodiales (14)
		14.	Geraniales (17)
Superorder 6	:		Celastriflorae
Order	:	15.	Celastrales (3)
Superorder 7	:		Santaliflorae
Orders	:	16.	Santalales (7)
		17.	Balanophorales (2)
Superorder 8	:		Violiflorae
Orders	:	18.	Violales (16)
		19.	Capparales (4)
Superorder 9	:		Malviflorae
Orders	:	20.	Malvales (12)
		21.	Urticales (3)
		22.	Rhamnales (2)
		23.	Euphorbiales (6)

Superorder 10	:		Rutiflorae
Order	:	24.	Rutales (25)
Superorder 11	:		Proteiflorae
Order	:	25.	Proteales (1)
Superorder 12	:		Hamamelidiflorae
Orders	:	26.	Hamamelidales (7)
		27.	Casuarinales (1)
		28.	Fagales (2)
Superorder 13	:		Rosiflorae
Orders	:	29.	Rosales (19)
		30.	Pittosporales (13)
Superorder 14	:		Loasiflorae
Order	:	31.	Loasales (1)
Superorder 15	:		Myrtiflorae
Order	:	32.	Myrtales (11)
Superorder 16	:		Gentianiflorae
Orders	:	33.	Oleales (2)
		34.	Gentianales (7)
		35.	Bignoniales (9)
		36.	Lamiales (3)
Superorder 17	:		Solaniflorae
Orders	:	37.	Solanales (7)
		38.	Campanulales (3)
Superorder 18	:		Corniflorae
Orders	:	39.	Cornales (10)
		40.	Araliales (3)
		41.	Dipsacales (5)
Superorder 19	:		Asteriflorae
Order	:	42.	Asterales (1)
Subclass 2	:		Monocotyledoneae (Liliidae)
Superorder 1	:		Liliiflorae
Orders	:	1.	Liliflorae
		1.	Liliales (10)
Superorder 2	:		Triuridiflorae
Order	:	2.	Triuridales (1)
Superorder 3	:		Alismatiflorae

Orders	:	3.	Alismatales (3)
		4.	Zosterales (7)
		5.	Najadales (1)
Superorder 4	:		Ariflorae
Order	:	6.	Arales (2)
Superorder 5	:		Cyclanthiflorae
Order	:	7.	Cyclanthales (1)
Superorder 6	:		Pandaniflorae
Order	:	8.	Pandanales (1)
Superorder	:	7.	Areciflorae
Order	:	9.	Arecales (1)
Superorder 8	:		Typhiflorae
Order	:	10.	Typhales (1)
Superorder 9	:		Commeliniflorae
Orders	:	11.	Commelinales (4)
		12.	Zingiberales (8)

(ii) Systems of Englerian School: Alexander Braun (1805-1876), Professor of Botany (Germany), is known for his significant contributions to various fields of plant science. He attempted at the arrangement of the plants of the Botanic Garden according to a natural system of his own. Braun contributed during 1859 – 1864 to a local German Flora (Ascherson's Flora der Provinz Brandenurg). A synopsis of his classification of the plant kingdom greatly influenced August Wilhem Eichler and Adlof Engler. Monocotyledons were classified by him so as to show a progression from the more simple to the more complex. He began with Lemanceae and ended with Orchidaceae. He considered naked *unisexual* flowers of Lemnaceae most primitive and arragned other families to show gradual complexity. In case of Dicotyledons, he commenced his classification with the Apetalae, plants with only one whorl of perianth members or with perianth undifferentiated into calyx and corolla. It was then followed by the *Sympetalae* (*Gamopetalae*) in which he started with the Lentibulariaceae. This group was then followed lastly by *Eleutheropetalae* (*Polypetalae*) which ended with the Leguminosae.

August Wilhem Eichler (1839-1887) published his system elaborately in the third edition of his 'Syllabus der Vorlesungen uber Specialle and Medicinisch- pharmaceutishe Botanik' in 1883. It is usually accounted as the earlist phylogentic classification of plants. During 1875 – 1878, he also proposed the rudiments of the first system based on a approach to genetical relationships between plants in his 'Blutendiagramme.' In his system, the Crytogamae and Phanerogamae were still retained as the two major groups of plants. The former were further divided into division i) Thallophyta (Algae and Fungi) ii) Hepaticae and Musci, Bryophyta and (iii) Pteridophyta (Equisetinae, Lycopodineae and Filicinae). The plant groups in parenthesis represent a class. The Phanerogamae were subdivided into division i) Gymmnospermae and ii) Angiospermae. It was for the first time these were classified into these two groups. Division Angiospermae contained class Monocotyleae and Dicotyleae. The latter was again recognsied as subclass Choripetalae and Sympetalae. It is to be noted that the gymnosperms preceded the angiosperms and the monocotyledons preceded the dicotyledons. It implies that earlier group in each pair was the more primitive or ancient. The dicotyledons were grouped into two as mentioned earlier, the old Monochlamydeae (Apetalae) was dispersed among them, and the catkin-bearing plants (Amentiferae) were considered at the beginning of the dicotyledons. This is suggestive of a fact that they were thought to be the most primitive in his Blutendiagramme (1875 - 78). The gamopetalous taxa were treated before the polypetalous ones but reversed this position in the 'Kiel Syllabus (1876) and his 'Berlin Syllabus' (1880). In the last, the Dicotyledons began with the Cupuliferae and ended with the Compositae.

H.G. Adolf Engler (1844 - 1930), the great German systematist, dominated the world of plant taxonomy at the turn of 20th century. He adopted Eichler's system with some modifications His system appeared first in a relatively minor publication 'Guide To Breslau Botanic Garden' in 1886. (Fuhrer durch den koniglich botanischen Gaerten der Universitat zu Breslau, 1886). It was then incorporated into his three series **(i)** Die Naturlichen Pflanzenfamilien (1887 - 1915) with Karl A.E. Prantl, second edition, 1924, incomplete). It comprised a revision down to the generic level of all plant groups (including subgeneric and sectional). **(ii)** Syllabus der Pflanzenfamilien (First edition, 1892, latest complete 12th edition by H. Melchior *et al.* in 1964 in two volumes). It comprised a revision of all plants down to family level (including subfamily and tribe), **(iii)** Das Pflanzenrich, 1936 (11th), edition by Engler and also written by many authors). It attempted to provide an up – to- date 'Species Plantarum' covering all plants but not completed.

Angiosperm families, in Englerian system, were arranged according to their increasing complexity of reproductive structures, the simplest of them being kept at the base. They were classified into two classes, the monocotyledons and Dicotyledons. The former was considered more primitive than the latter and hence placed before the latter. Angiosperms were classified into total 303 families. The monocotyledons included 11 orders and 45 families beginning with the *unisexual* and *apetalous* Pandanales which were then followed by the aquatic Helobiae and others. It ended with the highly specialized Microspemae. The dicotyledons were divided into two groups *viz.,* **(i)** Archichlamydeae, which included 33 orders of perianthless as well as polypetalous flowers, and **(ii)** Sympetalae which included 11 orders of gamopetalous flowers. The orders with perianthless unisexual flowers such as Verticillatae, Fagales Salicales, etc. were kept at the starting of the dicotyledons followed by *polypetalous* orders such as Centrospermae, Ranales, etc. and ended in the *sympetalous* orders like Contortae, Tubiflorae, etc. Compositae was regarded as a highly specialized family.

Several changes and rearrangements are made recently by Melchior (1964) in the 12th edition of Engler's Syllabus der Pflanzenfamilien'. For example, the dicotyledons are considered more primitive than the monocotyledons and hence kept before them. Various orders are splitted and therefore their number increased to 64 from earlier 55. The number of families similarly increased to 344 from earlier 303. Thus the system has been brought at par with the other contemporary classifications. The essence of Englerian concept of primitive flower is still maintained. Apart from morphological and embryological information, data from different fields is incorporated which changed the systematic position of some orders and families. In the following, synopsis of Melchior's scheme (1964), latest version of Engler's classification of angiosperms, is provided (Number of families in each category are indicated in paranthesis) :

(Divisions I to XVI include all other plants)

XVII Division	:	Angiospermae
A. Class	:	Dicotyledoneae (291)
1. Subclass	:	Archichlamydeae (227)
Order	:	Casuarinales (1)
		Juglandales (2)
		Balanopales (1)
		Leitnerales (2)
		Salicales (1)
		Fagales (2)
		Urticales (5)
		Proteales (1)
		Santalales (7)
Suborder	:	Santalineae (6)
		Loranthineae (1)
Order	:	Balanophorales (1)
		Medusandrales (1)
		Polygonales (1)
		Centrospermae (12+1)
Suborder	:	Phytolaccineae (6)
		Portulacineae (2)
		Caryophyllineae (1)
		Chenopodiineae (3)

Family of doubtful relationship-Didieriaceae

Order	:	Cactales (1)
		Magnoliales (22)
Suborder	:	Magnoliineae (8)
		Schisandrineae (2)

		Austrobaileyineae (3)
		Tetracentrineae (2)
		Eupteleineae (1)
		Cercidiphyllineae (1)
Order	:	Ranunculales (7)
Suborder	:	Ranunculineae (5)
		Nymphaeineae (2)
Order	:	Piperales (4)
		Aristolochiales (3)
		Guttiferales (16)
Suborder	:	Dilleniineae (6)
		Ochnineae (4)
		Theineae (5)
		Ancistrocladineae (1)
Order	:	Sarraceniales (3)
		Papaverales (6)
Suborder	:	Papaverineae (1)
		Capparineae (3)
		Resedineae (1)
		Moringineae (1)
Order	:	Batales (1)
		Rosales (19)
Suborder	:	Hamamelidineae (3)
		Saxifragineae (10)
		Rosineae (3)
		Leguminosineae (3)
Order	:	Hydrostahyales (1)
		Podostemales (1)
		Geraniales (9)
Suborder	:	Limnanthineae (1)
		Geraniineae (6)
		Euphorbiineae (2)
Order	:	Rutales (12)
Suborder	:	Rutineae (7)
		Malpighiineae (3)
		Polygalineae (2)

Order	:	Sapindales (10)
Suborder	:	Coriariineae (1)
		Anacardiineae (1)
		Sapindineae (7)
		Balsamineae (1)
Order	:	Juliniales (1)
		Celastrales (13)
Suborder	:	Celastrineae (10)
		Buxineae (1)
		Icacinineae (2)
Order	:	Rhamnales (3)
		Malvales (7)
Suborder	:	Elaeocarpineae (1)
		Sarcolaenineae (1)
		Malvineae (4)
		Scytopetalineae (1)
Order	:	Thymelaeales (5)
		Violales (20)
Suborder	:	Flacourtiineae (9)
		Cistineae (4)
		Tamaricineae (3)
		Caricineae (1)
		Loasineae (1)
		Begoniineae (2)
Order	:	Cucurbitales (1)
		Myrtiflorae (17)
Suborder	:	Myrtineae (15)
		Hippuridineae (1)
		Cynomoriineae (1)
Order	:	Umbelliflorae (7)
2. Subclass	:	Sympetalae (64)
Order	:	Diapensiales (1)
		Ericales (5)
		Primulales (3)
		Plumbaginales (1)
		Ebenales (7)

Suborder	:	Sapotineae (2)
		Ebenineae (5)
Order	:	Oleales (1)
		Gentianales (7)
		Tubiflorae (26)
Suborder	:	Convolvulineae (3)
		Boraginineae (3)
		Verbenineae (3)
		Solanineae (15)
		Myoporineae (1)
		Phrymineae (1)
Order	:	Plantaginales (1)
		Dipsacales (4)
		Campanulales (8)
B. Class	:	Monocotyledoneae (53)
Order	:	Helobiae (9)
Suborder	:	Alismatineae (2)
		Hydrocharitineae (1)
		Scheuchzerineae (1)
		Potamogetonineae (5)
Order	:	Triuridales (1)
Order	:	Liliiflorae (17)
Suborder	:	Liliineae (11)
		Pontederineae (1)
		Iridineae (2)
		Burmanniineae (2)
		Philydrineae (1)
Order	:	Juncales (2)
		Bromeliales (1)
		Commelinales (8)
Suborder	:	Commelinineae (4)
		Eriocaulineae (1)
		Restionineae (2)
		Flagellariineae (1)
Order	:	Graminales (1)
		Principes (1)

Synanthae (1)
Spathiflorae (2)
Pandanales (3)
Cyperales (1)
Scitamineae (5)
Microspermae (1)

A word about Engler and Prantl's system of classification as appeared in 'Die Naturalichen Pflanzenfamilien' (1897- 1915) appears pertinent at this juncture. It consisted of 20 volumes and provided keys and descriptions for all the then known genera of all plant groups. They divided plant kingdom into 14 divisions of which the first 11 divisions are devoted for algae, one division for fungi, one for bryophytes, pteridophytes and last one *viz.*, Embroyphyta Siphonogama included both seeded plants subdivision Gymnospermae and subdivision Angiosermae. Subdivision Angiospemaae was classified into classes *viz.*, Monocotyledonae and (ii) Dicotyledonae. The former included 11 orders and 45 families. It began with the order Pandanales and ended with the Microspermae. Likewise, Dicotyledonae were classified further into subclass *viz.*, (i) Archichlamydeae (ii) Metachlamydeae (Sympetalae). This subdivision consisted of 40 orders and 284 families. It began with the order Verticillatae followed by other *apetalous* orders and then by *polypetalous* and *gamopetalous* orders.

Alfred B. Rendle (1865- 1938) was the keeper of British Museum of Natural History, London, during 1906- 1930. He published a classification (1904, 1925, 1930, 1938). His system is basically Englerian, particularly of Engler and Prantl. It is one of convenience rather than one of modern phyogenetic significance. He treated the amentiferous and apetalous plants among the primitive Dicotyledons. He considered the grasses subordinate to the lilies. Also he did not accept the primitiveness of the Ranales to other dicotyledons. He considereed the monocotyledons as more primitive than the Dicotyledons. He treated the Palmae not as a separate order Principes but merged under Spathiflorae along with the Araceae. He thought the Palmae primitive among the Spathiflorae. After omission of some families, he retained only 32 families under nine orders.

He grouped the dicotyledons in three grades depending upon differentiation of floral structure. Plants in the first grade *viz.*, Monochlamydae included orders with a simple type of flower. In the second grade *viz.*, Dialypetalae, the orders were arranged in ascending sequence beginning with the Ranales. The third grade *viz.*, Sympetalae included orders that represented still higher grade of floral development which have originated from different dialypetalous groups. Woody habit, in his opinion, was primitive than the herbaceous. He laid emphasis on wind and insect pollination in his phyletic evaluation. Clarity and completeness of description, discussions of relationships and historical presentation of earlier classification are some of his significant features of two volume work 'The Classification of Flowering Plants'.

Richard von Wettstein (1862- 1931) was an Austrian botanist. He proposed system of plant classification in 1901 in his 'Handbuch der Systematscheme Botanik'. Its last version was published in two volumes (1930- 1935) in which he presented his latest view on the phylogeny of plants. He patterned his scheme basically on the Englerian concept of primitive flower. He rearranged the relative positions of many dicotyledonous families. His views were concurrent with the most contemporaneous phylogenists. It was a much better phylogenetic classification than that of Engler. This is particularly so because he based his system on the data from wider fields than any of the earlier botanists. He changed his view, unlike Engler, in relative position of two major groups of angiosperms, the Monocotyledons and Dicotyledons. He thought the Monocotyledons derived from Ranalian plexus and the Dicotyledons primitive than the monocotyledons. His system indicated arborescent habit to be more primitive than the herbaceous and many - flowered inflorescence to be more primitive than few - flowered or solitary flowered types. It also warranted the spiral arrangement of floral parts preceded the cyclic arrangement. He pointed out similarity between flowers of *Ephedra* and *Casuarina* and even regarded Casuarinales as the most primitive amongst the Amentiferae. He also sought origin of polypetalous families from two orders *viz.*, the Ranales and Euphorbiales. Polypetalae, accrording to him, gave rise to Sympetalae. Some features of his system, for instance, derivation of monocotyledons from the Ranales, consideration of Helobiae as primitive, placement of grasses and sedges in separate orders, consideration of Spathiflorae and Padanales as the most advanced, etc. are significant. These have been adopted in subsequent classifications.

August A. Pulle, Ultrect Botanical Museum, the Netherlands, proposed his system in 1938. It is a modification of Engler's system. His division Spermatophyta was further divided into four subdivisions *viz.*, (i) Pteridospermae (included two extinct families), (ii) Gymnospermae included except the Gnetales and Welwetschiales), (iii) Chlamydospermae included Gnetales and Welwetschiales, (iv) Angiospermae, the last subdivision, included Monocotyledonae and Dictoyledonae, the latter being divided into eight series. First series of the Dicotyledonae started with the Amentiferae and ended with the sympetalous orders. Similarly, Monocotyldonae began with the Spadiciflorae followed by the Pandanales and others and ended in the Gynandrae. His classification is in closer accord with the modern phyletic opinions. His classification especially about angiosperms is basically patterned after that of Wettstein. It appeared that the Monochlamydeae and the Sympetalae are unnatural taxa as he thought Sympetalae presumably polyphyletic in origin. He rearranged its constituent families.

Few other authors have also made their attempts to pattern their systems basically after Engler's system or followers of Englerian school. Mention may be made those of Car Skottsberg (1940) of Sweden, B. Hayeta (1921) and E. Warming (1895), etc.

Of late, Angiosperm Phylogency Group (APG) System has been patterned after cladistic analyses. Cladistic analyses divulged phylogeny particularly of angiosperms. Morphology of several major groups above family rank have lent better support for such studies. Presently, many elements of the major branching scheme of phylogeny of these groups have been established. The present state-of-art warrants a revised suprafamilial classification, nay it is desirable and feasible as well. Kate Bremer *et al.* (1998) complied the 'Angiosperm Phylogeny Group' system. They published it as 'An Ordinal Classification For The Families of Flowering Plants' in the Annals of the Missouri Botanical Garden. This system imbibed information revealed by many taxonomists working on this line.

APG system includes 462 families in 40 putatively *monophyletic* orders and some *monophyletic* informal higher groups. The groupings in this system are conceived as *monophyletic* Clades. It projects three most inclusive groups *viz.*, (I) Eudicots, (II) Unassigned (families of uncertain position for which no firm data available to place anywhere within this system) and (III) Monocots. It is to be remembered that Eudicots and Monocots are not formally named so or even ranked. An overview of the Eudicots and Monocots groups is as the following (Fig.7).

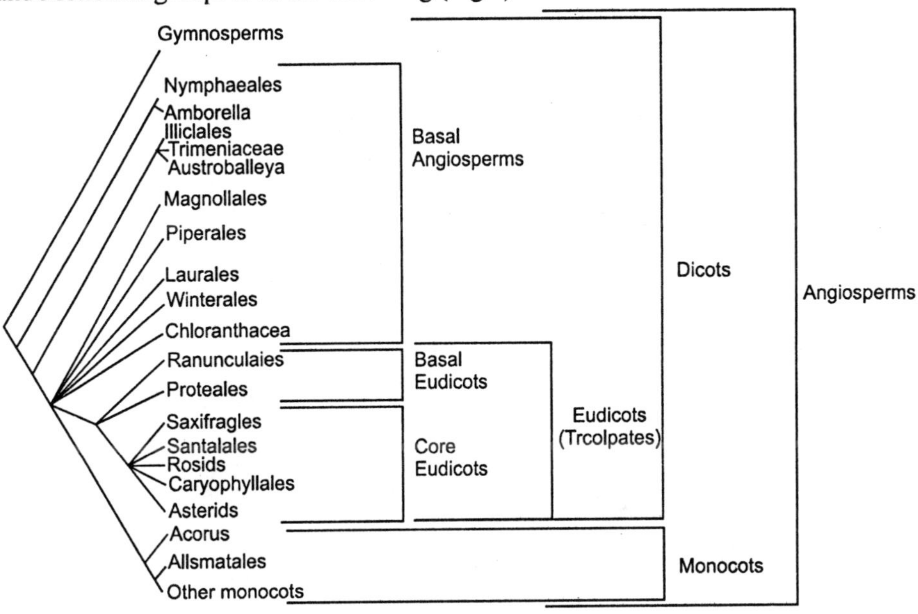

Fig. 7. An Overview of Angiosperm Phylogeny Group System (APG)

An outline of the three APG Groups mentioned above is presented below:

(I) APG Group: Eudicots

It is subdivided into five APG Groups *viz.*, (1) APG Group: Eudicots-Unassigned, (2) APG Group: Core Eudicots, (3) APG Group: Eudicots-Rosids, (4) APG Group: Eudicots-Asterids, (5) APG Group: Probabli Eudicots.

1. APG Group: Eudicot-Unassigned

   * Not Assigned To Order (04 Families)

   Order 1. Proteales (03 Families)

   2. Ranunculales (10 Families)

2. APG Group: Core Eudicots

   * Not Assigned To order (06 Families)

   Order 1. Caryophyllales (26 Families)

   2. Santalales (05 Families)

   3. Saxifragales (13 Families)

Rosids and Asterids contain large subgroups Rosides I & II, and Asterids I & II respectively. These are being supported as monophyletic (Fig.8).

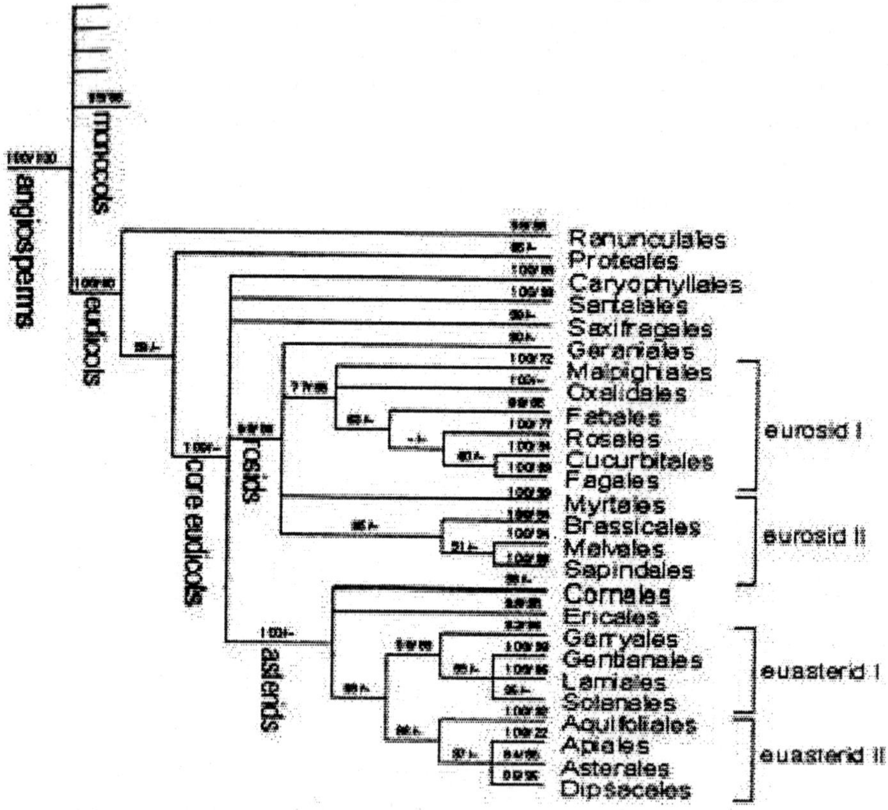

Fig. 8. Phylogenetic interrelationships of the orders of flowering plants, compiled from recent cladistics analyses. Jackknife support is given on the branches (a dash for values <50%), first jackknife values from analysis of 545 sequences of the rbcL, atpB, and 18S rDNA genes.

3. APG Group: Eudicots-Rosids.

   * Not assigned To Order (10 families)

   Order 1. Geraniales (07 Families)

   Subgroup 1. Eurosids I

   * Not assigned To Order (05 Families)

Order 1. Cucurbitales (07 Families)
2. Fabales (04 Families)
3. Fagales (08 Families)
4. Malpighiales (31 Families)
5. Oxalidates (06 Families)
6. Rosales (11 Families)

Subgroup 2. Eurosids II

* Not Assigned To Order (01 Families)

Order 1. Brassicales (16 Families)
2. Malvales (10 Families)
3. Myrtales (14 Families)
4. Sapindales (10 Families)

4. APG Group : Eudicots – Asterids

Order 1. Cornales (06 Families)
2. Ericales (24 Families)

Subgroup 1. Euasterids I

* Not Assigned To Order (03 Families)

Order 1. Garryales (04 Families)
2. Gentianales (05 Families)
3. Lamiales (21 Families)
4. Solanales (05 Families)

Subgroup 2. Eusterids II

* Not Assigned To Order (11 Families)

Order 1. Apiales (07 Families)
2. Aquifoliales (03 Families)
3. Asterales (14 Families)
4. Dipsacales (06 Families)

5. APG Group : Probabli Eudicot

* Not Assigned To Order (25 Families)

II. APG Group: Unassigned-Neither Dicots Nor Monocots

* Not Assigned To Order (11 Families)

Order 1. Ceratophyllales (01 Family)
2. Laurales (07 Families)
3. Magnoliates (06 Families)
4. Piperales (04 Families)

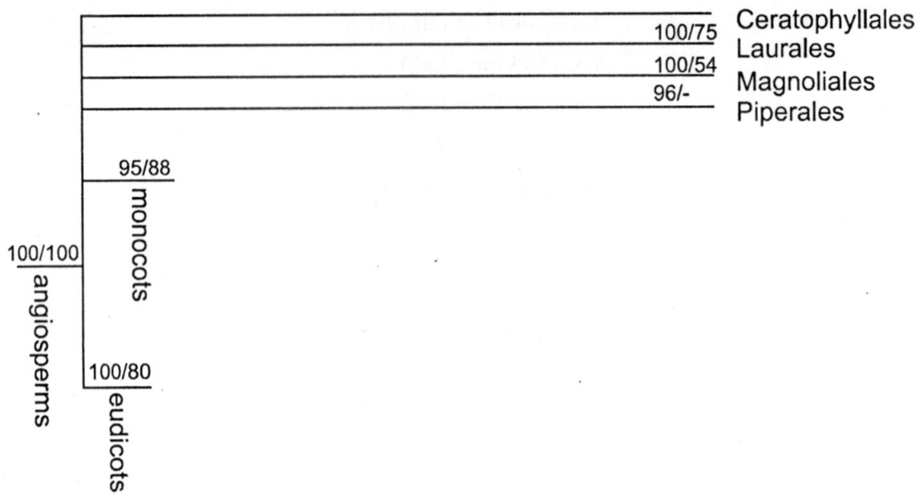

Fig. 9. Phylogenetic interrelationships of the orders of flowering plants, compiled from recent cladistics anslyses. Jackknife support is given on the branches (a dash for values <50%), first jackknife values from analysis of 545 sequence of the rbcL, atpB, and 18S rDNA genes.

III. APG Group: Monocots

Monocots, as already stated, are not named and ranked because it is still not clarified at what level they should be recognized. It is also the case of Commelinoids, a subgroup of monocots.

1. APG Group: Non-Commelinoid Monocots

   * Not Assigned To Order (05 Families)

   Order 1. Acorales (01 Family)
   2. Alismatales (13 Families)
   3. Asparagales (29 Families)
   4. Dioscoreales (05 Families)
   5. Liliales (09 Families)
   6. Pandanales (04 Families)

2. APG Group: Commelinoid Monocots

   * Not Assigned To Order (06 Families)

   Order 1. Arecales (01 Family)
   2. Commelinales (04 Families)
   3. Poales (06 Families)
   4. Zingiberales (08 Families)

Modern classifications are ideally founded on a very wide range of characters. Their incorporation is although gradual, it is continuous and hence it follows that the naturalness or predictivity of classification depends on the extent to which the plants have been investigated. It is also worth to note that classificatory systems are not permanent instead they are transient. They continue to be modified and improved upon the basis of new evidence. Again, there is no ultimate authority who will produce a really final system of classification. The problem of lumping and splitting exists in this arena of plant science. It is not important that how many taxonomic entities comprise the system. The acceptability and convenience of the users of a system matters better. All systems have past and are essentially modifications of the earlier ones. Study of only few modern systems would not suffice. One must be aware about developmental history of all systems and the principles on which they are based.

5

Sources of Taxonomic Information

Taxonomy: Interdisciplinary Approach

A.P. de Candolle (1813) introduced the term '*taxonomy*' which meant 'the theory of plant *classification*'. There have been again arguments about the two terms *viz.*, taxonomy and systematics with reference to their applicability and inclusiveness. In practice, we noted that they are used synonymously as well by many authors. It is but agreed that taxonomy primarily deals with three functions *viz.*, *identification*, *nomenclature* and *classification*. The review of pre-Darwinian and post- Darwinian systems of classification (Chapter-IV) indicate how all these three aspects (functions) of the science of taxonomy have steadily improvised with time, thoughts and concepts prevalent over a long past. Recently, Cronquist (1968) defined taxonomy as 'a study aimed at producing a system of classification of organisms which best reflects the totality of their similarities and differences'. The word 'totality' in modern time, is obviously suggestive of utility of all possible evidence or domains. Taxonomy has thus become transcendental. It comprehends and uses information from all other disciplines and sub-disciplines of science. It is an information system containing all pertinent knowledge relating to all plants. Nearly all systems of plant classification of modern time are therefore broad-scale based on contemporary information and conceptual thought. This has been possible only because the science of *taxonomy* changed over time. It first employed simply *exomorphic* features. Later, it started imbibing data from *endomorphic* features. Ultimately, taxonomy has become amoeba-like engulfing information from all sides. It is also pertinent to note the change in nature of systems. In earlier days, the systems were largely artificial. Gradually, they turned to be natural, phylogenetic or evolutionary. It is also important to notice the change in definitions/concepts of 'species' particularly a change from *typological* to *biological* concept of *species*. It is also imperative to consider phases of taxonomy. Taxonomy *vis-à-vis* classification has changed from one phase to another. Davis and Heywood (1963) recognised four phases *viz.*, the pioneer, the consolidation, the biosystematic and lastly the encyclopedic. It means the ultimate goal to be achieved is a perfected system based upon all available characters. This is what Turill (1938) expected in *Omega*-classification. All these trends or considerations directly demand information from all parts of a plant, from all populations of a species and all possible disciplines of plant science. The aim of this chapter is to survey the major sources of taxonomic evidence. It will be, however, desirable to place special emphasis on those features that tend to be ignored and on the 'newer' kinds which are increasingly employed today.

(I) Exomorphology

The type of data that has been employed most in plant classification is accrued from exomorphology and still will continue to reign taxonomically supreme for many more years (*cf.* Cronquist, 1975). Although the synthetic approach is putting long strides, yet the traditional method of plant taxonomy based on external forms is indispensable. This is so because, in general, there is a high degree of coincidence between the expressed phenotypic characters and the genotype of the taxon. In the ancient attempts to classification by Theophrastus (370-285 B.C.), the Elder Pliny (23-79 A.D.), Disoscorides (1st century A.D.) and later taxonomists such as Albertus Magnus (1193-1280), Brunfels (1464-1534), Bock (1458-1554), Fuchs (1501-1566), Caesalpino (1519-1603), Bauhin Brothers (16-17th centuries), Ray (1628-1705), Linnaeus (1707-1778), Adanson (1727-1806), de Candolle (1778-1841) and in the period of Bentham and Hooker (1862-1883) and so on to the systems of today like Englerian systems e.g. Engler and Prantl (1887-1915), Engler and Diels (1936), Wettstein(1901), and Ranalian systems such as Hutchinson (1973), Takhtajan (1969, 1980), Cronquist (1981,1988), Dahlgren (1975), Dahlgren and Clifford (1982) and Thorne (1992), etc., exomorphology has always predominated.

There have been many reasons for its reign over the field of *taxonomy*. First, the exonorphic features are readily observable. Moreover, they have innumerable variants and as such aid conveniently in delimitation and identification of various taxa. Second, one need not have elaborate or sophisticated laboratory facilities to analyse exomorphic features. A hand lens or dissecting microscope or at the most light microscope suffices to study them. Third, the effort (pain) and time required to tap information from other sources like cytology, embryology, anatomy, palynology, phytochemistry and serology, breeding behaviours and others ensure the advantage of *exomorphic* features over others in plant taxonomy. Lastly, the most systems of plant classification cited above, illustrated glossaries (Johnson, 1931; Core, 1955; Porter, 1967; Radford *et al.* 1974; Benson, 1979; Naik, 1984), books that deal with terminology of phytography (Feathorly, 1954; Harrington, 1957; Stearn, 1983), dictionaries of botanical terms (Jackson, 1928); as also general morphological texts (Troll, 1937- 1939; Bold, 1967; Bierhorst, 1971; Foster and Gifford, 1974, Sporne, 1975b, etc.) and few other sources are replete with well-knit terminology based on *exomorphic* features. All these aid greatly, nay suffice to describe morphological variations met within all plant groups. Even they are commonly available in various institutes, universities and like places. In brief, (exo) morphology continues to enjoy an edge over other sources of characters in plant classifications (*cf.* Holttum, 1968). Classifications based primarily on external forms are not necessarily regarded by the taxonomists to be ideal. Nevertheless, they are thought to be the most rapid and appropriate for the purpose of expressing diversity patterns prevalent in plants/ organisms. Moreover, they are useful as a reference system into which other kinds of data can be intercalated. In this attempt, the discussion is confined to those characters of plants that can be readily observed with the aid of a 10 X hand-lens or with naked eye.

Although there have been arguments in past in regard to the relative value of exomorphic and endomorphic features, it appears rightly concluded by Bailey (1951) that neither of these are more reliable in their comparison. It should also be remembered that exomorphic systems traditionally consider various populations of a species or taxon. *Endomorphic* features are often recorded on a single investigation, populations of a species or taxon are nearly ignored in our routine practice. Exomorphology may be regarded as old-fashioned by some, but it is still the foundation for solving any taxonomic problem. In past, use of *floral* and *vegetative features* has been much debated. The former are conceived more conservative and constant than the latter. In flowering plants, the sexual / reproductive parts are produced only for a brief period in majority of species and therefore they are subject to much less degree of evolutionary pressure. These parts are concerned only with one function, that·is, reproduction. The vegetative parts are subject to more degree of evolutionary pressure, as they have to withstand for nearly complete longer life of a plant species. Again they conduct various functions of plant body e.g. mechanical strength for support, conduction of water and food materials, storage of reserve food materials, etc. Thus the degree of constancy and conservation is ought to vary amongst the floral and vegetative characters.

(i) Vegetative Features: Vegetative features, particularly in higher plants, are often thought risky evidence since there are many cases where superficially similar morphological features are found in quite unrelated plants (Stace, 1989). Similar leaf-shapes and growth-habits are good examples. Leaf-shape in the species of *Acer* and *Platanus* are well known. They belong to two remote families. Distantly related or even unrelated taxa show cactus-like or broom-like plants within the same family, e.g. in species of two genera *viz.*, *Acquilegia* and *Thalictrum* of the Ranunculaceae, the leaves are very similar. This is evidenced by *A. thalictrifolia* and *T. aquilegifolium*. In flowering plant taxonomy, vegetative features have not been preferred to floral ones. This fact is evident from most classificatory systems. There are many superficial vegetative characters which are too plastic and hence not useful in classification. For instance, *Crotalaria* displays simple, *trifoliolate* and *digitately* compound leaves. They vary in respect of shape and size and cannot be employed for generic circumscription. The species *viz.*, *Lindernia anagallis* shows continuous variation in leaf character; certain forms were previously referred under *L.cordifolia*. Now all the forms have been brought under *L. anagallis*.

Although so, among the vegetative features analysis of the degree and pattern of variation is a promising field. Underground parts, stem, leaves, petioles, stipules and various other associated characters have been used in classification and even in phylogenetic consideration (Hickey and Wolfe, 1975). The neglected vegetative characters are being brought to fore-front and evaluated beneficially in identification. Underground parts are generally overlooked and represented in herbarium specimens. Some taxonomists paid attention on this line. For instance, two species in *Chlorophytum* (Liliaceae) *viz.*, *C. glaucum* Dalz. and *C. glaucoides* Blatt. are not easily indentifiable from other exomorphic features. The former has thick roots devoid of terminal tubers, whereas the latter possesses slender roots ending in ellipsoid tubers. Likewise, the two species, *viz.*, *C. tuberosum* (Roxb.) Baker and *C. borivilianum* Sant. and Fern. can be easily deciphered by their root characteristics. The former bears terminal roots tubers, whereas the latter has root tubers sessile. (*cf.* Naik, 1984). Davis and Khan (1961) recorded the shape of main roots,- ovoid, elongated, cylindrical, *fusiform* or *napiform* as constant feature useful to delimit various species of the genus *Aristolochia* (Aristolochiaceae). Morphology of root tubers is taxonomically significant e.g. in *Dioscorea* (Burkill and Prain, 1936). Likewise, form of root-structure is important in the taxonomy of the genus *Ranunculus* (Davis, 1960), and of growth form in relation to different habitats (Lam, 1960). Occurrence of *velaman* roots in some Orchidaceae and Liliaceae are important taxonomic feature. *Velamen* is single-layeral or multiple. It is also similarly significant (*cf.* Dahlgren and Clifford, 1982). Distribution of underground parts such as bulb, corm, tuber and rhizome are important at various taxonomic levels (*cf.* Dahlgren and Clifford, *loc. cit.*). These, however, still need further critical studies especially in petaloid monocotyledons and the scitaminean taxa.

Arborescent grasses are placed under the subfamily Bambusoideae of the family Poaceae. This subfamily is differentiated from other members of the Poaceae by the tree-like habit, hollow culms, rhizome, branch system, sheathing organs and petiolate leaf blades with tessellate venation. Bamboos are taxonomically one of the most difficult groups of grasses and have been highly problematic for field studies, collection and classification. Since the flowering in bamboos are erratic and at very long intervals, the vegetative characters are often given more emphasis to identify the specimens. For identification of the sterile specimens, vegetative characters are often given more emphasis at the genetic level and the culm-sheath morphology, especially of the new shoots at the specific level. Most of the new species under the *Thamnocalamus* group are known by their vegetative parts (*cf.* Saklani *et al.*, 2000).

Leaves exhibit multifarious characters e.g. shape, size, arrangement, type, form, duration, venation, apices, bases, margin, petiole features, etc. Some of the more neglected aspects will be considered briefly. The various features of leaves are extensively used for the recognition of species in many genera, e.g. *Alchemilla* (Walters, 1949); *Ulmus* (Melville, 1939, 1951, 1955) *Betula* (Natho, 1959). In plant group with reduced floral structures e.g. Gramineae, vegetative features play an important role in classification. Leaf shape in plant groups with divided or dissected leaves pose difficulty to describe e.g. *Ranunculus, Pulsatilla, Thalictrum* (Ranunculaceae),*Chaerophyllum, Caucalis, Oenanthe* (Umbelliferae). The position becomes often intricate due to variation in the degree of pinnation which may differ from plant to plant within a population, and even from year to year in biennial or perennial species. The notorious plasticity in some species may aggravate the problem. For example, leaf shapes with a wide range of variation from the base of the plant to the apex vary on different kinds of shoot. In such circumstances, it is necessary to ensure that comparable leaves are selected for taxonomic assessment. Basal leaves often exhibit peculiar features, e.g. assuming a rosulate appearance. Likewise, *ptyxis* in leaves most times is not carefully observed and has received less attention. It is easily observed in living material and generally only for a brief period of the season. *Ptyxis* involves the rolling or folding of individual organs. Terms have been coined to describe *ptyxis* e.g. *conduplicate*, *involute*, *revolute*, *convolute*, *plicate*, etc. The species of *Galanthus* are diagnosed by their ptyxis (Stearn, 1956). It is taxonomically important in *Primula*, especially at sectional level (Smith and Forrest, 1928). The Palmae have been divided into two groups, based on the way in which the plicately-folded prototype palm leaf has segmented (Tomlinson, 1960). In dried material, it can be misinterpreted. Wendelbo (1961a) considered leaf ptyxis in revising the genus *Dionysia* and also discussed its relationship to *Primula*. Melchior (1943) described sections *Dionysiopsis* and *Epicamptophyllum* as having involute ptyxis, whereas in section *Anacamptophyllum* it was revolute.

The occurrence of pulvinus in the leaf of the subfamily Papilionoideae (e.g. *Genista*) is studied by Dormer (1946). He thought it a primitive feature in the Leguminosae and of taxonomic and phylogenetic importance. Tomlinson (1962b) employed different leaf bases in the Palmae. They are mostly neglected in taxonomic characters in some genera of Leguminosae, Malvaceae, Rubiaceae and Euphorbiaceae, etc. They are hardly paid attention by taxonomists. They are simply mentioned as stipulate or exstipulate. The details are, however, ignored in taxonomic description of the plant and even so in the preparations of diagnostic keys. Their evolutionary significance is mostly overlooked. Thakur and Patil (2002, 2006) investigated their evolutionary, phylogenetic and taxonomic importance in some Euphorbiaceae, Shisode and Patil (2004), Patil and Shisode (2006) also studied them on similar line in some Rubiaceae and Malvaceae respectively. Similarly leaf venation, an aspect generally avoided in botanical researches, can be conveniently used in taxonomy. Corner (1958) discussed leaf venation in *Ficus*. Hickey (1973) has published terminology useful to describe venation. This feature of leaf, of late, is being investigated especially by the anatomists. Main types of venation were given by Kerner and Oliver long back in 1894, but ignored by the subsequent workers in angiosperm taxonomy. Such studies have been largely carried out by the paleobotanists. Plant pubescence has been also ignored in morpho-taxonomic studies. The degree of pubescence within the species is largely governed by ecological conditions. Johnson (1975) discussed plant pubescence in ecological perspective and Payne (1978) provided comprehensive terminology to describe them. Govindrajalu (1962) found that four sections of the genus *Alangium* (Alangiaceae) has their own specific combinations of pubescence. Lawrence (1951) described vesture types in his illustrated glossary of taxonomic terms, which are hardly paid attention by many. Discussion on types of trichomes and their taxonomic significance will be dilated later under role of anatomy in taxonomy.

Characters of plant pubescence are useful in tracing phylogeny. Raji (1971) used the stylar hair patterns of the tribe Viceae (Papilionaceae) to deduce phylogenetic pathways of its constituent genera. Similarly, *juvenile characters* e.g. type of germination, cotyledonary characters, leaf *dimorphism*, etc. have not received necessary attention. Sampathkumar (1982) and Gomez-Campo and Maria (1974) studied their taxonomic significance in Convolvulaceae and Brassicaceae respectively. Tomilnson (1984b) emphasised importance of shoot and seedling morphology in solving taxonomic problems.

(ii) Floral Features: In the classifications of angiosperms, it is the floral characters that caught all imaginations of mankind, and hence they have been most used in all times. The inflorescence, bracts and bracteoles, receptacle, hypanthium, calyx, corolla, nectaries, stamens, carpels and ovules find place in nearly all types of *systems* of classification, whether *artificial*, *natural*, *phylogenetic* or *evolutionary*. In plant descriptions, they occupy a prominent place and so also in the *diagnostic* keys. They are employed at all taxonomic levels. Thus the floral/ reproductive features acquired a remarkable reliance and this is thought, as stated earlier, because of their more conservative nature than the other features. Inspite the privileged role in taxonomy of the floral features, some characters may, in certain instances, be ignored or sometimes they may not be duly appreciated. Consequently, their taxonomic significance is not properly evaluated. Inadequate studies of these features may even lead to their less reliability.

Sharsmith (1961) brought to light features like adaxial petal appendages of considerable complexity and the attachment of the petal to the rim of the cup in the genus *Hesperolinon* (Linaceae) which are distinct from the genus *Linum*. Cullen (1962) pointed out calyx type which separated two main groups in the *Anthyllis Vullneraria* complex, a feature which was neglected for considerably long period. Zohary (1956) while revising the genus *Tamarix* (Tamaricaceae) in Israel recorded floral disc constant in certain species. He negated the erroneous observations made earlier. The *staminal nectaries* e.g. in Cruciferae are employed, together with other features, for generic and tribal classifications (Schulz, 1936). The *staminal disc* and associated nectaries are widely employed in the taxonomy of many plant groups (*cf.* Fahn, 1953, Frei, 1955). Staminal appendages are traditionally used in the Ericaceae (Palser, 1961). *Staminodes* are also significant taxonomically in certain cases e.g. *Scrophularia*. Filament hairs provided key features in the genus *Verbascum* (Murabect, 1933). The number of parts of the flower is correctly recorded. They are simply described as many, several or numerous. The lower limit is frequently noted. The number of stamens documented in case of *Cleome simplicifolia* (Capparidaceae) is 12 (Cooke, 1958). Recent floras however noted 10-25 stamens (Patil, 2003). Similarly, Haskell (1949) showed stamens in *Stellaria media* to be 3 and 4. Previously, many floras reported 3, 5 to 10 stamens. Flower colour also poses problems. They change during ontogeny e.g. in some Boraginaceae, *Gloriosa superba*. They change after anthesis e.g. in some Cruciferae; *Eranthemum roseum*, etc. Few species express more than one flower color e.g. *Wahlenbergia flexuosa* (Patil, 2003). Such expressions also pose difficulty in identification. Useful characters are derived from inflorescence, apart from general types. Difference in posture of the pedicels during and after anthesis, the angle and degree of branching and the manner of opening and development, etc. are some of the features useful for identification, but have been mostly unnoticed. For example, in the genus *Silene* groupings are based on type of inflorescence. In the perennial sections *paniculate* inflorescences are found with more or less opposite branches bearing groups of 3-flowered *dichasia* (Chowdhuri, 1957).

Smaller fruits receive cursory treatment in most taxonomic works. Of late, the situation is redeemed. For instance, in the Umbelliferae details of fruit structure have been emphasised (Baumann, 1946; Klan, 1947). In the Chrysantheminae (Compositae) fruit characters, along with other embryological features provide basis for subdivision of the group (Heywood, 1954a, 1959a). Fruit characters in the order Rhoeadales are useful for different families.

Differences between seeds are often constant and valuable taxonomically. Seeds vary externally in shape, size, colour, surface and markings, appendages like *arils*, *caruncles* and *elaisomes*. Studies using low-power compound microscope by Cole (1961) in the genus *Chenopodium* revealed marking patterns on the testa. These aided to separate species. Duke (1961) found shape, colour and sculpturing providing critical indication of the taxonomic position of species in *Drymaria* (Caryophyllaceae). Obermeyer (1962) opined that the only reliable distinction between the two liliacecous genera *viz.*, *Chlorophytum* and *Anthericum* are the number and shape of seeds. Hairy outgrowths on the testa, their colour and length also provide distinctive features for genera and / or species in some families e.g. Malvaceae, Convolvulaceae, Acanthaceae, Asclepiadaceae, etc. Seeds are so unique that they are very easily identified or they are very characteristic of them e.g. *Entada pursaetha* DC. (very large seeds), *Lodoecia maldivica* (Gmel.) Pers. (double coconut), etc. *Glinus* Linn. and *Mollugo* Linn. can be separated, the former having appendaged seeds, whereas the latter have seeds without appendage. Corner (1951a) found usefulness of seed characters: (i) to separate the leguminous seed from the rosaceous seed, (ii) to link the mimosoid and caesalpinoid seeds, (iii) to recongnise the Swartzieae as a fourth group (subfamily) linking the other three, as opined by de Candolle (1925).

One must also be cautious about using seed characters while using them in taxonomy. They must be adequately evaluated. Certain environmental variations are also noted in certain cases of plant species. For example, features like *papillae* on the seeds *in Spergula*, presence of mucilage on testa in *Alyssum*, two kinds of seeds *in Atriplex*, etc.

(II) Anatomy

Systematic anatomy has long history. It fascinated people since the early days of microscopy that saw range of variations in plant anatomy. Workers in anatomy then began to recognize patterns of similarity and eventually recognized that plants sharing many common anatomical features were probably closely related. The use of anatomical features in taxonomic descriptions can be traced back to Caesalpin (1583). Since then plant anatomists became concerned with the taxonomic significance of internal structures of plants. Grew (1682) in his 'The Anatomy of Plants', Malpighi (1675-1679) in his 'Anatome Plantarum' and Hill (1770) in his book 'The Construction of Timber' revealed role of anatomy in taxonomy. Later, Mohl (1851) also pointed out their taxonomic value in his 'Grundazuge der Anatomie und Physiologie der Vegetabilischen Zelle'. Bureau (1864) employed anatomical characteristics in delimitations of taxa at different ranks within the Bignoniaceae. Radlkofer (1895) followed similarly treating the Sapindaceae for Engler and Plantl's 'Die naturlichen Planzenfamilien' This viewpoint was taken up by his student Solereder in 1885 in his classic 'Systmatische Anatomie Werth der Holzstruckur den Dikotyledonen'. It was later translated into English (1908). The trend of presentations of comparative anatomical data culminated into the compilation by Metcalfe and Chalk (1950).

Anatomists, in general, are /were more interested in plant structure in relation to function than to classification. It also appeared that anatomists have not always realized the limitations of their mode of investigation and have sometimes drawn conclusions that are improbable to a taxonomist. Conversely, skilled taxonomists have sometimes been unable to evaluate the importance of anatomical investigations (*cf.* Metcalfe and Chalk, 1950). However, in recent years sufficient data have accumulated to help draw better conclusions (*cf.* Davis and Heywood, 1963). It has become easier and more convenient now-a-days to tap anatomical data from 'Plant Micromorphology Bibliorgraphy Database' on the website of the Royal Botanic Gardens, Kew (England). It goes back into the literature and is kept upto date and thus has greatly simplified the task of the botanists. Anatomical data can be used for two purposes : (i) for identification, group making, etc. (ii) for studying phylogenetic trends and degrees of relationship. Bailey (1951) opined that the use of anatomical characters depends on the objectives in mind. For example, Naik and Nirgude (1981a) employed anatomy at species level in *Chlorophytum*. It is essential to obtain reliable evidence regarding salient trends of evolutionary specialization in the various organs and internal structures of plants. It is against this backdrop that we should employ the contribution of anatomical studies to classification and phylogeny. Anatomical data can be viewed as consisting of two types *viz.*, endomorphic (as contrasted with exomorphic or morphological data), and ultrastructural. The former are observable largely with the light microscope, whereas the latter by use of the TEM. Only the role of former is dilated in the following. The role of latter is explained elsewhere in this book.

Trichomes: Trichomes are epidermal emergences and known well for their diversity in Angiosperms. They are first divided into *glandular* and *eglanular* (*non-glandular*) categories. They may be unicellular or multicelluler, simple or branched, terete or *dendric*, *stellate*, *whiplike* or *T-shaped*, etc. The length, size and density are liable to vary with environment and hence not taxonomically useful. Only distinctive types of trichomes find place in taxonomic thinking. They have been classified by anatomists in their own way e.g. Ramayya (1972) and Uphof (1962) categorized them on different bases. Payne (1978) also provided comprehensive terminology to describe them.

Metcalfe and Chalk (1950) summarized family-wise trichome types in the dicotyledons. He remarked that some kinds of trichome types possess diagnostic value because of their restricated occurrence. He also opined that the combination of trichome types is valuable in families where several kinds occur. He reached to these conclusions on the basis of literature then available. The cruciferous genus *Arabis* (Rolling, 1941) have trichomes unicellular and *eglanduler*. In some species, they are simple, whereas in others they are branched. The latter ones are dendritic or star-like. Still in other species, they are terete, flattened or swollen. Trichomes thus help identify the various species in this genus Govindarajalu (1962) noted trichomes useful at sectional lavel within the genus *Alangium*, for example, Sect. *Angolum* and Sect. *Rhytidandra* show erect, reclinate and articulate, *unicellular* trichome, whereas Sect. *Marlea* possess only erect and reclinate types. The Sect. *Conostigma* is featured out by *stellate* and uncinate unicellular and *uniseriate mulicellular* trichomes besides the erect and reclinate types. Ramayya (1969) proposed a key based on trichome features for the Indian Compositae. Cowan (1950) used trichomes in *Rhododendron* (Ericaceae) for specific and infrageneric distinction. Heintzelman and Howard (1948) found trichomes types and their distribution useful to distinguish members of the family Icacinaceae. Their study lent support to separation of *Ottoshulzia* from *Poraquiba.* Presence of *sessile glandular* trichomes in Typhaceae and Sparganiaceae as well provides a link between them (Solerader and Meyr, 1933). Patil and Patil (2009) and Patil and Patil (2009) investigated trichome morphology in some Rubiaceae and Acanthaceae respectively and found them very useful to indentify them.

Trichomes, in general, are much less frequently found in the monotyledonous taxa than in the dicotyledonous ones. But still they exhibit great and various categories of structures. Some families of monocotyledons entirely or prevailingly glabrous e.g. Araceae, Typhaceae, Taccaceae, etc. Others have a variety of trichome types e.g. Commelinaceae, Eriocaulaceae Dioscoreaceae, etc. Dahlgren and Clifford (1982) reviewed briefly trichomes in the monocotyledons. He concluded that trichomes mostly belong to types such as (i) *unicellular*, (ii) *mulitcellular* with a unicellular base and (iii) multicellular with a multicellular or cushion-like base. Stellate trichomes are observed in Dioscoreacae, Hypoxidacae, Bromeliaceae and some Alismataceae. *Unicellular* trichomes have a wide distribution and probably of little taxonomic significance. The mulitcelluar trichomes are uniseriate, either glandular or eglandular. Branched trichomes are found in *Hanguana*, other Restionales being glabrous. Likewise, *Gaimardia* have branched trichomes, other members of Centrolepidaceae have uniseriate ones. The Haemodoraceae have branched trichomes but mostly with multicelluar base, whereas the Philydraceae have *uniseriate* trichomes with long end cell and sunken *glandular* trichomes in combination. In the Poaceae unicellular trichomes are relatively common. They also show 'microhairs' with two to several cells. Dahlgren and Clifford (*loc. cit.*) thought trichomes in the monocotyledons taxonomically interesting.

Colleters (dendroid trichomes) occur on the adaxial surface of stipules on shoot tips of leaf nodulated rubiaceous species. Such *colletors* are reported in *Pavetta* (Faber and von, 1912; Lersten, 1974), *Psychotria* (Horner and Lersten, 1968; Miller *et al.*, 1983), *Gleasomia* (Rogers, 1985), *Neorosea* (*Lersten,* 1974), *Tricalysia* (Robberecht, 1988; Lersten, 1974).This interesting structure, however, needs thorough studies in the Rubiaceae and its related families to reach sound taxonomic conclusions.

Stomata: Structure and development of stomata gained increasing importance in systematics. They also give clues to the different evolutionary trends particularly in angiosperms. Well illustrated terminology has been developed by Vesque (1889),

Metcalfe and Chalk (1950), Van Cotthem (1970), Pant (1965), Stebbins and Khush (1961), Dilcher (1974) and Payne (1979). Metcalfe and Chalk (1950) enlisted families showing a consistent type of stomata in angiosperms, for example, *anomocytic* (Ranunculaceae, Nymphaeaceae, Rosaceae, etc.), anisocytic (Brassicaceae), *diacytic* (Acanthaceae, Lamiaceae, Caryophyllceae), *paracytic* (Rubiaceae, Convolvulaceae, Magnoliaceae). Prat (1932) showed consistency of *gramineous* stomata in the Gramineae and suggested removal of the genus *Jouvea* from the tribe Hordeae (Gramineae).

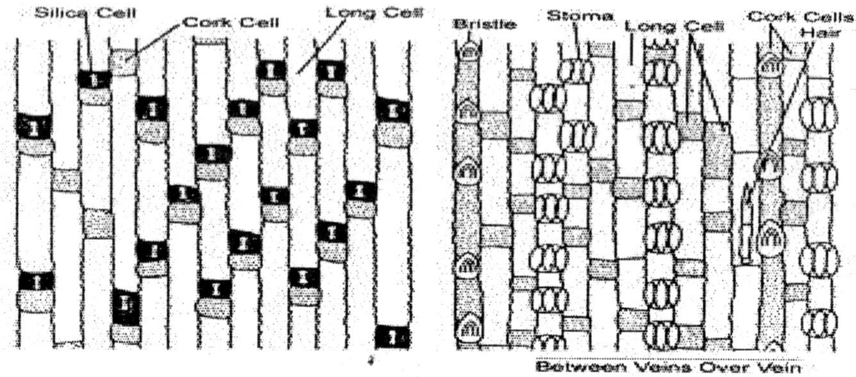

Plate-I (A): Gramineous Epidermis Showing Silica, Cork, Long Cells and Stomata.

Stace (1965) employed distribution, structure, shape, size, degree and pattern of thickening of walls of guard cells, etc. in taxonomy of Combretaceae. Verdcourt (1958) recognized three subfamilies. *viz.,* Rubioideae, Cinchoideae and Guettaroideae. This treatment has been substantiated on the basis of structure and development of stomata in the Rubiaceae. Similarly, Cronquist (1968a) grouped the families *viz.,* Alangiaceae and Cornaceae in the order Cornales, and Araliacceae and Umbelliferae in the order Umbellales. The former two families show *aperigenous* development of stomata, whereas the other two families poss *paramesogenous* development (Kannabiran and Krishnamurthy, 1979). Bremekamp (1953) suggested transfer of the genus *Elytraria* from Acanthaceae to Scrophulariaceae. The genus, however, showed *diacytic* stomata and *syndetohelic* development and hence should be retained in the Acanthaceae. The Scrphulariaceae have *anomocytic* stomata and haplochelic development (Paliwal, 1966). Thakur and Patil (2010) noted *cycloyctic* stomata in *Simmondsia* which is not encounted elsewhere in the Euphorbiaceae. These authors in consideration with other endomorphic features are inclined to refer the genus outside the Euphorbiaceae in a family of its own-the Simmondsiaceae, Shisode and Patil (2000, 2008a) investigated predominantly anomocytic stomata both in the Celastraceae and Hippocrateaceae (*sensu stricto*).They opined that the latter should by merged with the former.

Epidermis: Epidemal cells except those associated with stomata, trichomes, veins, etc. have been found to differ in some plant groups.

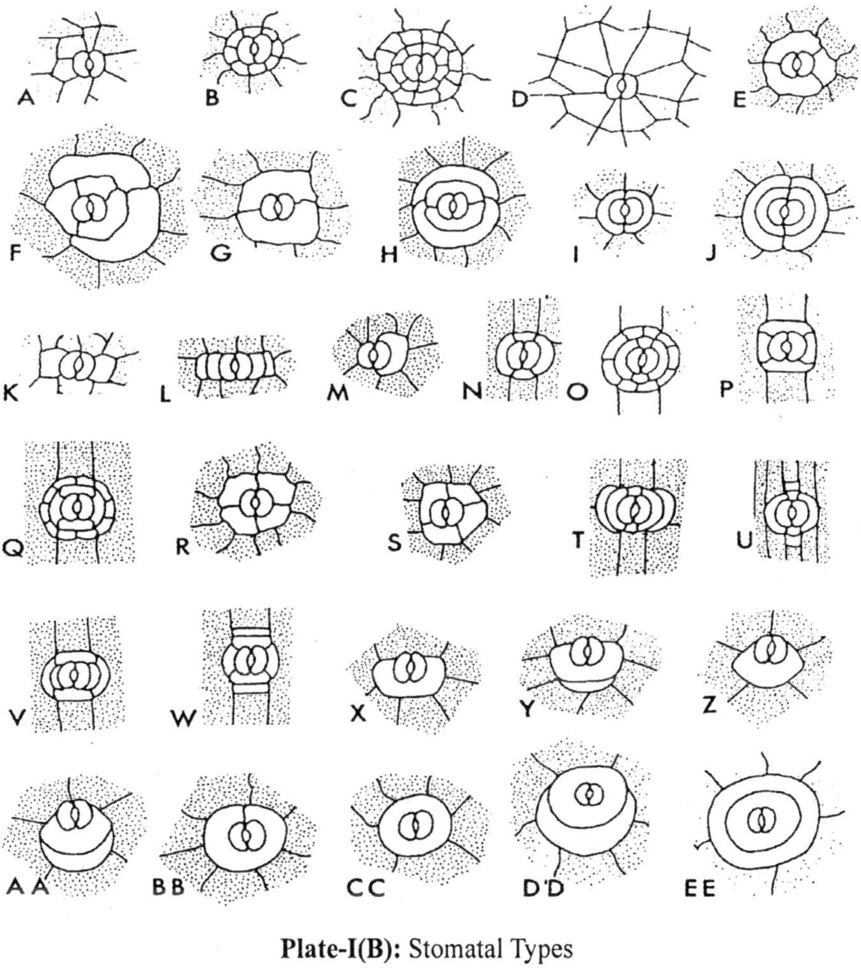

Plate-I(B): Stomatal Types

A: Anomocytic; **B:** Cyclocytic; **C:** Amphicyclocytic; **D:** Actinocytic; **E:** Anisocytic; **F:** Amphianisocytic; **G:** Diacytic; **H:** Amphidiacytic; **I:** Paracytic; **J:** Amphiparacytic; **K:** Brachyparacytic; **L:** Amphibrachyparacytic; **M:** Hemiparacytic; **N:** Paratetracytic; **O:** Amphiparatetracytic; **P:** Brachyparatetracytic; **Q:** Amphibrachyparatetracytic; **R:** Staurocytic; **S:** Anomotetracytic; **T:** Parahexacyticmono polar; **U:** Parahexacytic-dipolar; **V:** Brachyparahexacytic-monopolar; **W:** Brachyparahexacytic-dipolar; **X:** Polocytic; **Y:** Copolocytic; **Z:** Axillocytic; **AA:** Coaxillocytic; **BB:** Desmocytic; **CC:** Pericytic; **DD:** Copericytic; **EE:** Amphipericytic. (Adapted After Dilcher, 1974).

They vary in size, shape, orientation, undulation of anticlinal wall, curvature of periclinal wall, *papillae*, etc. The systematic expression of the distribution of cell types is termed as dermogram. The *dermograms* can be presented for each species as a *dermotype* (Prat, 1932), for example, Ramayya and Rajgopal (1968, 1971) proposed a generic key for the Indian Aizoaceae and Portulacaceae based on their dermotypes. Ahmad (1976) supported to the treatment of *Asteracantha longifolia as Hygrophila auriculata* by Heine (1962) based on foliar epidermal contours. Singh *et al.* (1974) justified *Berberis* and *Mahonia* as distinct genera, having many epidermal features. Tayade and Patil (2003) in their foliar epidermal study in some species of the genus *Argyreia* (Convolvulaceae) pointed out subtle differences in cutcular striations and cell wall contours. They also proposed a key to indentify species of the genus. Shisode and Patil (2004b) also studied foliar epidermal features of some speices of the genus *Boerhavia* (Nyctaginaceae).They employed epidermal features in a key to decipher the species studied. They also supported distinct taxonomic status of *B. diffusa* L. and *B. repens* L. based on epidermal characteristics.

Leaf Anatomy: Systematic importance of foliar anatomy was rightly appreciated by Carlquist (1961) who stated that 'the leaf is perhaps anatomically the most varied organ of angiosperms and its anatomical variations often concur closely with generic and specific and occasionally familial lines'. It affords most reliable characters useful in taxonomy e.g. size and shape of epidermal cells, epidermal *papillae*, nature and thickness of *hypodermis*, structure of mesophyll and vein sheaths pattern of sclerenchyma, *scleríeds*, crystals, secretory cavities, *foliar nectaries*, *stomatal crypt*, occurrence of *hydathodes*, etc. Cutler (1965) investigated several anatomical features and inverted vascular bundles, a unique type in the leaves of the genus *Thurnia.* The genus has been placed either in the family Juncaceae or Rapateaceae. The anatomical features supported for distinct family of its own, the Thurniaceae. Paliwal and Kakkar (1970) lend support to the erection of independent family Garryaceae for the genus *Garrya* based on scleried type and other anatomical features. Webster (1967) used chloroplast structure together with other morphological chararcters in the delimitation of the genus *Chamaesyce* from *Euphorbia.* In case of Gramineae, differences have been noticed with respect to endodermal sheath, parenchyma sheath, epecially arrangement of cells outside the parenchyma sheath, mesophyll, etc. (Brown, 1958). Brown, (*loc. cit.*) recognized six types on the basis of tissue arrangement in grass leaves. He discussed evolutionary tendencies in the various groups within the grasses. The vascular bundles are partially surrounded by sclerenchyma adaxially in the form of crescent form or V-shaped and abaxially by either u-or-y-shaped. Ayensu (1974) revealed that *Vellozia* and *Barbacenia* (Velloziaceae) can be distinguished from one another by the form of sclerenchyma in the leaves. The structure of the bundles can also vary especially in Gramineae with C4 photosynthesis (Kranz anatomy) in which bundle sheath cells have chloroplasts centrifugally localized and without grana. Such cases can be conveniently pointed out taxonomically (Johnson and Brown, 1973; Brown, 1977). Govindarajalu (1969) proposed a key based on leaf anatomical features to distinguish the various species of the genus *Cyperus.* Benson (1962) noted considerable difference in the *palisade* layer and the spongy mesophyll tissue in different species of *Ranunculus.* Thakur and Patil (2011a) noted the mesophyll in *Simmondsia* (Euphorbiaceae) not distinguishable into the palisade and spongy tissue. Cells of the mesohyll are polygonal, irregular or elongated but without intercellular spaces. These features favour its treatment in a family of its own, the Simmondsiaceae as stated earlier. Shisode and Patil (2010) studied species of the genus *Ziziphus* and keyed out them using the features of stomatal distribution, *hypodermis* and vasculature of the leaves.

Petiole: Petiole is of considerable taxonomic importance since it is little affected by environmental change (Metcalfe and Chalk, 1950). It presents various vascular patterns, different shapes of vasucular arcs or bundles, different shapes of outline, extent of hypodermal development, cell contents, extent of development of *sclerenchyma* lining the vascular arc or bundles, etc. The principal and special types of vascular system in dicotyledons are summarised by Metcalfe and Chalk (1950). Soladoye (1982) found petiolar features such as petiole outline, parenchymatous layers of cortex, distribution of perivascular fibres and vascular system, etc. in 64 species of *Baphia,* a legumious genus. These features are helpful in infrageneric classification of the genus and correspond to systematic groupings identifield on other parameters. Rathnakumari (2002) investigated petiolar anatomy of different varieties of *Morinda pubescens* and found it taxoniomically significant. Shisode and Patil (2011) studied petiolar anatomy in 13 species of the Rhamnaceae In their opinion, petiole outline, epidermal features, number of layers of parenchyma, cells in the cortex, pattern of vascular system, occurrence of crystals, etc. are useful to distinguish taxa studied. Similar studies have been carried out by Thakur and Patil (2016) in Euphorbiaceae, by Shisode and Patil (2008b) in some members of the Celastrales, by Tayade and Patil (2008) in Convolvulaceae. All these investerations employed petiolar features to distinguish taxa studied. Dehgan (1982) and Dehay (1935) found resemblance between the Euphorbiaceae on the one hand and Malvaceae, Tiliaceae and related families on the other hand on petiolar structure. They pointed out close relationship between Buxaceae and Euphorbiaceae on similar petiolar structure.

Venation: Since very early times, venation patterns particularly of leaves have been source of taxonomic data. Foliar venation pattern is one of the characters used for primary division of angiosperms into dicotyledons and monocotyledons. The former possess generally reticulate venation, whereas the latter are characterised by parallel veins. However, these are not without exceptions, for example, reticulate venation occurs in some monocotyledons such as Dioscoreaceae, Taccaceae, Araceae Smilacaceae, Petermanniaceae, Funkiaceae, Convallariaceae, Philesiaceae and genera like *Tricyrtis* and *Veratrum.* This dicotyledonous character previously caught attention of phylogenists. Hallier (1912) emphasised it as one of the characters held in common by the Araceae (Monocotyledon) and Piperaceae (Dicotyhledon). Huber (1969) also pointed out that this character is also possessed in common by the Dioscoreaceae (Monocotyledon) and woody Polycarpiceae (Dicotyledon). Hickey (1973), for the first time, classified the architecture of dicotyledonous leaves. Later Hickey and Wolfe (1975) also provided systematic summary of architectural features of dicotyledon leaves. They indicated taxonomic significance of foliar venation patterns, the number of primary and secondary veins, angle of the latter with the former, nature of tertiary and higher order of venation, characteristics of leaf margin, elevation and presence of grooves on midvein adaxially, vein-endings, etc. These are being employed in leaf architectural investigation in view of their use in systematics.

Dede (1962) provided a key to identify members of the Rutaceae based on foliar venation pattern. Tucker (1964) indicated use of vein islets while deciphering *Liriodendron chinense* and *L. tulipifera.* Handro (1964, 1967) employed venation pattern alongwith other leaf anatomical features to distinguish Amaranthaceae at generic and specific levels. Singh *et al.* (1978) showed distinction in leaf architecture of two genera *viz.,Podophyllum* and *Holboellia* from the family Berberidaceae and suggested separate families for them. Leaf architecture is also useful taxonomically at higher taxonomic hierarchy. Such data support Takhtajan's (1969) placement of the Euphorbiales to the Dilleniidae. It also justifies Cronquist's (1968) assignment of the Lecythidaceae and allied families to the Dillenidae rather than to the Rosidae. Melville (1969) studied venation patterns in angiosperms and *Glossopteris* and pointed out similarities in between them. However, Doyle and Hickey (1976) refuted glossopterid ancestry of angiosperms on the basis of developmental patterns of veins in either of these plant groups.

Nodal Anatomy: The fundamental nodal types *viz., unilacunar, trilacunar* and *multilacunar* with equal number vascular traces are recognized depending upon the leaf gaps in angiosperms (Sinnott, 1914). A fourth type, *uniclacunar* with *two-traced* was added by Marsden and Bailey (1955).

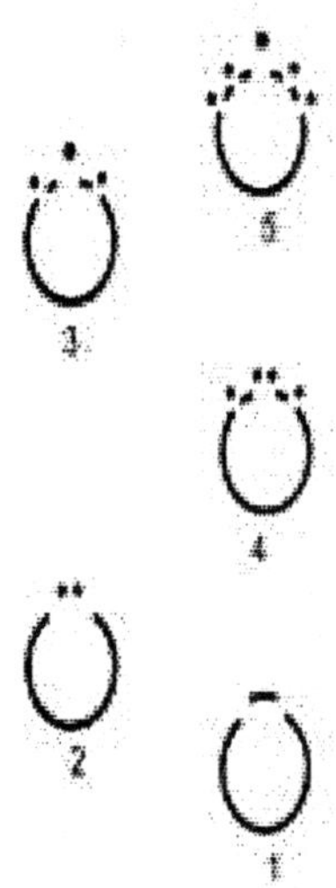

Plate-II: Types of Nodes (Dicotyledons)

1: 1-lacunar, 1-traced, **2:** 1-lacunar, 2-traced, **3:** 3-lacunar, 3-traced, **4:** 3-lacunar, 3-traced (medium trace double), **5:** Multi-lacunar, multi-traced.

A regressive to the unilacunar and *vice-versa* progressive series is thought operated in angiosperms. Conflicting views although have been advanced about primitive types, comparative study of nodal anatomy is useful at various taxonomic levels. The sufamily

Icacinoidcae of the family Icacinaceae may be categorized into two separate sections – one with *trilacunar* nodes and another with *unilacunar* nodes (Bailey and Howard, 1941). Particular type may be characteristic for a taxonomic entity, for example, node is unilacunar in whole order Centrospermae (Sinnott, 1914). It is unilacunar in the family Cireasteraceae (Cronquist, 1968a). Different nodal types may occur in the same family e.g. Euphorbiaceae (Patil, 1988; Thakur and Patil, 2002, 2006), Malvaceae (Pati and Shisode, 2006), Rubiaceae (Shisode and Patil, 2004a). Amongst the members of Euphorbiaceae, *Ricinus* with *multilacnar* node of seedling can be marked out from other Euphorbiaceae (Patil, 1988). Likewise, *Macaranga* is distinctive in its mature nodes amongst the Euphorbiaceae (Thakur and Patil, 2006), However, the genera *viz.,Bombax, Ceiba* and *Adansonia* which are sometimes separated from the Malvaceae into an independent family Bombacaceae (Cronquist, 1968) have similar trilacunar nodes as the other Malvaceae (Patil and Shisode, 2006).

Stem Anatomy: Alike the other plant organs, stem affords many characters of taxonomic value. The features which can be employed for characterization are such as shape of outline and ridges, distribution and abundance of collenchyma, pattern of collenchyma thickenings, distribution of fibres, structure of endodermis, shape and size of sclerenchyma girders, transfusion tissue, cell morphology of pith, arrangement and types of vascular bundles, etc. Metcalfe and Chalk (1950) illustrated several examples of this kind. Scattered vascular bundles are generally thought characteristic of monocotyledons and used for monocotyledon-dicotyledon distinction. Although so, according to Metcalfe and Chalk (1950) the frequent occurrence of several rings of vascular bundles or scattered vascular bundles in the herbaceous Ranunculaceae shows close affinity with the monocotyledons. The genus *Punica* is included under the family Lythraceae by Bentham and Hooker (1862-1883). The occurrence of *intraxylary phloem* in stem-axis but shows their affinity (Metcalfe and Chalk, 1950). The possession of intraxylary phloem by the Loganoideae sharply distinguishes this tribe from the Buddleiodeae in which it is wanting (Metcalfe and Chalk,*loc. cit.*). Vasculature is noted to be systematically significant at higher levels e.g. *medullary bundles* in Nyctaginaceae, Chenopodiaceae, Amaranthaceae, Begoniaceae; cortical bundles in Oleaceae, Crassulaceae, Araliaceae, Buxaceae; *intraxylary phloem* in Apocynaceae, Asclepiadaceae Lythraceae, Myrtaceae, interxylary (included) phloem in Salvadoraceae, Onagraceae, Loranthaceae, etc, (Metalfe and Chalk, 1950). The species of the subgenus *Genuini* of the genus *Juncus* (Juncaceae) are identifed on the basis of size and shape of sclerenchyma girders, degree of elevation of stem ridges and characteristic of pith (Stace, 1970). Occurrence of *bicollateral* vascular bundles in two alternate rings help characterize the family Cucurbitaceae (Metcalfe and Chalk,*loc. cit.*).

Root Anatomy: Traditionally, roots have been mortly neglected by research workers. Being underground, attempts are rarely made to take out and extend observations on anatomical grounds. The data revealed so far is not most times sufficient for comparative studies; hence roots are regarded as less valuable from taxonomic standpoint. The roots with economic significance or with curious structure have been studied in detail. Others have generally been overlooked. It is doubtless that variations in case of mycorrhizal or bacterial associations, meristem orgnisation might afford more information if investigated. The encyclopedic compilation of plant anatomy by Metcalfe and Chalk (1950) provided meagre data. The plant growing in marshy places provide easily intact roots. They may be taken under consideration in future. It is also worth to note that edaphic factors may also cause vaiations in root structure, for example, *Silene vulgaris* Linn, (Caryophyllaceae) differ in the amount of periderm formed and have more numerous vessels when grown in sand than in clay (*cf.* Millner, 1934). Few interesting structure or inclusion may be noted for taxonomic purpose. e.g. (i) *tyloses* in vessel of *Aucoumea* (Burseraceae) (Metcalfe and Chalk, 1950). (ii) occurence and variations in sclerenchimatous idioblasts e.g. *Bruguiera, Rhizophora* (Rhizophoraceae) (Metcafe and Chalk, 1950). Murrey (1934) noted that the genus *Primula* (Primulaceae) can be divided into groups based on differences in root structure. These groups agree closely with the taxonomic sections of the genus based on external plant features. Root anatomy help indentify eight genera of Caprifoliaceae found in the British Isles (Gasson, 1979). These instances although few, provide positive example of use root anatomy in systematics. Thus root is one organ in need of more anatomical reaserch.

Velamen is another interesting tissue found in the roots of some monocotyledons e.g. some members of Liliaceae, Amaryllidaceae, Araceae, Taccaceae and Orchidaceae. It is specialized by usually multiseriate or occasionally uniseriate tissue comprised of dead suberised cells arranged compactly around the core of roots. The exodermis is situated at the inner side of the velamen and interrupted by passage cells. They help absorb water and protect the roots especially when they are aerial. However, they are found in aerial as well as subterranean roots. The number of layers and cell contours may provide some evidence for taxonomy. Still, velamen has remained neglected from taxonomic standpoint. Their detailed structure, occurrence and development await further study in various taxa. Dahlgren and Clifford (1982) summarized its taxonomic distribution in the monocotyledons. However, information is still far from satisfaction to employ in systematics.

Wood Anatomy: The information from wood anatomy has been used to assess taxonomic and phylognetic status. Many features of woods are useful for the said purposes, for example, (i) axial parenchyma (*apotracheal* and *paratracheal* and their subtypes), (ii) vascular rays, their abundance, dimesions, thickness, pitting whether homogeneous or heterogeneous etc., besides cartain inclusions and special structures, (iii) vessels of different types, (iv) woods *ring-porous* or *diffuse-porous*, storied or non-storied, *homoxylous* or *heteroxylous*, etc.

Wood anatomy plays a great role in providing evidence for treating taxa of uncertain affinity. For example, (a) Bentham and Hooker (1862-1883) included the genus *Bixa* under the Family Flacourtiaceae. The absence of septate fibres and the presence of diffuse parenchyma distinguish the genus from the Flacourtiaceae (Metalfe and Chalk, 1950), (b) *Balanites* is placed under the family Zygophyllaceae (Bentham and Hooker, 1862-1983) or treated as an independent family the Balanitaceae (Hutchinson, 1973). Heimsch (1942) showed some wood anatomical similarity with the Zygophyllaceae. However, he also stated that the high and wide rays of *Balanites* may be evidence from a structural point of view for excluding this genus from the Zygophyllaceae.Metcalfe and Chalk (*loc. cit.*) although noted some exceptional wood anatomical features, have not showed importance to them e.g. intervascular pitting absent, presence of crystals, tracheids non-storied, etc. Record (1921) stated that the wood structure of *Balanites* bears no resemblance to that of genera of Zygophyllaceae. In a nutshell, *Balanites* should be removed from the Zygophyllaceae and may be placed in a family of its own, the Balanitaceae. (c) The family Lecythidaceae is submerged under the Myrtaceae (Bentham and Hooker, 1862-1883). The latter differs from the former principally in having mostly solitary pores, fibres with distinctly bordered pits and wood parenchyma paratracheal or diffuse instead of distinctly *metatracheal*. The Lecythidaceae may be also separated into groups on the basis of the presence or absence of crystals, which is a unique and distinctive character (Metcalfe and Chalk, 1950). The genus *Garrya* although differs from the family Cornaceae in having superior ovary and shows affinity with Ameniferae (Wangerin, 1910). The wood is very similar to that of the Cornaceae and no wood anatomical characters ally it with the Amentiferae.Separation of *Leucobalanus* (white oaks) and *Erythrobalanus* (red oaks) is substantiated by differences in wood anatomical features such as late-wood pores, *multiseriate* rays, ray-vessel pitting, *tyloses* and also colour (*cf.*, Barefoot and Hankins, 1982).

Besides the taxonomic importance, wood anatomy has played significant role in phylogenetic considerations. For example, it is a well known fact that botanists of Englerian school consider the Amentiferae as the most primitive, whereas the Ranalian–magnolian complex is conceived primitive by the botanists of Ranalian school. The Ranalian–magnolian taxa show some vesselless wood, vessels with *scalariform perforation* plates, vessels long with small diameter, oblique end walls and *scalariform* inter-vascular pitting, besides diffuse porous wood, diffuse parenchyma, etc. All these features indicate their primitiveness. On the contrary, the wood anatomical features revealed by the Amentiferae are clearly advanced and can not be thought primitive (*cf.*, Tippo, 1938).

In general, secondary wood behaves as a very conservative feature in evolution. Certain features of anatomy of the wood have evolved at a considerable slower rate than the external features. This is why the wood anatomical features are conceived more reliable while using them in taxonomy.

Floral Anatomy: Even a bird's eyeview on major plant classifications indicate that reproductive characters have been greatly employed in systematics. This is so because these characters exhibit great conservation and hence more stable or reliable.

Classifications based on them appear to have been reasonably successful as judged by their almost universal use. Since floral exomorphic characters are slow to change, it is natural to suppose that their vascular supply is likewise conservative. The vascular system of the flower has been extesively studied and the results are thought of great value in understanding the nature and trends of evolution of the floral parts and in deducing systematic relationships.

Although the nature of primitive flower has been debated, it is now agreed by most morphologists that the ancestral flower was bisexual with numerous stamens and carpels with or without a bract-like perianth. All the foral parts were spirally arranged and the flower was symmetrical and without fusion among its parts. Puri (1952, 1962) pointed out that, during evolution and specialisation of the flower, the vascular plan undergoes modification through major evolutionary forces *viz.*, (i) *adnation*, (ii) *connation*, (iii) *reduction* and (iv) *amplification.* These principles have provided much fuel to the systematists. Eames (1953) opined that the floral anatomical characters of families and genera are generally well marked and even specific characters may be quite clear in some cases. Puri (1951) has brought out the significance of floral anatomy in the solutions of morphological problems. The role of floral anatomy in relation to taxonomy and phylogeny has been much disussed by Eames (1953), Puri (1952, 1958, 1962), Murthy and Puri (1980), Govil (1995) and Pai (1992).

The typical angiospermic flowers possess sepals, petals, stamens and carpels. Generally, sepals receive three vascular traces, petals and stamens one and the each carpel is supplied with three vascular traces. Obviously, the monocotyledonous flowers exhibit floral vasculature in conformity with the *trimerous symmetry*, whereas the dicotyledonous ones with 4 or 5 - merous symmetry. The above mentioned four evolutionary forces have brought up changes and accordingly the flowers in angiosperms exhibit variations which are of taxonomic and phylogenetic importance. It appears worth to take help of some examples in support of the role played by floral anatomy.

Ramadevi and Narayana (1989) studied floral anatomy of the genus *Hydrocera* of the Balsaminaceae. It exhibits no connation or adnation between perianth parts and their vascular traces as in other members of the family. The carpels are 5-traced in *Hydrocera* and 3-traced in other Balsaminaceae. *Placentation* is anatomically axile and not parietal in it. It contains three pendulous ovules in each ovarian locule which is not encounted in other Balsaminaceae. These authors, therefore, suggested a separate subfamilial status the Hydroceroideae for the genus *Hydrocera* under the family Balsaminaceae.

Patil and Pai (1981b) investigated floral morphology of the genus *Doryanthes* and also other genera of tribe Agaveae (Patil and Pai, 1981a, 1985). It shows *basifixed extrorse* anthers and double vascular supply to the perianth members. These features are not encounted in the members of the tribe Agaveae (*sensu* Hutchinson, 1973). It is thus suggested to refer *Doryanthes* in a separate tribe Doryantheae. Likewise, the genus *Polianthes* is placed under the tribe Polyantheae (Hutchinson, 1973). It shows similar floral morphological features such as 3-traced tepals, *placentation axile*, anther introrse and *dorsifixed* as in the Agaveae, (Patil and Pai, 1986, 1981a). It is, therefore, thought better placed under the tribe Agaveae (*sensu* Hutchinson *loc. cit.*). Govil and Lavania (1980) investigated floral structure in *Cuscuta*. Singh and Pai (1977) studied the family Convolvulaceae on similar line, to which the genus is sometimes referred (*cf.* Bentham and Hooker, 1862 - 1883). In *Cuscuta*, the sepals, petals and stamans are 1-traced, whereas carpels have only *ventral bundles*. The gynoecium is bicarpellary and *placenation* is anatomically parietal. The sepals and carpels are 3-traced in Convolvulaceae and the *placentation* is parietal or basal. Vascular supply to carpels and sepals in *Cuscuta* is reduced but in general there is similarity with convolvulceous taxa. Therefore, it can not be separated from the Convovulaceae. The families *viz.*, Celastraceae, Rhamnaceae and Vitaceae are sometimes included either under one order or in different orders (*cf.* Bessey, 1915; Bentham and Hooker, 1862 – 1883; Rendle, 1959; Hutchinson, 1973, etc.). Floral anatomy of these have been studied (*cf.* Berkeley, 1953; Nair and Sharma, 1961; Prichard, 1955). It is concluded by these authors that the Rhamnaceae and Celastraceae arose from a hypothetical ancestor with *obdipostemous* condition. The loss of *antisepalous* stamens resulted in the former, whereas the disappearance of the *antipetalous* stamens culminated in the latter. The families Rhamnaceae and Vitaceae showed common features e.g. *antipetalous* stamens, conjoint petal-stamen trace, conspicous *interstaminal* disc and basically *basal placentation* derived from axile condition. These hold them together in one taxonomic entity.

Subramanyam and Nair (1975) imvestigated floral anatomy of the controversial genus *Pentharum* sometimes included under Crassulaceae or Saxifragaceae. Rydberg (1905) recognized separate family of its own. Britton and Brown (1913) and Quimby (1939) also supported their treatment. It differs floral anatomically from the Crassulaceae (Subramanyam and Nair, *loc. cit.*) in considerable fusion of carpels to which other floral organs are adherent basally, devoid of secondary marginal carpellary traces and different course of ventral traces, etc. The vascular supply is derivable from Group IV of *Sedum* (Crassulaceae) but is thought to be a climax of speciation evolved from the Crassulaceae. These authors, therefore, supported a separate familial status, the Penthoraceae to the said genus. Tiagi and Kshetrapal (1975) observed tetracarpellary gynoecium in the Orobanchaceae and remarked that this ancestral condition rules out any possibility of their derivation from the parasitic forms seen among the Scrophulariaceae. They suggested to search out their ancestors among the primitive Tubiflorae such as the Solanaceae wherein 5 - carpellary gynoecia occasionally occur. The family Olacaceae is thought allied either to the Proteales (Rendle, 1959) or the Santalales (Benson, 1957; Hutchinson, 1959; Cronquist, 1968a; Takhtajan, 1980). Floral anatomical features of *Olax* (Olacaceae) (Patil and Pai, 1984) such as type of *placentation*, vascular supply to the petals, vascular supply to a *calyculus*, reduction of calyx to a *calyculus* and reduction of the androecium ally it with to the Santalales (Smith, 1942) instead to the Proteales (Patil and Pai, 1984).

Fruit and Seed Anatomy: Characters divulged by fruit and seed anatomical studies have also been found useful to taxonomic thinking in many cases. Taxonomists are ever hardput ot treat two lythraceous genera *viz.,Rotala* and *Ammania.* Van Leeuwen (1979) pointed out different orientation of the cells of pericarp. In *Rotala*, the inner layers of cells are linear, strongly lignified and transverse to the outer. In *Ammania*, the cells of both layers are nearly similar. Shah *et al.* (1975) investigated pericarp structure in some species of the genus *Tephrosia* (Papilionaceae). They found anatomical structure of pericarp alongwith exomorphic features useful to distinguish the species. Hayden and Dwyer (1969) investigated seed coat morphology in the tribe Morindeae (Rubiaceae). They noted sclerenclymatous elements of the seed and thereby suggested a strong affinity between the tribe Morindeae and the tribe Coussareae. Rout and Deb (1999) noted seed coat parenchymatous in *Ixora* and different from the allied genus *Pavetta.* Corner (1976) described seed anatomy of some celastraceous genera. He regarded the genus *Perrotetra* entirely anomalous in the Celastraceae. He pointed out that the meliaceous genus *Turraea* has the celastraceous fruits and seeds. Based on seed structure, he lent no support to the affinity of the Celastrales with the Santalales. He reviewed seed anatomy of the Rhamnaceae and showed no affinity with the Vitaceae. Coner's study (1976) on the seeds of the family Convolvuaceae are interesting taxonomically. The seeds of Convolvulaceae differ from the other unitegmic seeds of *Sympetalae* by the special differentiation of the three outer cell-layers of the integument. The innermost of these layers form a *palisade* of narrow prismatic cells with thickened and sometimes lignified walls. The outer epidermis multiplies into the distinctive and persistent layers of seed coat, and of these, the innermost becomes the mechanical *palisade*.

Laticifers : Latex is a suspension or emulsion with varying refractive index. It may be white (milky) e.g. *Euphorbia, Asclepias, Lactuca;* yellow – brown e.g. *Cannabis;* yellow-orange e.g. *Chelidonium* or colourless e.g. *Morus*. It diffes in chemical compostion. It may contain rubber particles, waxes, resins, proteins, essential oils, mucilages and starch grains. It may also contain salts, organic acids, sugars, alkaloids, proteolytic enzyme (papain). This liquid may occur either in series of cells (articulated latifcifers) or in single long cells (non-articulatal laticifers). Both of these are called '*laticifers*' Laticifers may occur in dicotyledons, monocotyledons and even pteridophytes. Articulated laticifers are characteristic of some species of Compositae, Convolvulaceae, Papaveraceae, Euphorbiacae, Caricaceae, Sapotaceae, Liliaceae and Musaceae. Simple *non-articated* latifiers occur in *Musa, Allium, Convolunulus, Ipomoea,* etc. *Articulated laticifers* are found in *Sonchus, Lactuca, Carica, Papaver, Manihot, Hevea,* etc. Thus presence or absence, their types, chemical contents, etc. have taxonomic utility. For example, Papaveraceae have well developed latex system, whereas Fumariaceae lack it and should be treated distinct (Sperlich, 1939). Treub (1975) considered *Allium* and related taxa possessing latex cells under the order Alliales, distinct from the other Liliales.

Dahlgren and Clifford (1982) mentioned laticifers in some members of monocotyledonous families' *viz.,* Araceae, Alismataceae, Musaceae, Alliaceae and Cyclanthaceae. They considered origin of latcifers independent in all angiosperms, whether dicotyledons or monocotyledons. Rudall (1994) observed *articulated* and *non-artuculated laticifers* common in the Crotonoideae and Euphorbioideae of the Euphorbiaceae, besides the Acalyphoideae. He discussed evolution of laticifers in relation to phylogeny of the family Euphorbiaceae.

Sclereids: Sclereids are sclerenchymatous cells of various shapes but generally not much elongated. They have thick lignified secondary wall which often shows presence of pits. Tschirch (1889) classified them basically into four types : (i) *brachysclereids* (or stone cells) which are nearly isodiametric in form, (ii) *macrosclereids* which are rod-shaped, (iii) *osteosclereids* which are bone-shaped or spool-shaped and ends of which are enlarged, lobed or branched, (iv) astrosclereids which are variously branched and often star-shaped and trichosclereids which are trichome-like.

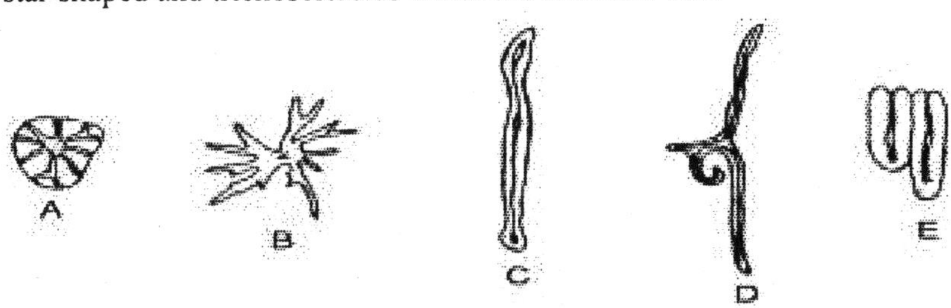

Plate-III: Types of Sclereids

A: Brachysclereid; **B:** Astrosclereid; **C:** Macrosclereid; **D:** Trichosclereid; **E:** Osteosclereid.

They occur in many different places in the plant body e.g. shell of walnuts, seed coats, leaves, phloem, cortex, bark, etc.

The different types of sclereids may be characterstic of species and of systematic significance at various levels. For example, they have been employed at generic level in case of *Camellia, Capparis, Garrya, Rhizophora, Jasminum,* etc.; at sectional level in case of Nelsonieae (Acanthaceae) Abutoideae (Ericaceae), Monsteriodieae (Araceae); at familial level especially the Nymphaeacae, Oleaceae, Connaraceae, Hamamelidaceae, Theaceae, etc. (*cf.* Rao, 1980). Barua and Wight (1958) also employed them at species level in the genus *Camellia.* Their taxonomic significance is utilized also for species of *Memecylon* (Rao, 1957), *Limonium* (Rao and Das, 1981) and palms (Tomlinson, 1959). In monocots, however, they are rare and mostly found in Arecaceae (Palmae,) Agavaceae, Araceae, etc, Carlquist (1961) used them at tribal level. He noted elongate sclereids in the ovular epidermis of various species of the tribe Cynaceae.

Vessels: When the vascular elements have perforated end walls and the cell contents are non-living, they are defined as vessels. They differ in length, obliquity of end walls, number and shape of perforations. The vessel elements which are long and narrow with oblique walls perforated by numerous slits i.e. scalariform perforation plates are thought the most primitive. The vessel elements which are short and broad, and having transverse end walls with a single large circular perforation i.e. simple perforation plate are considered the most advanced. There are intermadiate ones in between these two extremities.

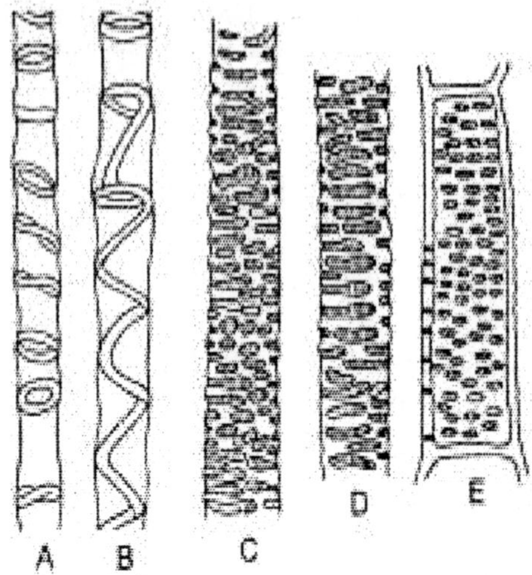

Plate-IV(A): Secondary Wall Thickening In Vessels

A: Annular; **B:** Partly annular, partly helical; **C:** Reticulate; **D:** Scalariform reticulate; **E:** Pitted.

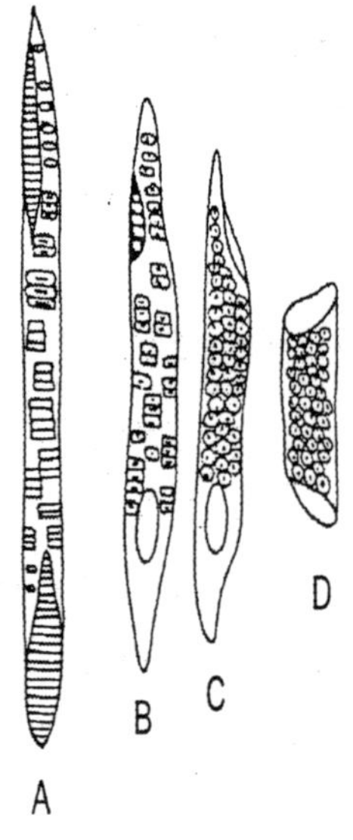

Plate-IV(B): Types of Vessels.

Vessel members are generally thought originated independently in pteridophytes, gymnosperms and angiosperms. In case of angiosperms, they first arose in the secondary xylem and proceeded gradually into the *primary xylem*. In monocotyledons, they appeared in the roots and then in the stem and leaves. Comparisons of vessels amongst these groups and also in angiosperms should be cautiously used particularly for drawing phylogenetic conclusions. Only a few angiosperms like Trochodendraceae, Tetracentraceae Winteraceae Amborellaceae and *Sarcandra* in Chloranthaceae are informed primitively vesseless. Their absence in aquatic families (Ceratophyllaceae, Lemnaceae, Nymphaeaceae, etc.) is considered a reflection of secondary loss of vessels owing to aquatic habitat. Nearly all characteristics of vessel elements are of taxonomic and phylogenetic significance.

Families exclusively with scalariform vessels are much less numerous and most of them (e.g.Austrobaileyacae, Cannellaceae, Eupomatiaceae, Illicinaceae) belong to the Magnoliidae. Vessels in herbs in all families of dicotyledons except few, show *simple performation*. This is an evidence to believe that herbs are derived from woody ancestors (Cronquist, 1968a; Fahn, 1982). In case of monocotyledons, vessels lack in leaves and stems of many families, whereas they are generally present in roots. Secondly, when vessels are present in all three organs, the one present in roots are more advanced. For example, leaves and stems in Alismatiflorae and Ariflorae lack vessels. However, vessels occur in their roots. Vessels in roots of Alismataceae and Butomaceae are simple perforated and in Hydrocharitaceae they show *scalariform perfomration* plates. In Liliiflorae, vessels are present in roots, except *Petrosavia*. Some large families have root vessels with *scalariform* perforation plates e.g. Dioscoreaceae, Taccaceae, and most Liliaceae, and Hypoxidaceae, Root vessels with *simple perforation* plates occur in Velloziaceae, Asparagaceae, Dracaenaceae, Agavaceae, Xanthorrhoeaceae, etc. Roots vessels with *simple perforation* plates are also found in the families *viz.*, Commelinaceae, Restionaceae, Xyridaceae, Gramineae and Cyperaceae (Dahlgren and Clifford, 1982; Fahn, 1982). Several families with particular type of vessel elements have been documented by Metcalfe and Chalk (1950). Since then, many researchers have added data on vessel morphology and their utility in taxonomy. Research work carried out in monocotyledons by Naik and Nirgude (1981b), Cheadle (1942,1943), Cheadle and Kosaki (1971, 1972), Cheadle and Tucker (1961) and Cheadle and Uhl (1948) supported a well-worn generality that vessel specialization is higher in late *metaxylem* than in early *metaxylem* and the root is the organ in which much difference of specialization of vessels occurred in monocotyledons.

A literature resume indicates that some curious or interesting features have not been mostly utilized in taxonomic treatments. For example, different types of nectaries and velamen structure in monocotyledons, colleters, domatia, fruit structure, etc. in all angiosperms have not been paid considerable attention. Future focus on these is obviously needed.

(III) Micromorphology and Ultrastructure

Morphology and anatomy are inseparable aspects of structure. Based on this relation, alternative pairs of terms such as macromorphology, exomorphology (external) and endomorphology (internal) are being used for them in recent time. The former of these two pairs has become useful in the context of the scanning electron microscope (SEM), the term micromorphology generally being applied to electron microscopic features. The term ultrastructure is more usually used for structure seen with the transmission electron microscope (TEM). This advanced technology in microscopy has benefitted greatly in understanding morphology and anatomy. Consequently, it has also benefitted taxonomy. Electron microscope has shown up new details of plant surface and structure. Data is now fairly available for taxonomic and phylogenetic assessments of plants. Leaf surface, trichome surface, fruit and seed surface, pollen grain wall features, stigmas and minute plant bodies of lower plants, etc. are being examined critically. Ultrastructural details cover structure of pollen/ spore walls, sieve tube elements, vessels, few embryological structures and cell inclusions, etc. A subdiscipline called '*Ultrastructural Systematics*' has emerged to aid *synthetic taxonomy* (*cf.* Blair and Turner, 1972) and the gains received from studies are being reviewed and incorporated in various books and literary sources (Heywood, 1967, 1968, 1969; Heywood and Dakhshini, 1971, Cole and Behnke, 1975; Naik, 1984; Sivarajan, 1984; Stace 1989; Stuessy, 1990). A few cases of such studies are highlighted in the following.

The seed surface patterns have yielded very fruitful taxonomic information. Chaung and Heckard (1972) in their SEM studies investigated seed coat micromorphology of the Scrophulariaceae, while Mussleman and Maun (1976) extended similar observations on the Scrophulariaceae and Orobanchaceae. Seavey *et al.* (1977) also studied tribe Epilobieae of Onagraceae. The family Caryophyllaceåe is investigated on similar line by Crow (1979). The genera like *Lupinus* (Heyn and Herrnstadt, 1977). *Mentzelia* (Hill, 1976) and *Lindernia* (Sivarajan, 1981a) have been also studied. The generic distinction of *Mollugo* Linn. and *Glinus* Linn. is validated by seed characters. The former have seeds non-appendaged, whereas the latter showed seeds appendaged. Light microscopy cannot help to distinguish *G. oppositifolius* and *G. lotoides.* SEM studies, however, divulged different shapes of tubercles in these two species. Sivarajan and Usha (1983) supported to reinstate the *Linnean species M. stricta* which has been hitherto considered as *conspecific* with *M. pentaphylla* Linn. A distinct variety, *viz.,M. pentaphylla* var. *rupestris* is described by Cooke (1903) from the then Bombay Presidency. It is distinguished from *M. pentaphylla* by its linear leaves. SEM studies revealed seeds of *M. pentaphylla* areolate covered with a microreticulum, which is lacking in others. Apart from this, the seeds of *M. pentaphylla* var. *rupestris* are similar to those of *M. stricta* rather than those of *M. pentaphylla.* The taxon is, therefore, merged with the latter (Sivarajan and Usha, 1983). Newell and Hymowitz (1978) studied the *interspecific* variations in the genus *Glycine*. Subodh *et al.* (1977) also revealed *intraspecific* variations in the 'Phaseolus mungo-radiatus, sublobatus' complex. These two latter instances suggested that adequate sampling of seeds should be scanned so that a complete range of variation patterns of seed coat microsculpture will be available for taxonomic and phylogenetic analysis. Researches based on single seed samples may lead to erroneous generalizations.

Achenes of *Cyperus compressus* L. and *C. difformis* L. are examined by using scanning electron microscopy to ascertain micromorphological characters to delimit *conspecific* cytotypes by Wujek *et al.* (1997). The achene micromorphology correlated with the vegetative morphology in that the number of nodulated cells increase with chromosome number. All the cells of the n=64 *cytotype* bear a central nodule, while the n = 56 *cytotype* possesses both nodulated and non-nodulated cells. Accordingly, the euploid cytotypes of *C. compressus* bear diagnostic micromorhpological achene characters. However, no diagrostic micromorphological characters were noted between the achenes of the euploid or aneuploid cytotypes of *C. difformis*. These results supported the utilization of achene microphology as a potential tool for the taxonomic recognition of euploid *Cyperus* cytotypes. Darok (2000) surveyed 120 species of the Rubiaceae for their cuticle ornamentations based on SEM studies. He proposed terms of the principal primary and secondary ornamentation of the cuticle present in the family. These terms are of descriptive-diagnostic value.

SEM studies are found useful taxonomically in some asteraceous taxa. Achenes present a variety of features in various asteraceous groups. Kynclova-Petrocyora (1970) found achene characters such as shape, presence or absence of wings, carinae, ribs and border, colour, etc. stable and reliable distinguishing characters in the tribe Anthemidae. Dittrich (1968) examined the achenes of subtribe Cardueae-Centaureinae and used the characters like shape of hilum, detachment of fruits, presence of epicarp, hairiness, etc. Saklani *et al.* (2000) dealt with SEM-characterisation of achenes of 23 species of *Saussurea* DC. from India. The characters such as the base, surface and apex of achenes are found useful in differentiating the species and as such a key to all species is devised.

Stigmatic characters in angiosperms using SEM is also helpful in taxonomy. Heslop-Harrison and Shivanna (1977) and Heslop-Harrison (1981) investigated about a thousand species of 900 genera belonging to 250 families using SEM. They recorded many families homogeneous. For example, Boraginaceae is a family in which this character is, in their opinion, of potential taxonomic value. They also noted correlations between stigmatic types and type of incompatibility systems.

Palynological studies generally involve observations of the number of apertures, exine ornamentation and its internal structure. Light microscopy is generally extended in past. With SEM and TEM techniques, further details of the pollen features and also intine characteristics can be discerned.

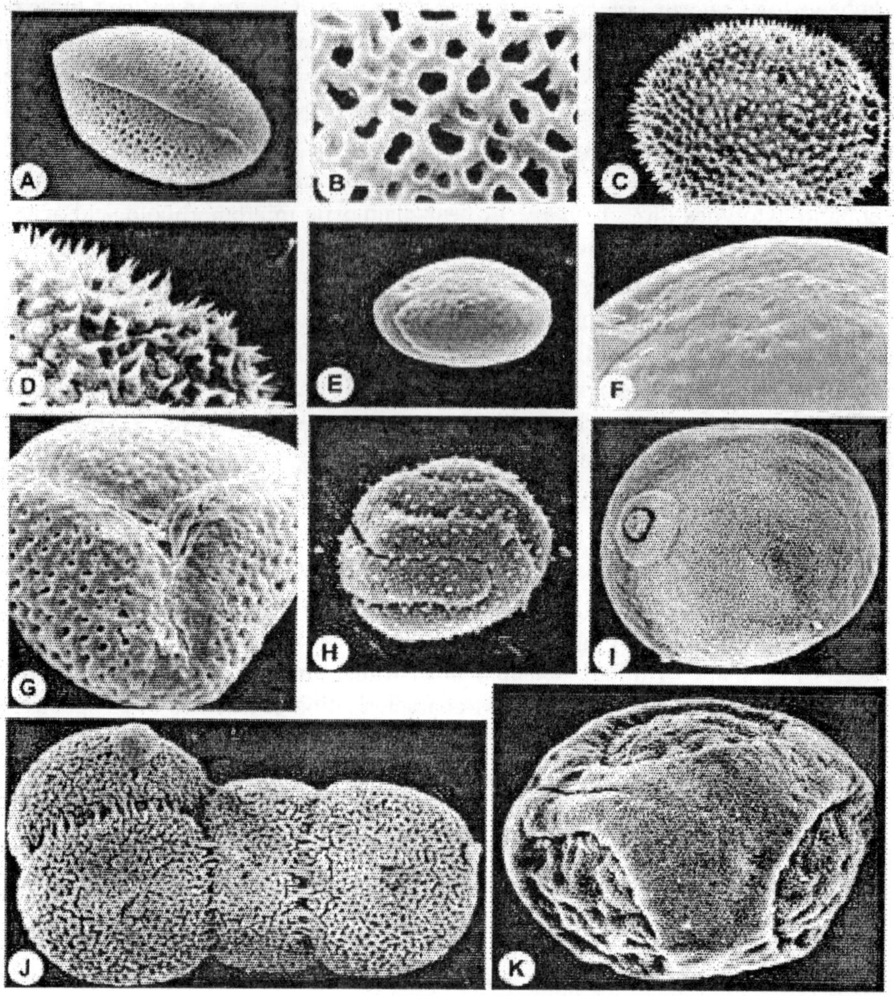

Plate-V: Pollen Grains : SEM

(A) *Butomus umbellatus*: 1-Sulcus at distal face; **(B)** Reticulate exine; **(C)** *Ottelia ulvifolia*: Inaparturate pollen; **(D)** Spinulose exine; **(E)** *Acorus calamus*: 1-sulcoidate pollen; **(F)** Smooth Perforated exine; **(G)** *Houttuynia cordata*: 3-chotomosulcate pollen; **(H)** *Eriocaulon tagawe*: Spireaperturate pollen with spinulose exine; **(I)** *Triticum aestivum*: Ulcerate, Operculate pollen; **(J)** *Typha latifolia*: T-tetrad; **(K)** *Luzula pilosa*: Tetrahedral tetrad.

These have taxonomic and evolutionary significance. The details of surface sculpturing provide an amazing wealth of details of taxonomic value. For instance, three types of ektexine *viz*., columellar, alveolar and granular are recognised. Similarly, two types of endexine are known, *viz*., laminated and nonlaminated. It has been possible to distinguish angiosperms from fossil gymnosperms of doubtful angiosperm affimities. Angiosperms are characterised by granular or columellar ektexine and nonlaminated endexine or its absence in the nonapertural regions of the pollen grains. Gymnosperns are characterised by granular or alveolar *ektexine* and laminated *endexine*. Doyle *et al.* (1975) opined, after a critical review of the available data, that the typical endexine was lost during the transition from gymnosperms to angiosperms, and then re-elaborated with different structure within the angiosperms. Walker (1974b) recongnised three basic types of exine structure *viz.*, tectate, semitectate and intectate. These types are based on their LM and TEM techniques. The *tectate* pollen may be tectate-perforate and tectate-imperforate. In their survey, they noted majority of families having pollen grain with tectate exine. They noted evolutionary trend from tectate-imperforate to tectate- perforate to *semitectate* and more rarely to intectate pollen. More specialied genera exhibit *semitectate* and *intectate* pollen. Although, angiosperm pollen show well developed exine and thin extine, there are certain families, especially of monocotyledons, wherein a highly reduced exine and an elaborate *intine* are observed, e.g. families Musaceae, Zingiberaceae, Costaceac, Marantaceae, Heliconiaceae, Strelitziaceae, Cannaceae and Iridaceae. The details of these are being studied in recent years. For example, in *Heliconia*, a very thin layer of exine is present with few spinules. The wall is almost entirely intine (Kress, Stone and Sellars, 1978). A similar situation is known in *Canna* (Rowley and Skarla, 1986).

Palynodiversity of 21 species of the genus *Berberis* (Berberidaceae) is studied using SEM technique by Rao *et al.* (1999). They found the genus eurypalynous and the diversity with the extremes such as inaperturate to *spireaperturate* apertural system; *psilate*-smooth to foveolate surface ornamentation, and other extremes of shape and size. They proposed subdivision of the genus and also provided palynological species key for indentification. They concluded that the evolution of surface ornamentation within the taxa is also parallel to the apertural evolution.Skvarla and Nowicke (1976) and Nowicke and Skvarla (1977, 1979) investigated the order Caryophyllales using SEM technique. It is characterised by 'centrospermous type' of pollen with a spinulose, punctuate-perforate tectum. The genus *Cordia* (Boraginaceae) with 250 species is studied using SEM technique by Nowicke and Ridgeway (1973). They correlated exomorphology and pollen features such as aperture and exine ornamentation in the different sections of the genus. They suggested distinet generic status for the three groups.

'*Dilated Cisternae*' (DC) of the endoplasmic reticulum are found in the root cells of *Raphanus sativus*. This was revealed for the first time by Bennet and NewComb (1965) using TEM technique. They are similarly noted in phloem parenchyma of foliar veins in case of *Brassica chinensis*. This fact was noted particularly by Favali and Gerola in 1968. Later, these have been also noticed in *Capparis cynophallophora* by Jorgensen (1981). '*Dilated Cisternae*' of the *ER* occur only in families of the order Capparales, where they appear to be functionally related to the presence of glucosinolates, a group of Carpparales-specific secondary plant products. They are equally restricted to myrosin-cells which contain the enzyme myrosinase. Thus the DC are characteristic of the order Capparales and hence of taxonomic significance.

The most interesting and useful taxonomic features revealed particularly during 1965 to 1983 on *seieve-tube* plastids in a series of papers by Behnke and his co-workers. These plastids were first detected in phloem in case of Dioscoreaceae (Behnke, 1965, 1967). Since then more than 1500 species belonging to 380 families have been searched out (*cf.* Behnke and Turner, 1971; Behnke and Dahlgren, 1976; Mabry and Behnke, 1976; Mabry, Behnke and Eifert, 1976; Behnke and Mabry, 1977; Behnke, Pop and Sivarajan, 1983, etc.). The occurrence of many basic plastid types with differences relating to the inclusions and shapes have been investigated using TEM technique. Their great potentiality for use in taxonomy has been well illustrated and is being documented.

In a nutshell, these plastids are basically of three principal types based on their inclusion of starch and protein: (i) *S-type*: These accumulate starch. (ii) *P-type*: These accumulate protein and (iii) *So-type*: These accumulate neither starch nor protein.

S-type have been recorded for about 65% of the flowering plants (Behnke, 1977). They occur in the subclasses Ranunculidae, Dilleniidae and Hamamelidae, many orders in the Rosidae, nearly 50% in the Magnoliidae, a few in order Caryophyllidae and Rhamnaceae (Behnke 1974b). P-type occur only in Pinaceae among gymnosperms (Behnke, 1974a).They occur among the monocotyledons in total 21 families and among the dicotyledons in Vitaceae and Leeaceae (Behnke, 1974b) and nearly 50% members of the Magnoliideae. So-type are observed in the families e.g. Rafflesiaceae, Crassulaceae and in some taxa of Urticaceae, Ulmaceae and Moraceae.

Occurance of P-type is thought to be primitine condition in flowering plants. Loss of protein in the P-type plastid resulted in the development of S-type, which is thought an advanced condition. The Magnoliidae in the light of this study is considered heterogenous. This is attributed to its basal position in the evolution of flowering plants (Behnke, 1975). The occurrence of P-type in all groups of monocotyledons and predominance of *S-type* in the basal groups of dictotyledons is conceived as an additional witness for the independent origin of dicotyledons and monocotylelons (*cf.* Meeuse, 1970). They are also divided into subtypes and are characteristic of some plant groups e.g. monocotydons show P II, Centrospermae PIII, Magnoliales/ Laurales PI and Fabales PIV. Some taxonomic realignment have been also possible in flowering plants. For instance, (i) *Dysphania* (Mabry and Behnke, 1976) and *Halophytum* (Hunziker *et al.*, 1974) have been confirmed in the Caryophyllales based on these plastids. (ii) *Theligonium*, often placed under Phytolaccaceae or Chenopodiaceae, is found containing S-type plastid (and also andhocyanin). Its affinity with Caryophyllales (Centrospermae) is ruled out (Mabry *et al.*, 1975).

The ultrastructural features, hand in hand, the chemical data, have revolutionised taxonomic concepts. For example, the features exhibited by the blue-green algae such as their cell-wall structure, ribosome characteristics and lack of endomembranes are more closely related to bacteria than to algae. In recent time, these two groups are generally kept under one taxonomic entity, the Procaryota. Even within the true algae, ultrastructural characters of plastids have strengthened and clarified our views on the basic differences between the major divisions such as Chlorophyta, Chrysophyta,

Rhodophyta and Phaeophyta. The ultrastructural studies alongwith biochemic studies, also suggested that the Charales are less closely related to most of the Chlorophyta (*cf.* Raven, 1977; Stewart and Mattex, 1975).

(IV) Embryology

Hofmeister and Strasburger, as far back as the second and third quarter of the 19th century, indicated the possibility of using embryological features in taxonomy. Much higher standard were set in descriptive embryology with the commencement of the 20th century. Particularly in Germany and Sweden, many embryologists started to discuss the usefulness of embryological data in view of the interrelationships of the families and genera they studied. Since then various opinions regarding the applicability and status of embryological features in relation to taxonomy have been expressed. It appears pertinent to review them briefly to which the reader is referred.

Corner (1951b) commented on the different classifications for various kinds of embryo sac, endosperm and *embryogenesis*. He further stated that the different types seem to have a little or no taxonomic significance and that they are either examples of polyphyletic convergence from diverse families or the classifications are not critical enough and apparent similarities have been confused. Davis and Heywond (1963) remarked on the list of embryological features of taxonomic value provided by Cave (1959). In their opinion, the list may be impressive but the data contributed by embryology to taxonomy are rather scarce and are generally not well known. They further stated that there is much in the subject that is difficult to interpret and technical problems add to the complications. According to Kapil and Bhatnagar (1980), the embryological characters are less prone to adaptive stresses and they are relatively stable and have acquired great significance in plant taxonomy, especially when external morphological characters are inconclusive and misleading due to convergence. Herr (1984) evaluated the use of embryological characters in plant taxonomy. He stated that these characters have not yet reached its potential in past due to the great number of embryological features that potentially can be documented and of which only a few usually are described in any one study. He further remarked that the analysis of embryological characters in comparative studies, while sufficiently broad, often lacks in-depth and insufficient attention to detail. Nevertheless, we cannot ignore the concluding remarks made by Maheshwari (1950) in his chapter on 'Embryology In Relation To Taxonomy'. He stated that what has already been written is sufficient evidence of the value of embryological data in an elucidation of the interrelationships of families and genera. Although it is not claimed that these data will always prove important, they should form a part of any thorough taxonomic analysis. He further remarked that every contribution, however small, will be one step nearer to the final goal, and occasionally there will emerge fresh ideas and new viewpoints which will more than compensate for the effort extended upon the work.

Coulter and Chamberlain (1903) summarised the basic understanding of the reproductive processes in plants and signalled the readiness of the discipline for comparative studies for purpose of assessing relationships. Schnarf (1929, 1931) gave the earliest compendium of comparative data and the most recent by Johansen (1950) and Davis (1966). The potential role of embryology in taxonomy has been excellently discussed at length by various workers (*cf.* Maheshwari, 1958, 1964; Johri, 1963, 1967; Kapil and Bhatnagaer, *loc. cit.*, Kapil, 1962; Cave, 1962, etc.). Additional data from various taxa have steadily accumulated over later period. The most significant insights in structure and function of embryology have come recently with the use of Electron Microscope.

Any particular character varies enormously from one plant group to another in its taxonomic value. It is quite impossible to predict this value in a group in which that character has not been previously investigated. Too much reliance, as indicators of taxonomic relationship should not be placed on characters which have been shown to have frequently resulted from evolutionary convergence. There can be a score of features of taxonomic value. These have been well summarized by Maheshwari (1950), Cave (1959), Johri (1963), Palser (1975), Bhojwani and Bhatnager (1983) and Herr (1984). These have been provided in many other texts (*cf.* Naik 1984; Mondal, 2005, etc.). Their repetition is avoided here for precision.

Maheshwari (*loc. cit.*) interpreted embryology in a broader sense so as to include those events that take place in the ovule prior to fertilization as well as afterwards, including the development of the male gametophyte and fertilization. Cave (*loc. cit.*) also adopted a similar view. An assessment of the various embryological features has been of considerable service in deciphering the proper placement of several difficult plant groups and subgroups. Sometimes, it has given a new orientation to our idea of their affinities. It is, however, impossible to review the vast amount of literature on this subject matter put on record over the last few decades. The following selection is offered to assess the value of the different embryological characters in plant taxonomy and relationships.

Applications of Embryological Data:

- ***Paeonia***: The genus *Paeonia* has greatly attracted the attention of morphologists and workers in various disciplines. Most systematists included the genus in the family Ranunculaceae (Bentham and Hooker, 1862-1883; Benson, 1957; Engler and Diels, 1936). There are others who referred it to a separate family of its own-the Paeoniaceae (Hutchinson, 1973; Takhtajan, 1969). Exomorphologically it is different from other members of the family Ranunculaceae in highly undifferentiated perianth, persistent sepals bending down while in fruit, gaudy corolla, and slightly concave receptable, broad lobed surrounding the carpels, centrifugal stamens, and fruit an etaerio of massive follicles. The genus is investigated anatomically (Ozenda, 1949), floral anatomically (Eames, 1961), karyologically (Gregory, 1941) and palynologically (Erdtman, 1952; Wodehouse, 1935) etc. All these studies favour its separation from the Ranunculaceae. The genus *Paeonia* has been well investigated embryologically (Murgai, 1959, 1962; Yakovlev, 1951; Yakovlev and Yoffe (1957a, 1957b,

1959, 1961, 1965). It differs from the Ranunculaceae in anther tapetum, ovule arillate and with massive outer integument overtoping the inner one, nucellus ephemeral, presence of vasculature in the integument, embryo sac enclosed by the inner integument, coenocytic free nuclear phase during embryogenesis and hypogeal germination. The coenocytic phase in the embryogenesis is common in the gymnosperms but unknown elsewhere within the angiosperms. The above embryological dissimilarities strongly support the exclusion of *Paeonia* from the Ranunculaceae to a separate family, the Paeoniaceae. Taktajan (*loc. cit.*) erected an order of its own, the Paeoniales. It is, however, taxonomist's acrobatics.

- ***Trapa*:** This genus has received various treatments based on exomorphology. It is included in the Onagraceae (Bentham and Hooker, 1862-1883). Hutchinson (1959, 1973), Engler and Diels (1936), Takhtajan (1969), Cronquist (1968a) and Benson (1957) include it in the *unigeneric* family, the Trapaceae (*cf.* Pulle, 1938). The family Onagraceae exhibits characteristic features such as ovary inferior and 4-locular, ovules numerous on axile placentae and fruit generally being loculicidal capsule. On the contrary, in *Trapa,* the ovary is semi-inferior and 2-locular, ovule solitary per locule and the fruit being drupe. The family Onagraceae and the genus *Trapa* are investigated embryologically by Ram *nee* Ghosh (1956). Earlier Maheshwari (1950) also called attention to those features in which the embroylogy of *Trapa* differs from that of the Onagraceae. Later, these differences were also assessed by Johri (1963) and Seshvataram (1970). The members of the Onagraceae are characterised embryologically by 4-nucleate embryo sac of the Oenothera-type derived from micropylar megaspore, nuclear endosperm, embryogeny Onagrad type and embryo with typically two cotyledons. The genus *Trapa* is quite distinct in having 8-nucleate embryo sac of Polygonum-type derived from chalazal microspore, endosperm wanting (embryo develops at the expense of the nucellus), embryogeny Solanad type and embryo with one cotyledon suppressed. These embryological differences fully support the disbanding of the genus *Trapa* from the Onagraceae and its *monotypic* family Trapaceae. Eames (1951) also lent support to such a conclusion on anatomical basis.
- ***Actinidia*:** The genus is variously treated in different systems of classification. Hutchinson (1959, 1973) and Cronquist (1968) recognised it as a distinct family from the Dilleniaceae and placed in the Theales. Benson (1957) also treated it similary. Solereder (1898) assigned it to the Theaceae. Bentham and Hooker (1862-1883), Baillon (1867-1895) and Chesnais (1941) placed *Actinidia* in the Dilleniaceae. Engler and Diels (1936) recognised Actinidiacae but placed along with the Dilleniaceae and Theaceae in his Theineae of the order Parietales. Takhtajan (1969) kept the Actinidiaceae under his order Ericales. Lawrence (1951) separated it from the Dilleniaceae by stamens distinct or adnate to the petals, pistils distinct, *bitegmic* seeds, and small embryo. Van Tiegham

(1899) on the basis of *unitegmic* ovules recongnised the Actinidiaceae distinct from Dilleniaceae wherein the ovules are *bitegmic*. *Actinidia* is investigated by Vijayraghavan (1963). The families *viz*., Dilleniaceae and Theaceae have been studied by a number of embryologists (*cf.* Johri, 1963; Vijayraghavan, 1970). The genus *Actinidia* has ovules *unitegmic*, *hypostase* and *endothelium* present, endosperm cellular, embryo Solanad type and non-arillate seeds. All these are, however, unknown in the Dilleniaccae. On the contrary, the genus *Actinidia* is similar to the Theaceae in secretory tapetum, ephemeral nucellus, degenerating antipodals, Solanad *embryogeny*, *polyembryony* prevalent and seeds with persistent outer epidermis of the integuments. It also resembles to the another thealean family, the Marcgraviaceae. The creation of separate family Actinidiaceae is justified. Furthermore, it should be assigned to the Theales rather than the Dilleniales.

- ***Cuscuta*:** Bentham and Hooker (1862-1883) included this genus in the Convolvulaceae in their tribe Cuscuteae. Engler and Diels (1936) placed it in their subfamily Cuscutoideae under the Convolvulaceae. Thorne (1992) maintained similar treatment. Rendle (1979) and Benson (1957) retained it similarly under the Convolvulaceae. It reçeived a separate family status, the Cuscutaceae e.g. Wettstein (1935), Hutchinson (1959, 1973), Takhtajan (1969, 1980) and Cronquist (1968). Floral parts in *Cuscuta* and other members of Convolvulaceae receive similar vascular supply except calyx and gynoecium. (Singh and Pai 1977; Tiagi and Gupta, 1963; Govil and Lavania, 1980). Embryo sac development is followed that of Allium type, which is however Polygonum type in the Convolvulaceae (Davis, 1966; Govil, 1970; Govil and Lavania, 1980; Tiagi 1951, 1965). The adult embryo in *Cuscuta* is undifferentiated, acotyledonous, terete and coiled, which is not observed in the other Convolvulaceae. Moreover, Tiagi (1951) stressed separation of *Cuscuta* from the Convolvulaceae on the basis of absence of parietal cell, bisporic embryo sac, coenocytic and vesicular suspensor, besides the much coiled hypocotyl. A sum total of evidence indicate its placement in a distinct taxonomic entity, the family Cuscutaceae.
- ***Martynia*:** The genus is a part of the family Pedaliaceae in some taxonomic treatments (Bentham and Hooker, 1862- 1883; Cronquist, 1968). On the basis of parietal placentation in *Martynia* and *axile placentation* in the Pedaliaceae, Engler and Prantl (1897) separated the former in a distinct family alongwith few other genera. It has also been accorded similar position by majority of taxonomists (Engler and Diels, 1936; Takhtajan, 1969; Huthinson, 1973; Benson, 1957). Palynologically it is distinct from other Pedaliaceae in *inaparturate* pollen grains (Erdtman, 1952). Embryological investigation is at par with above evidence. The pollen grains are shed at 2-celled stage in *Martynia* and at 3-celled stage in the Pedaliaceae. *Martynia* differs characteristically from the Pedaliaceae in having two layers of palisade cells in the cotyledons

in contrast to single layer in the Pedaliaceae. Catalpa variation of Onagrad type *embryogenesis* observed in *Martynia* is not encountered in the Pedaliaceae. Likewise, *hypostase* is absent in the former as noted in the latter (Singh, 1970, 1970b, 1960, 1963, 1964). Embryologically, *Martynia* shows some distinctive features and amply supports its disbanding from the Pedaliaceae and institution of a family-the Martyniaceae alongwith few others.

Loranthoideae-Viscoideae (Loranthaceae): The family Loranthaceae comprises two subfamilies *viz*., Loranthoideae and Viscoideae (Danser, 1929, 1933a, b). Lawrence (1951) also accepted these as natural groups within the Loranthaceae and characterised them as *calyculus* present in the Loranthoideae and its absence in the Viscoideae. Takhtajan (1969), Benson (1957), Cronquist (1968a) and Hutchinson (1973) are silent regarding its division. Miers (1851) suggested the erection of a new family, the Viscaceae. Engler and Prantl (1897) and Schultze-Motel (1964) also recongnised two subfamilies within the Loranthaceae. Bentham and Hooker (1862- 1883) treats these subfamilies as two separate tribes. Barlow (1964) and Van Tiegham (1896), however, accorded them familial ranks. Van Tiegham (*loc.cit.*) even raised them to the ordinal rank, the Loranthales and the Viscales. Maheshwari (1954) pointed out a need for two separate families based on embryological characteristics. Embryological features have been revealed by some workers (*cf.* Bhandari and Kanan, 1968a, 1968b; Bhandari and Indira, 1969; Correa, 1958; Dixit, 1958a, 1958b, 1961; Schaeppi and Steindl, 1942, 1945, etc.). Their embryological status is evaluated (*cf.* Raj, 1970; Johri, 1963; Maheshwari and Dixit, 1957). The Loranthoideae have mostly shrubby plant body and bear bisexual flowers with a calyculus. While the Viscoideae have a reduced plant body with a tendency towards complete parasitism, the flowers being unisexual and without a calyculus. The Loranthoideae show several embryological features peculiar to it such as monosporic Polygonum type of development of embryo sac and its elongation into the style, composite endosperm, vertical division of the zygote and downward growth of the biseriate proembryo. In contrast, the Viscoideae have a bisporic Allium type of embryo sac development with a characteristic curvature, the endosperm not composite, the zygote dividing transversely and the embryo without a suspensor. There are good reasons for elevating both the subfamilies to the familial status *viz*., Loranthaceae and Viscaceae (*cf.* Johri and Bhatnager, 1960, 1972).

Menyanthoideae-Gentianoideae (Gentianaceae): The family Gentianaceae is subdivided into two subfamilies *viz*., Gentianoideae and Menyanthoideae in some treatments (Bentham and Hooker, 1862-1883, Rendle, 1952), while there are others who are silent in regard to their such a splitting (Benson, 1957, Engler and Diels, 1936). Lawrence (1951) although adhere to such a treatment, but discussed about this splitting. Hutchinson (1973), Cronquist (1968a) and Takhtajan (1969) raised these subfamilies to the level of independent families *viz*., the Gentianaceae and the Menyanthaceae. Hutchinson (*loc.cit.*) and Takhtajan (*loc.cit.*) included both of them under the same order, the Gentianales,. However, Cronquist (*loc.cit.*) referred Gentianaceae to the Gentianales and Menyanthaceae to the Polemoniales. He opined that the Menyanthaceae would be a highly discordant element in the Gentianales. Lawrence (*loc.cit.*) characterised the Gentianoideae by leaves opposite, corolla involute or imbricate and plants terrestrial. The Menyanthoideae have, in his opinion, leaves alternate, corolla induplicate-valvate and plants marshy or aquatic. This is also supported by vegetative anatomy (Perrot, 1897), floral anatomy (Lindsey, 1938) and chemistry (Cronquist, *loc.cit.*). Cronquist (*loc.cit.*) also remarked that these disciplines render it at home in the Polemoniales. Embryological investigations are also at par and at least favour their separation from each other (Souges, 1943b, Crete, 1949a, 1949b, 1955, 1956a) and cirtically evaluated be Johri (1963). The Menyanthoideae have characteristic embryological features such as tapetal cells bi-to multinucleate and show nuclear fusion commonly, suppressed *parietal* placentation, *endothelium* present, endosperm cellular and *embryogeny* of Asterad type. While the Gentianoideae lack all these characteristics. These fully justify their disbanding from each other and lend support to institute independent families, the Gentianaceae (*sensu stricto*) and the Menyanthaceae.

Limnanthaceae: This family contains only two genera *viz.*, *Limnanthes* and *Floerkea.* It is referred either to the order Sapindales (Engler and Diels, 1936; Engler and Prantl, 1897) or to the Geraniales by the most taxonomists (Hutchinson, 1926, 1973; Melchior, 1964; Takhtajan, 1969; Cronquist, 1968a; Benson, 1957). The Limnanthaceae, however, show distinctive exomorphic features such as regular flowers, stamens two-whorled and alternating with petals, ovaries separate from each other, style *gynobasic*, ovule solitary and *basal-parietal* and fruits generally schizocarpic, one-seeded indehiscent achenes (nutlets). These exomorphic features are, however, not encountered either in the Sapindales or the Geraniales, and appear to have been overlooked in past by the systematists. This alliance has been well searched for their embryology (*cf.* Johri 1970).

The Limnanthaceae differ embryologically from both the Sapindales and the Geraniales in unitegmic, *tenuinucellate* ovules and unusual type of tetrasporic embryo sac. The exomorphic features pointed out above are characteristic of the Limnanthaceae. They are suggestive of raising the Limnanthaceae to the level of order, the Limnanthales. Embryological evidence is also at par to such a conclusion. Maheshwari and Johri (1956) and Johri (1970) also suggested an order of its own, distinct from the Geraniales and Sapindales.

The construction of classificastion depends upon a careful comparison of attributes of the plants (organisms). It does not suffice by simply stating that plant 'A' is more similar to plant 'B' than either is to 'C' without expressing the particular reasons for such a conclusion. Such expressions of relationship are based upon a comparison of features of plants. Data for analysis come in many forms and the plant taxonomist must be able and ready to handle many different types. A taxonomist usually finds it necessary to generate original data so that the relationships among the considered plants can be understood more clearly. In the following, is discussed type of embryological data useful in assessing relationships amongst the various plant groups.

Moringaceae: The systematic position and relationship of this unigeneric family have been a matter of controversy and debate. Bentham and Hooker (1862-1883) placed it in their 'Ordines anomali' which is in between their series Disciflorae and Calyciflorae, although have hypogynous disc and calyx-tube. Jussieu (1789) and Hallier (1905, 1912) included it in the Leguminosae. Lindley (1830) assigned to the order Violales. Benson (1957) included in his order Papaverales alongwith capparalean families. Hutchinson (1972), Cronquist (1968a) and Takhtajan (1969) treat it under their order Capparales. This resume is quite suggestive of its relationships based on exomophic features. It is generally, characterised by 5-*merous* bisexual flowers, tricarpellary syncarpous gynoecium, *parietal* placentation and a short *gynophore*. Embryologocially it is investigated by Hedaytullah (1935), Narayana (1962), Puri (1934), Rutgers (1923) and later also assessed by Narayana (1970). Moringaceae have little in common with the Violaceae wood anatomically (Metcalfe and Chalk, 1950) and also palynologically (Erdtman, 1952). Persistent *hypostase* and Asterad type of embryogeny in Moringaceae also lessen the possibility of its relationship with the Violaceae. Its placement in the family Leguminosae is also untenable (Dnyansagar, 1955). The Leguminosae are characterised by the presence of monocarpellary gynoecium, *marginal* placentation, coenocytic chalazal endosperm haustorium and diverse type of *embryogeny*. Moringaceae cannot be related to Polygalaceae as it lacks '*endosperm process*' found in the latter (Srinivasachar, 1942). Moringaceae embryologically resemble Capparidaceae in glandular tapetum, *simultaneous cytokinesis*, Polygonum type of embryo sac and *nuclear endosperm* (Narayana, 1970a. 1970b). Narayana (*loc.cit.*) also thought derivation of Moringaceae from Capparidaceae. Palynologically, with 3-colpate pollen grains (Narayana, *loc. cit.*; Erdtman, *loc.cit.*) and anatomically, especially in the presence of '*myrosin cells*'. Moringaceae can be also allied with the Capparidaceae.

Podostemaceae: It is an exclusively tropical family of obscure affinities. Even their identification is misled by the beginners because of their exomorphic features. They are generally described as submerged herb-like mosses, hepatics or algae (Lawrence, 1951; Hutchinson, 1973). It is the common experience of many that during field studies they are usually collected under the name of the aforesaid plant groups. This is because of their habitat similarity and being attached to stones or rocks, often thallus-like with photosynthetic creeping roots. Engler and Diels (1936), Cronquist (1968a) and Takhtajan (1969) included only the Podostemaceae in their order Podostemales. Takhtajan (*loc. cit.*) thought it derived from the Crassulaceae-like ancestors. Hutchinson (1926, 1973) and Benson (1957) placed it alonghwith the Hydrostachyaceae under their order Podostemales. Hutchinson (*loc.cit.*) derives the Podostemales from the Saxifragales, and Benson (*loc.cit.*) relates them to the Crassulaceae and Saxifragaceae. The relationship of Podostemaceae has been also suspected with the Saxifragaceae by Warming (1891), Engler (1930) and Rendle (1952) and with the Crassulaceae by Mauritzon (1933). This alliance has been investigated embryologically (Chiarugi, 1933; Chopra and Mukkada, 1966; Magnus, 1915; Mukkada, 1962, 1964; Razi, 1949a, 1955; Subramnyam, 1962; Went, 1908, 1910, 1912, 1926). The Hydrostachyaceae differ from the Podostemaceae in *parietal* placentation, pollen grains united in tetrad, unitegmic ovules, monosporic embryo sac and *cellular endosperm*. It is notable that *pseudo-embryo* sac, generally observed in the Podostemaceae, is absent in the Hydrostachyaceae. The Saxifragaceae also differ from the Podostemaceae in structure of tapetum, pollen grain single (not in pairs), ovule *crassinucellate*, embryo sac monosporic Polygonum-type and endosperm present (cellular, nuclear or helobial). All these are wanting in the Podostemaceae (Rau, 1970; Kapil, 1970). There are some similarities in between the Podostemaceae and the Crassulaceae, except absence of bisporic development and *pseudo-embryo* sac. Mauritzon (1933) suggested closer relationship amongst them based on the highly developed suspensor haustorium. Reduced endosperm and mode of life of *Crassula aquatica* is thought a connecting link between these families. Although so, the family is still distinctive embryologically. The set of embryological features such as double pollen grains, pseudo-embryo sac, bitegmic *tenuinucellate* ovules, bisporic development of embryo sac, Solanad type of *embryogenesis*; and also absence of *triple fusion* and endosperm, etc. is unmatched and does not find parallel amongst the allied families.

***Exocarpos* (*Exocarpus*):** The genus was suggested for assigning a separate family, the Exocarpaceae by Gagnepain and Boureau (1946, 1947). Interestingly, it was further suggested for placement near the Taxaceae under the gymnosperms. These authors justified their viewpoint on the basis of 'naked' ovule and articulated pedicel, similar to the gymnospermic genera *Podocarpus* and *Acmopyle*, and presence of a *pollen chamber*. It was included by Lam (1948) in the Protangisopermae alongwith *Casuarina* (Casuarinaceae) and *Salix* (Salicaceae). Smith and Smith (1942) thought *Exocarpus* a specialised genus of the family Santalaceae. Studies on the anatomy of the stem, node and pollen morphology lend further support to the view expressed by Smith and Smith (*loc.cit.*). Sussenguth (1954) pointed out a need for embryological study of this genus before any such conclusion. Later, Ram (1959) investigated three species of the genus. The features such as the usual angiospermous structure of flower and anther, pollen grains 2-celled at shedding, Polygonum type of embryo sac development, the primary nucleus giving rise to a micropylar and a chalazal chamber (micropylar chamber produces the main body of endosperm zygote) dividing transversely and the suspensor proliferating into additional embryos, the integument consumed and endosperm surrounded by the pericarp (the seed not 'naked'), etc. are all angispermic. This set of embryological features also lends support to its placement within the Santalaceae. It differs from other genera of the Santalaceae only in its straight embryo. (*cf.* Bhatnagar, 1970).

***Elytraria*:** Lindau (1895) constituted four subfamilies within the family Acanthaceae, of which the subfamily Nelsonioideae has been doubted for its validity under the Acanthacae by Bhaduri (1944). Bremekamp (1953) suggested its transfer to the family Scrophulariaceae. Johri and Singh (1959) investigated embryology of *Elytraria acaulis* (Nelsonioideae). It differs from rest of the Acanthacae in having placentation parietal, integumentary tapetum, persistent antipodals, *ruminate* endosperm, and absence of conspicuous testa. *Elytraria* are rosette herbs with alternate leaves and scapigerous racemose inflorescence, whereas most Acanthaceae are erect herbs or shrubs with opposite leaves and cymose inflorescence. There are certain embryological resemblances amongst the genus *Elytraria,* tribe Rhinantheae (Scrophalariaceae) and family

Acanthaceae. They share common features such as *unitegmic tenuinucellate* ovules, Polygonum type of embryo sac development, *cellular* endosperm with chalazal and micropylar *haustoria. Elytraria* and Rhinantheae both show early segmentation of the endosperm corresponding to the Pedicularis type (Crete, 1951). In later development, however, they differ fundamentally (Banerjee, 1961; Crete, *loc.cit.*; Berg, 1954). This feature is not known in the other allied families (Mauritzon, 1934). Interestingly, seeds in *Elytraria* possess *jaculators*, a feature unknown in Rhinantheae. Arekal (1963) on the basis of embryological features opined that the Rhinantheae are not related to the Nelsonioideae. In a nutshell, embryologically *Elytraria* is better at home in the Acanthaceae rather than the Scrophulariaceae (*cf.* Johri, 1963).

There is a score of embryological characters that would help taxonomists in their job. For instance, the delimitation of angiosperms as a whole largely depends on embryological characters. Enclosed mostly anatropous ovules, Polygonum type of embryo sac development, double fertilisation and triple fusion, and post-fertilisation development of polyploid endosperm are some of the features that not only characterise the entire group of angiosperms but also lend support to its probable monophyletic origin. The phenomenon of '*double fertilisation*' is exclusively the most distinctive characteristic of flowering plants in contrast to lower groups. The pollen grains caught in the stigmatic exudate germinate under the influence of the phenolic compounds in the secretion only if the pollen wall and the pistillate proteins are compatible (Kapil and Bhatnagar, 1975). The absence of endothelium and presence of *helobial* endosperm characterise the monocotyledons from the dicotyledons. The most remarkable dichotomy-mono-dicot in angiosperms is also characterised by embryo characters. At the time of anthesis, angiosperm pollen is 2-nucleate or 3-nucleate. Generally, monocotylelons, polypetalous and monochalmydeous dicotyledons are 2 nucleate, whereas *gamopetalous* dicotyledons are 3-nucleate (Dufou, 1961).

Equally good examples of the efficacy of embryological data exist at the various taxonomic levels. The anther contains many such features. For example, Buss and Lersten (1975) employed tapetal nuclear number in the Leguminosae. Variations in endothecial cell wall patterns in the Compositae are valuable (Dormer, 1962; Vincent and Getliffe, 1988) and also in the Araceae (French, 1985a, b). Prosperi and Cocucci (1979) and Cocucci (1983) thought variation in presence or absence of *callose* in pollen tubes valuable diagnostically for the Polemoniales. Many aspects of ovules have been examined taxonomically. The situation in the parasitic Loranthaceae is especially fascinating in that the ovules are sometimes without integuments and even disappear as distinct units (Maheshwari, Johri and Dixit, 1957; Cocucci, 1983). Apart from usual Polygonum type embryo sac, the Poaceae variant of it is interesting. Anton and Cocucci (1984) found the antipodal dividing to form 6-300 cells. Fish (1970) discovered two different embryo types in *Clematis* (Ranunculaceae) in fertile (monosporic) and sterile (tetrasporic) ovules which led to inferences on evolutionary directionality within the genus. Anderson (1970) used variation amongst the species of the genus *Chrysanthemum* (Compositae) especially in the number of antipodal cells. In the Gramineae, six major patterns are noted *viz.*, festucoid, bambusoid, centothecoid, arundinoid-danthonioid, chloridoid-eragrostoid, and panicoid (Reeder, 1957, 1962). These are characterised by the course of vascular system, presence of epiblast, fusion of scutellum to the coleorhiza, and the cross-sectional appearance of the embryonic leaves. The different embryo types correlate with other features to help define tribal limits. Philipson (1977) recognised four types of ovules in the dicotyledons by integument number and nucellar thickness *viz.*, bitegumentary- crassinucellate, bitegumentary- tenuinucellate, unitegmic-crassinucellate, unitegmic-tenuinucellate. He discussed the value of these characters in defining the major subclasses in angiosperms. It is inferred that iridoid compounds are confined to plant groups which exhibit *unitegmic, tenuinucellate* ovules, even though it does not occur in all of them (Jensen *et al.*, 1975).

A mention may be made finally of cases of taxa which are clearly marked out by their embryological features. It is doubless that double fertilisation is characteristic of group of angiosperms, as also the endosperm formation during post-fertilisation period. The family Cyperaceae is demarcated from all other angiosperms in the development of the male gametophyte. Although four microspores are formed, only one microspore enlarges and develops a pollen grain (*pseudomonad*) (a point explained separately elsewhere). This feature is sufficiently distinctive to provide a ready and dependable method of indentifying members of the Cyperaceae from their pollen grains alone. Similary, the members of the family Onagraceae (except *Trapa)* show a monosporic 4-nucleate embryo sac. The Ericales are a homogenous group of sympetalous families in which most members share a long series of embryological features in common, including, notably, the production of pollen in tetrads. Similarly the Centrospermae or the Caryophyllales are embryologcially very distinctive. They show characteristically trinucleate pollen, bitegmic crassinucellate ovules, inner integument longer than the outer, embryo curved, presence of perisperm in seed, etc. No plants outside this order are known to present this combination of characters (*cf.* Cronquist, 1968a). Similar is the case of the family Podostemaceae explained earlier. The monocotyledonous order Helobiae (*sensu.* Engler) is characterised by the helobial type of endosperm. Farinosae (*sensu* Engler) usually have a mealy or starchy endosperm. The highly specialised family Orchidaceae (or the Orchidales) is characterised by undifferentiated embryo and no or very little endosperm. The parietal cell is present in the archesporium of the Cyperaceae, while it is wanting in the Poaceae.

Enormous literature on embryology is now available, but it is not essential to cite all of them here. Although, it is not claimed that these data will always be useful, they should, at least, form a part of any thorough systematic analysis. The importance of embryology in taxonomy was realised much later. This has been mainly because it is time –consuming and painstaking laborious work. Even to study a few representatives of a family, it may take years. Mohan Ram (1959) and Harling (1954) cautioned about the interpretation of embryological data. It is a matter of common knowledge that external appearance of vegetative parts may sometimes lead to wholly incorrect notions of relationship. It is, therefore, important to also lay emphasis like embryological features which are regarded more conservative (Mahashwari, 1963). Davis and Heywood (1963) concur: "Clearly, a strict check of the taxonomic identification of material used for embryological studies is vital, as indeed it is for any other systematic studies. It is exceedingly difficult to assess to what extent early results may be suspect due to such reasons, in the same way that many of the earlier chromosome counts have had to be rejected".

(V) Palynology

The knowledge of pollen grain structure and function has obviously depended on advances in microscopy especially in the earliest years. The rock crystal lens dating as far back as 200 B.C. was recovered from the ruins of Nineveh in Greece. Since then various magnifying appliances, whether simple or compound microscopes, were fabricated. Fern spores were perhaps examined as early as 1628. This is evidenced in the record of a letter from John Faber of Banberg. A botanist, Cesi, was commissioned to sketch spores (*cf.* Singer, 1953). Robert Hooke (1665) discovered cells in cork and following him many biological objects like pollen grains were examined by the early naturalists. Marcello Malpighi (1628-1694) in Italy and Grew (1680) in England observed and reported pollen grains in the 17th century using microscopes developed prior to 1665 by Robert Hooke. However, the biological significance of pollen in reproduction was unknown. Martinet (1777-1779) in 'Katechismus der Nature' (Catechism of Nature) wrote about pollen morphology, pollination, pollen dispersal and some aspects of floral biology. A botanical artist at Kew Gardens, Francis (Franz)Bauer, compiled a series of drawing of pollen grains of about 175 species belonging to 57 families during 1758-1840(*cf.* Lack, 2008). These remained unpublished. This volume is deposited in the British Museum. He was aware of the taxonomic importance of this data. Johannes Evangesta Purkinje of Moravia (present in Czechoslovakia) also observed pollen grains, developed terminology and published in 1830. However, some purposeful investigations were carried out by some earlier workers. Lindley, an Englishman (1830-1840) correlated pollen features with other features to establish tribes of the Orchidaceae, Hugo Von Mohl (1851), a German botanist, proposed a classification based on pollen data in 1835 of angiosperms and gymnosperms of about 211 families. He included features such as number of pores, furrows and surface contours. Similarly, Fritzsche Carl Julius (1808-1871) again German worker, published his most notable "Ueber den Pollen' (1837). The terms *viz.*, intine and exine was coined by him. He also developed a classification of pollen grains of flowering plants. Pollen of the Mimosoideae was studied by Sergius Rosanoff (1866). Hugo Carl Albert Fischer(1890) worked for his doctoral thesis on 'Beitrage zur vergleichenden Morphologie der Pollenkorner'. He observed 2200 species and concluded that related species are generally similar. There are also records of pollen features in Engler and Prantl's 'Die Naturlichen Pflanzenfamilien' (1887-1915) and Hooker's 'Flora of British India (1872-1897).

The study of pollen morphology was being made till the close of the 19th century to advance knowledge in morphological botany or to supplement the taxonomic descriptions of plants. The science of palynology mushroomed enormously in the

20th century. Notable works are those of Wodehouse and Erdtman. Wodehouse (1935) studied pollen morphology as a part of his aerobiological research. He thought the origin and development of the pollen *morphoform* related to phyletic and the factors such as external and internal environment. His understanding about the arrangement, distribution and function of furrows on the surface of the pollen grain is indicative of his depth and vastness of his thoughts. He developed a simplest terminology to describe pollen morphology. He also enunciated his principles of pollen evolution. The work of Professor Gunnar Erdtman is another landmark. He published 'Pollen Morphology and Plant Taxonomy' (1952) and also later in 1969 the 'Handbook of Palynology'. The former book marked the beginning of a new phase in the history of pollen morphology. He laid down the principles and methods for the study of pollen morphology and also proposed an elaborated code of terminology for describing pollen and spore. He also proposed terms like '*sexine*' and '*nexine*' in place of his earlier terms '*ektexine*' and '*endexine*' respectively.

Faegri (1956) said: "does the taxonomist take the pollen grains into account when describing or revising a group of plants, and when descriptions are given of the pollen grains they tend to be superficial. As a result of the lack of data, palynological characters scarcely enter into current taxonomic thinking". Later, Keating (1979) remarked that "the usefulness of palynology has become so obvious that it is now routinely incorporated into most systematic and evolutionary studies". It seems that after the time of Faegri (*loc.cit.*) the value of palynological data was greatly appreciated. Later, pollen as an aid in taxonomy had become more widely acknowledged mainly because of an increase in analytical studies of pollen morphology and wall structure. Its application also in phylogeny becomes apparent. Stern (1962) was probably the first to give pollen variation for a group of species with a proposed phylogeny. Palynological data have been demonstrated to be of value in both taxonomic and phylogenetic research (*cf.* Davis and Heywood, 1963; Cerceau-Larrival, 1971). TEM became available is 1952 and made possible the study of pollen wall ultrastructure using thin section preparations. Obviously, the science of palynology has entered into taxonomic thinking.

Pollens were studied as a part of other disciplines, such as embryology, cytology, floral or pollination biology, reproductive biology, aerobiology, etc. Palynology was recongnised as an independent discipline of biology only when it was coined by Hyde and Williams in 1945. The genesis and progress of palynology was thus intertwined with other disciplines. It was also largely dependent on the inventions and advancement in optical technology. Concomitantly, pollen morphological terminology was also developed. Both TEM and SEM approaches provided the best understanding of outer and inner features of pollen grains.

The major characters of pollen grains employed in taxonomy hail from (i) association, shape and size of grains, (ii) number, shape and position of aperture, (iii) external and internal wall layers (stratification and sculpture/ ornamentation) and (iv) nuclear number and protoplasmic characters. Different schools of terminology have evolved.

Glimpses of these are presented in the following with a view to expose complexity of pollen characters:

Wodehouse (1935) prepared a glossary of the simplest terms for describing the pollen morphology. Some such terms are aspidate, colpate, lophate, operculate, psilate, reticulate and zonate, etc. Selling (1947) recognised five morphological categories *viz.*, *monads*, *pseudomonads*, tetrads, *massulae* and *pollinia*. He also categorised them further on the basis of exine ornamentation, apertural number, size and distribution. He also introduced two terms *viz.*, *colpodiporate* and *demicolporate*. Iversen and Trels- Smith (1950) proposed 24 pollen classes based on apertural forms and their number (e.g. *vesiculate, polyplicate, inaperturate, monocolpate, dicolpate, tricolpate, stephanocolpate, pericolpate, dicolporate, tricolporate, stephanocolporate, pericolporate, diporate* etc.). Erdtman (1952) described pollen using terms such as *monolete, trilete, hilate, sulcate, ulcerate colpate, porate, rugate* and *forate*,. Later, Erdtman and Straka (1961) proposed the main morphotypes e.g. *atreme, monotreme, tritreme, tetratreme, pentatreme, hexatreme, polytreme* and *anomotreme*. This is known as *NPC-system* based on the number (N), position (P), and character (C) of apertures of pollen. Nair (1974, a, b) gave morphotypes based on aperture form, number, distribution and position. These morphotypes and the major groups of plant forms correspond so strikingly that he proposed a palynological classification of the whole plant kingdom into: (i) Amorphosporophyta (thallophytes without apertural markings, (ii) Trimorphosporophyta (archegoniatae with *monolete*, *trilete* or *alete* spores), (iii) Polymorphosporophyta (angiosperms with diverse forms of apertures). Good glossaries of pollen morphological terms are also made available by Kremp (1965), Kapp (1969) and Moore and Webb (1978). A computer-based numerical coding system is also developed for adequate description of pollen grains (*cf.* Germeraad and Mullar, 1970). A numerical approach to pollen grain description involving geometrical location in 3-dimensional space, similarity and set theory, as applied in the Saxifragaceae. (*cf.* Hideux and Ferguson, 1975, 1976). Nair (1968) proposed terms to describe exine layers as (i) ectine (subdivided into *perectine, exectine, intectine* and *basectine*) and (ii) endine (subdivided into *intendine* and *exendine*). This system is an elaboration of his earlier one in 1961. Walker (1974) recognised basically three types of pollen on the characteristic of tectum using both LM and TEM *viz.*, tectate, semitectate and intectate. The former one is again categorised into (i) *tectate perforate* and (ii) *tectate – imperforate*. Meyer (1975) recognised three major types of ektexine structure *viz.*, *columellar*, *alveolar* and *granular*. He also proposed two types of endexine *viz.*, *laminated* and *non-laminated.*

In recent time, few other pollen features, which may be considered as embryological data, have emerged useful. For instance, wall development e.g. Austrobaileyaceae (Zavada, 1984), synchronous vs. asynchronous formation of tetrads e.g. Winteraceae; (Sampson, 1981), pore development e.g. *Helianthus*; (Horner and Pearson, 1978), spatial orientation in the anther e.g. *Sorghum*; (Christensen and Horner, 1974), etc. These are taxonomically useful and hence workers should pay attention to such emerging features.

Applications of Palynological Data

Palynological data are known to be useful at all levels of the taxonomic hierarchy. This is illustrated well by the resume given above. Some taxa are said to be *stenopalynous* or *unipalynous* (with little or no variation), and may be pollen feature/s exclusive of the group e.g. Poaceae, Lamiaceae, Asclepiadaceae, Brassicaceae, etc. Some taxa, are said to be eurypalynous or multipalynous (containing much variation), and are thought heterogeneous e.g. Verbenaceae, Acanthaceae, Rubiaceae, etc., Certain genera may be *stenopalynous* or *eurypalynous*. For instance, the genus *Ambrosia* (Asteraceae, Payne and Skvarla, 1970) is stenopalynous. The genera like *Justicia, Strobilanthus* (Acanthaceae), (Bremekamp, 1994) are *eurypalynous*.

Broadly, pollen grains in angiosperms can be categorised into two types *viz*., *uniaperturate* and *triaperturate*, from these other less common types are thought originated. Their apertures, in general, may be proximal or distal in position. The proximal aperture is sometimes observed in least specialised angiosperms (Nair, 1968), whereas the distal position of aperture is predominant in them. The proximal position of aperture is generally noted in Pteridophyta. It is both proximal or distal in Pteridosperms (Potomie, 1967). *Triaperturate* condition is observed at the distal position. Other types of aperture are situated zonally or globally. The *triaperturate* or triaperturate-derived pollen grains are observed in majority of dicotyledenous taxa. The *multiaperturate* condition is derived from either *uniaperturate* or *triaperturate* ones. The *inaperturate* type results from progressive shortening and eventual elimination of the apertural furrows.

Pollen grains in angiosperms are liberated at anthesis and contain either two or three nuclei. (*cf.* Schurhoff, 1924, 1926). There is a definite pattern of distribution of this nuclear condition among angiospermous taxa and are significant taxonomically and phylogenetically (Dufou, 1961; Dahlgren, 1975; Dalhgren and Clifford, 1982). Brewbaker (1967) also reached to the similar conclusion. Webster and Rupert (1973) recongnised this condition as 'Schurhoff- Brewbaker Law'. According to them, binucleate condition is more primitive in angiosperms as a whole and within individual taxa. Trinucleate condition is derived from the binucleate one and this change is irreversible. Binucleate condition dominates in the monocotyledons as they do in the dicotyledons. Davis and Heywood (1963) mentioned broadly that the *polypetalous* dicotyledons are 2-nucleate, *gamopetalous* dicotyledons 3-nulceate, and the *monochlamydeous* dicotyledons and the monocotyledons 2-nucleate. The order Centrospermae, however, exhibits uniformly 3-nucleate condition, although they are monochlamydeous taxon. Brewbaker's (*loc. cit.*) study revealed most families homogeneous with respect to nuclear number. Among monocotyledons, the 3-nucleate pollen grains are observed in case of the highly reduced wind-pollinated groups, aquatics and achlorophyllous saprophytes. The Poales and Juncales show 3-nucleate condition. The pollen grains are mostly 2-nucleate, except the Burmanniales wherein both types prevail. The Arales exhibit both conditions. The Zingiberiflorae and Areciflorae have mostly 2-nucleate pollen grains (Dahlgren and Clifford, 1982). Amongst the dicotyledons, the orders such as Caryophyllales, Plumbaginales, Polygonales and Asterales show uniformly 3-nucleate condition (Dahgren, 1975). Brewbaker (*loc.cit.*) noted 2-nuclate condition throughout the orders *viz.*, Magnoliales, Piperales and Nymphaeales. An interacting correlation has been divulged by Heslop-Harrison *et al.* (1975). Plant groups, in their opinion, with 3-nucleate pollen grains have dry stigmas and those with wet stigmas have 2-nucleate pollen grains. Another interesting correlation has been studied by Kooiman (1972) in the family Lamiaceae. Binucleate pollen grains are generally 3-colpate and trinucleate ones exhibit 6-colpate condition of pollen aperture in this family.

Nair (1970) enlisted certain families for their pollen associations. At maturity, pollen grains usually remain isolated from each other. But in few plant groups, they occur as *dyads, tetrads, polyads* or *pollinia*, for example, (i) dyads in Podostemaceae, (ii) tetrads in Annonaceae, Apocynaceae, some Araceae, Begoniaceae, Droseraceae, Ericaceae, Gentianaceae, Juncaceae Onagraceae, Malvaceae, etc., (iii) *tetrads* and *polyads* (*massulae*) Mimosaceae, (iv) *pollinia* in Asclepiadaceae and Orchidaceae.

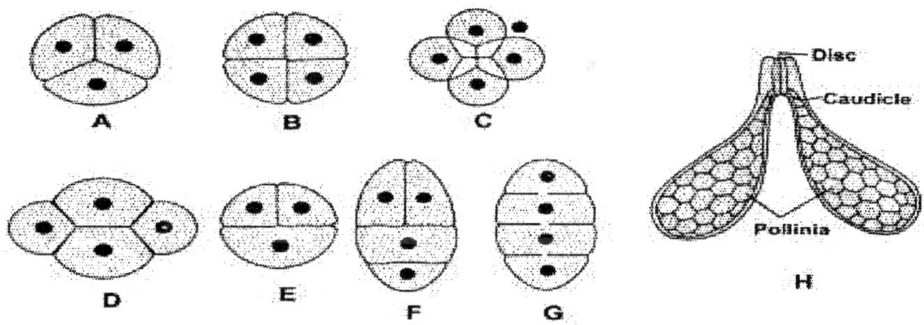

Plate-VI: Types of Pollen Tetrads and Pollinia.

Pseudomonads are characteristic of the family Cyperaceae. Nair (*loc.cit.*) also gave systematic account of pollen features in various orders of angiosperms (*sensu* Engler) based on the information then available. The account is available for eleven orders of monocotyledons and forty orders of dicotyledons. Taxonomic significance of the families is indicated, apart from the generalisation of their respective order. Readers are advised for the detailed account in his 'Pollen Morphology of Angiosperms' (1970).

The shape and size of pollen grains are characteristic of some taxa. In the subfamilies *viz.*, Thallasioideae and Halophiloideae of the Hydrocharitaceae, the pollen grains are threadlike (Nair, 1970). One of the largest and the most unusually shaped pollen grains is found in *Crossandra stenostachya* (Acanthaceae) which measures 250 x 19 mm (*cf.* Brummitt *et al.* 1980). Another such extremely large spherical grain is observed in *Cymbopetalum odoratissimum* (Annonaceae) which measures 350 mm in diameter (*cf.* Walker, 1971b). Walker and Doyle (1975) recognised six size classes based on the length: minute (<10mm), small (10-24mm, medium-sized (25-49mm), large (55-99mm), very large (100-199mm) and gigantic grains (>200mm). The pollen are either radially or bilaterally symmetrical. They show various shapes-circular, tringular in polar view; and *peroblate, spheroidal, subspheroidal, sublobate, oblate, oblate-spheroidal, prolate, prolate- spheriodal, subprolate, perprolate*, etc. in equatorial view. All these expressions with respect to shape, size and symmetry are characteristic of certain taxa at various levels and hence useful taxonomically.

There are certain taxa which are referred to various taxonomic entities. Palynological evidence throws more light on their taxonomy. Engler and Prantl (1857) separated the genus *Martynia* from the family Pedaliaceae based on parietal placentation in the former and axile in the latter. The pollen grains of the *Martynia* exhibit features such as *exine* with *reticulate* thickening, 2 nucleate at anthesis and *polyrugoidate* condition. Thus its disbanding from the family Pedaliaceae is supported and justifiable (*cf.* Signh, 1970a, b).

The genus *Paeonia* is referred to the Ranunculaceae in most taxonomic treatments (*cf.* Bentham and Hooker, 1862-1883; Benson, 1957; Engler and Diels, 1936). Its separation on various grounds is already discussed elsewhere in this book. Palynologically it is also distinctive in having reticulately pitted exine (*cf.* Tiagi, 1970). Palynological evidence support separation of Podophyllaceae from Berberidaceae, Bombacaceae from Malvaceae, Fumariaceae from Papaveraceae, Cuscutaceae from Convolvulaceae. Nelumbonaceae from Nymphaeaceae, etc.

Several families of angiosperms show distinctive pollen grains. Malvaceae, Convolvulaceae and Compositae have consistently *spinulose exine*. In Cruciferae, pollen grains are 3-*colpate*. Plumbaginaceae have *verrucate*. Pollen grains are 1-colpate in the Piperaceae, Chloranthaceae and Saururaceae. They are 3-porate (*pororate*) in the Myricaceae. Pandanaceae show 3-colporate condition. The pollen in Gramineae are 1-porate *operculate*. There are certain orders which are also distinctive palynologically. The order Urticales (porate pollen), Malvales, Rhamnales and Ericales (3-colporate) are characteristic for their pollen features (Nair, 1970).

Viscin threads help adhere the pollen surfaces. They are formed of ektexine. Their number, dimensions and sculpturing have been found helpful in delimiting the tribes and genera of the Onagraceae. Smooth threads are observed in *Ludwigia* and *Circea.* Compound threads occur in the tribe Epilobeae, whereas segmented threads appear in the tribe Onagreae and genera like *Lopezia* and *Fuschia* (*cf.* Skvarla *et al.*, 1978). The position of viscin threads is also taxonomically significant, e.g. they are present on the distal polar surface in the Ericaceae and on the proximal polar position in the Onagraceae (*cf.* Faegri, 1981).

Venkata Rao (1971) threw more light on the pollen grains in his monographic study of the family Proteaceae. He noted three distinct pollen types in Proteaceae *viz.*, the tringular (3-porate), spherical (3-porate) and ellipsoidal (2-porate). These types are characteristic of tribes and genera. The cytokinesis in case of 3-porate type is *simultaneous* and *successive* in case of 2-porate type. Type of *cytokinesis* itself has taxonomic significance (*cf.* Maheshwari, 1950). On the basis of ellipsoidal pollen the genera *Austromnellera* and *Musgravea* are included in the tribe Musgravecae, and also retaining only one genus *Embothrium* in the tribe Embrothrieae, and other ten genera into a new tribe the Telopeceae. The ellipsoidal pollen is restricted to the subfamily Grevilleoideae. The spherical pollen, in his opinion, evolved as its counterpart in the subfamily Proteoideae.

Venkat Rao (*loc.cit.*) also proposed a key based on pollen features in conjunction with exomorphic features for delimiting the two subfamilies and the tribes included in them.

Individual genera and species may also be characterised by certain pollen features. For example, *Salix* has long and narrow 3-furrowed (i.e. 3-colpate) pollen grains compared with *Populus* which has spherical pollen grains without distinct apertures (i.e. inaperturate). Also, pollens are reticulate in *Salix* and granulose in *Populus* (*cf.* Davis and Heywood, 1963; Nair, 1970).

Hanks and Fairbrothers (1976) described 41 species of two fagaceous genera viz., *Fagus* and *Nothofagus* using light and scanning electron microscopy. They prepared keys on the basis of 29 character ranges. They used palynological data to produce diagnostic keys. They hypothesised the origin, migration and systematic relationships of the two genera. The *Fagus* species not distinguishable with the light microscope could be separated easily with the SEM. Nowicke and Ridgeway (1973) in their LM and SEM studies of 250 species of *Cordia* (Boraginaceae) revealed significant palynological differences and also correlated with other exomorphic features between the different sections of the genus. The sect. *Varronia* has 3-porate grains with a reticulate exine. The sect. *Cordia* shows 3-colpate or 3-colporate pollen with striato-reticulate exine, while sect. *Gerascanthus* and sect. *myxa* have 3-colpate or 3-colporate grains with spinulose exine. These authors also suggested distinct generic status to these three groups. Exine patterns are useful to identify different species of a genus. For example, in *Bauhinia*, *B. acuminata* pollen are psilate, those of *B. krugii* striate, *B. retusa* verrucate, *B. racemosa* reticulate, *B. purpurea* reticulate-tuberculate and *B. malabarica* spinulate (*cf.* Nair, 1974b). In some cases pollen features are taxonomically significant at varietal level e.g. the two varieties of *Caltha palustris viz.*, var. *normalis* and var. *alba* are identifiable by 3-colpate pollen and pantoporate pollen respectively (Nair, 1974b).

Palynological investigations also contributed to the elucidation of phylogenetic relationships. They are helpful in shedding light on the problems of the primitive Ranalian dicotyledons and their relationships to the monocotyledons. The pollen features do not lend support to the existence of a sharp line of demarcation between the Dicotyledons and Monocotyledons. Among the widely accepted primitive taxa of the present-day dicotyledons the monocolpate and trichotomocolpate forms alone are found in the Magnoliaceae, as in pre-angiosperms, while an array of apertural forms dominated by the tricolpate and devoid of the monocolpate and trichotomocolpate, is found in the Ranunculaceae. There are two clear phylogenetic stocks among the primitive dicotylalons *viz.*, (i) the dominantly *monocolpate* and *unipalynous* 'magnolian' stock, and (ii) the dominantly *tricolpate* and *multipalynous* 'ranalian' stock, represented by the Magnoliaceae and the Ranunculaceae respectively. The characteristic monocolpate form of the 'magnolian' families is paralleled in the monocotyledons as a whole. This is indicative of a fact that the monocotyledons are more closely related to the 'magnolian' dicotyledons and less to the 'ranalian' dicotyledons. It is concluded, following Campbell (1930), that the 'ranalian', 'magnolian' and the 'monocot' stocks might represent three independent phylogenetic lines with their phyletic roots in protangiosperms. It is further assumed that angiosperms originated possibly in the pre-angiospermous pteridosperms which are characterised by trichotomosulcate pollen grains. Accordingly, the earliest angiosperms must have possessed trichotomocolpate pollen grains alone (Nair, 1970).

Nair (1972) also used pollen morphological features in assessing the systems of plant classification especially of primitive angiosperms. In his opinion, the systems of Bessey and Cronquist confirm with the facts of pollen phylogeny, with regard to primitive angiosperms. Nair (1970) threw light on plant evolution on the basis of pollen data. The plant kingdom is composed of the Thallophyta, Bryophyta, Pteridophyta, Gymnospermae and Angiospermae. It is in the bryophytes that the beginning of organised morphological evolution has been set in motion. Further evolution within the group produced the *monolete* and inaperturate *morphoforms*. The trend of spore evolution established in the bryophytes has been repeated in the pteridophytes, and the gymnosperms. The aperture is trichotomously divided (*trichotomocolpate*) in the primitive gymnosperms. The earliest taxon in the angiosperms must have been one with *trichotomous* aperture. From such a primitive protangiosperm taxon, the hypothetical monocolpate forms represented by the 'Proto-Monocots' on the one hand and the 'Proto-Magnolian' dicots on the other perhaps diverged (*cf.* Nair, 1970).

Prior to application of palynoloy in solving taxonomic and phylogenetic problems, certain cautions must be kept in mind. The chemical processing of the pollen grains can charge their size and different treatments bring about different size alterations. (Reitsma, 1969). Nutritional factors can also affect their sizes (Bell, 1959). The correct identification of the voucher specimens is very necessary. With these cautions in mind, we can sum up that palynology can contribute useful types of data for taxonomic purpose at various levels, and also in origins and phylogeny of plant groups.

(VI) Cytology

Broadly cytology encompasses all aspects of cells. In practice, especially in taxonomic consideration, the chromosomes and their various attributes have remained in focus (*cf.* Lewis, 1969). Chromosomal data has two-pronged importance for the purpose of taxonomy: (i) they are strictly anatomical information – the number of chromosomes is as important as the number of carpels or stamens in a flower or type of chromosomes is to be considered in the same way as the shape of leaves, sepals or petals, or the type of alkaloids present, (ii) they also constitute a special type of information, for chromosomal number and homology largely determine pairing behavior at meiosis which partially controls the level of fertility of hybrids and hence the breeding behaviour and pattern of variation of populations. Both of these standpoints are thought valid. While interpreting chromosomal information both should be borne in mind. The former standpoint may be true for taxonomic (phenetic) purposes, whereas the latter assumes greater importance in especially biosystematic (phylogenetic) studies. Lewis (1957) stressed that chromosomes derive their prominence as a tool in taxonomy from their direct relation to the genetic system of which they are an integral part. Davis and Heywood (1963) also claimed a special role for cytological (chromosomal) data in taxonomy for the same reason given by

Lewis (1957). The discovery of chromosomes in the last quarter of the 19th century and the subsequent demonstration that they are the carriers of genetic information provided an impetus to, or at least a physical foundation for, the experimental evolutionary studies on which are based our knowledge of such taxonomically important matters as variation within plant species and the significance of the breeding structure of populations and of the barriers to gene exchange between them (*cf.* Moore, 1978). Earlier Moore (1968) also opined that the genetical tools made available to the taxonomist and the data provided by the chromosome cytologists have gained the widest currency.

Prior to appreciation of significance of chromosomes in taxonomy, it appears necessary to know genomic organisation in two different sectors of organisms, the prokaryotes and eukaryotes. The prokaryotes lack membrane bounded cell organelles, chromosomes with histone, proteins and meiotic and mitotic mechanisms. The 'chromosomes' in prokaryotes are called '*genophores*'. Complexity in chromosome structure of eukaryotes gradually evolved through different stages from a prokaryotic system (Sharma, 1976). The chromosomes of Dinophyceae with a well-defined chromosome structure (except for the histones characteristic to eukaryotic chromosomes) represent a connecting link. They, however, yet to gain the perfection of eukaryotic chromosomes since they lack the mitotic mechanism. Soyer and Happala (1974) therefore referred the family Dinophyceae to the 'Mesokaryota', an intermediate group of organisms between prokaryotes and eukaryotes. Chromosomes represent a stage of packing of the genetic material rather than a direct reflection of it. It is hence important to keep in mind as we consider the question of variation in chromosome number and its taxonomic significance. Secondly, only the 'chromosomes' of the eukaryotes will be dealt with here.

Plant cytology, with a direct bearing on taxonomic issues, gained momentum with the contribution by East (1928) on *Nicotiana* (Solanaceae), by Cleland (1923, 1936) on *Oenothera* (Onagraceae) and by long series of papers on *Crepis* (Compositae) by Babcock and collaborators (*cf.* Babcock and Cameron, 1934; Babcock, Stebbins and Jenkins 1937; Babcock and Jenkins, 1943, etc.). The last series of papers was particularly most influential in showing the position value of cytological data in plant taxonomy. The first 30 years of 20th century saw an increasing impact of cytogenetical concept and procedures on at least some taxonomists. The increased output of chromosome data from a wide variety of plants, facilitated by the development of the squash technique, and the experimental demonstration, in genera such as *Nicotiana*, of hybridisation followed by *polyploidy* as a method of speciation, provided some of the fuel for the '*New Systematics*' (*cf.* Huxley, 1940). During these years, the important role of the chromosomes in aiding taxonomic understanding became evident through classical studies on Cruciferae, Gramineae, Agavaceae and monographs on some genera such as *Crepis*, *Datura*, *Nicotiana*, etc. *Biosystematics* or *experimental* taxonomy became then as now the growing point of taxonomy. Stebbin's (1950) masterly 'Variation and Evolution In Plants' was also a powerful stimulus to many workers. Moreover, taxonomy has always, amoeba-like, ingested those concepts and techniques which have proved useful and the increasing development of a broadly based '*general purpose taxonomy*', drawing as many sources as possible (*cf.* Davis and Heywood, 1963). Modern taxonomy is expected (i) to provide a convenient method of identification and communication, (ii) to provide a classification which as far as possible expresses the natural relationships of organisms, and (iii) to detect evolution at work, discovering its process and interpreting its results (*cf.* Davis and Heywood, (*loc.cit.*). These three aims of taxonomy constitute both the phylogenetic and phenetic approach to taxonomy. Chromosome data play an important role in both.

Different types of chromosomal data have been employed in taxonomy. The subject matter can be usefully considered under as: chromosome number, (including base numbers), *B-chromosomes*, chromosome morphology, size, their behavior at meiosis and banding patterns. The sources in the plant are most commonly mitosis and meiosis and can be observed with proper techniques in two regions of the plant (i) the root, shoot and lateral meristems (mitotic divisions), and (ii) the sporogenous tissue. In conventional practice, most mitotic data come from root tips and almost all meiotic observations from microsporogenesis in young anthers.

(i) Chromosome Number: Generally, all individuals within a species have the same chromosome number in each cell (although there are few exceptions). Further, it is hold that the more closely related species are, the more likely they are to have the same chromosome number; and the more distantly related the more likely they are to have a different number. Thus the relative conservativeness renders the chromosome number as important and most oftenly used taxonomic character. Organisms may be haploid, diploid or polyploid. The diploid condition itself presumably arose at some stage with a phylogenetic advance from a prior haploid genome (*cf.* Crowson, 1970). The lower forms of organism have predominantly haploid body e.g. bacteria, aquatic forms of plant-life. Diploidisation set in and has been restricted to the zygotic stage during the course of evolution in sexuality. The evolution and elaboration of diploid phase in the life cycle and the gradual suppression and ultimate elimination of the haploid (gametophytic) phase in the higher forms is now quite known.

Usually the information on chromosome number is provided in the form of the diploid number (2n). It is customary to quote new chromosome counts in diploid number (2n) when the count is based on mitosis in sporophytic tissue and the haploid number (n) when it is based on mitosis in gametophytic material or on meiosis. In somatic tissue, whether haploid or diploid, counts are generally made on rapidly and synchronously dividing cells e.g. cells of embryos, young sporogenous tissue and meristems. Chromosome numbers have varied in plant kingdom and show wide range. In *Haplopappus gracilis* (Compositae) n=2 (Jackson, 1957) and *Brachyscome* leariloba (Compositae) n=2 (Smith-White, 1968; Watanabe and Smith-White, 1987). In *Colpodium versicolor* (Gramineae) n=2 (Sokovskaya and Probatova, 1977); in *Kalanchoe* sp. (Crassulaceae) n = ca 250 (Baldwin, 1938); and in *Ophioglosssum reticulatum* (pteridophyte) n=620 (Love and Kapoor, 1967b). This great diversity of chromosome numbers and their relative constancy within populations and species provide an important character for taxonomic purpose. There are three following broad classes of chromosome number relationship with taxonomic groups (*cf.* Warburg, 1949):

1. **Constant Number:** No deviation in chromosome number is exhibited throughout the whole group, whether as a gènus or family. For example, in *Quercus, Fagus* and most other genera (except *Nothofagus*) of the Fagaceae, all species have the same number, n = 12. *Nothofagus* species have 13 chromosomes (Wardle, 1967; Hunks and Fairbrothers, 1976). In these and several other examples chromosome number is obviously no significance for taxonomic discrimination within the group, although the constant number may be useful criterion to distinguish the group from related others.
2. **Euploidy:** Individuals whose somatic chromosome number is an exact multiple of the basic number are termed '*euploid*' and the series is described as 'euploidy'. For example, a remarkable range of polyploid numbers is found in the family Malvaceae (Skovsted, 1941), (i) x=5,2*x*(10), 3*x*(15), 4*x*(20), 5*x*(25), 6*x*(30), 7*x*(35) and 8*x*(40); (ii) *x*=7, 2x(14), 3x(21). 4*x*(28), 5*x*(35) and 6*x*(42). The numbers in parenthesis are chromosome numbers. Various members of such an euploid series may be unified by a basic number or *x* (conventionally so designated) which is the genetic number of a diploid species and other species can be described at triploids (3*x*), tetraploids (4*x*), hexaploids (6*x*) and so one. Thus one of series in Malvaceae has x=5, while the other series have x=6,7, 11 and 13. In the groups where the diploid species are unknown the basic number is inferred. The accuracy in such cases however relies upon the chromosome data available for the group. In certain groups more than one basic number may be present. In case, living diploid members are wanting, the inferred basic numbers ranging between 2 and 13 may be conveniently referred to as the 'primary basic numbers, whereas the remainder ones are termed 'secondary basic numbers'. There are numbers in some taxa, for example (i) *Chlorophytum*, x = 7 and 8 (Naik, 1976) (ii) *Salix*, x = 12 and 19 (Grant, 1976), (iii) *Cochlearia*, x = 6 and 7 (Davis and Heywood, 1963). There can be an array of basic numbers in a family e.g. (i) Amaryllidaceae, x = 7, 8, 9, 10 and 11 (Flory, 1977), (ii) Himalayan Anacardiaceae, x = 7, 10, 12, 14, 15, 16, 24 and 30 (Khosla *et al.*, 1973). On the contrary, there are certain families cytologically more or less homogeneous e.g. Apocynaceae, x = 11; Cactaceae, x = 11; and Magnoliaceae, x = 7. Earlier taxonomists largely agreed with generic status based on basic numbers. Favarger (1960) revealed an interesting case. Two species viz., *Cicendia filiformis* and *Mircrocala pusilla* belonging to Gentianaceae, once kept under the same genus *Cicendia*, have different basic numbers, n = 13 and n = 10 respectively. This generic separation is also supported by morphology of their pistil. However, this can not be universal rule.

   Hybridisation plays a dominant role in evolution. It involves crossing of any two genetically unlike individuals, which have different basic numbers. Such hybrids undergo polyploidy. This type of polyploidy is termed 'dibasic polyploidy'. Karpechenko (1927, 1928) provided an excellent example. He artificially synthesised *Raphanobrassica,* which is a hybrid between *Brassica oleracea* (2n = 18) and *Raphanus sativus* (2n = 20), and has 2n = 38 chromosomes. Different inferred dibasic numbers may characterise subgenera, as in *Veronica* (x= 7, 8, 9; x2 = 15, 17; x3 = 26) (*cf.* Davis and Heywood, 1963). In *Spartina*, the new basic number, x = 61, has originated from parents with base numbers x = 30 and x = 31 (*cf.* Stace, 1989). Several such examples are recorded by the Stebbins (1950) and Wagner (1968). In instances like the Brassicas conclusions about dibasic polyploidy can be safely drawn, but in larger groups it is easy to fall into errors in drawing inferences about the origin of apparent dibasic numbers, and the temptation to play chromosome arithmatic and accept the results must be resisted unless there is strong supporting evidence from other sources (*cf.* Davis and Heywood, 1963).

3. ***Aneuploidy***: When chromosome numbers found within a group bear no simple numerical relationship to each other or chromosome numbers are not in the proportion of exact multiples, then the individuals of the group are termed '*aneuploid*' and the series is described as '*aneuploidy*'. There is gain or loss of one or more regular chromosomes in aneuploid series. Anequploidy covers two distinct kinds of phenomena: first, there is an increase or decrease in number whereby the same genetic material becomes distributed in a different number of chromosomes, leading to a change in the basic number; second, the basic number is unaltered but the genetic material present is changed through the addition or loss of chromosomes. Increase or decrease in chromosome number may be brought about by some irregularities during cell division or changes in the chromosomes themselves e.g. fragmentation, misdivision of the centromere, etc. Sometimes, the term 'dysploid' is applied when the aneuploid sequence does not apparently form a particular series. Frequently a species has one chromosome number throughout its range, and sometimes this number differs by one from that of its closest known relative. The difference is obviously a taxonomic character. Such a relationship also can be suggestive of an evolutionary directionality between the two taxa. In most groups, the loss of a single chromosome is more common event than a single gain. In nature, various species of *Carex*, a cyperaceous genus, show a wide range of chromosome numbers from n= 6 to 112 with multiple of 5, 6, 7 and 8. In the genus *Vicia* (Papilionaceae), there are chromosome numbers of 2n=10, 12, 14, 24 and 28, obviously consisting of aneuploids clustered around the diploid and tetraploid levels. In *Crepis* (Compositae), there is a progressive decrease in basic number from 6 to 3, step by step, one chromosome at a time. A decrease from n=4 in *C. neglecta* or some related species to n=3 in *C.fuliginosa* is thought arisen through an unequal reciprocal translocation (*cf.* Togby, 1943; Babcock, 1947a). A similar decrease is noted in *C. kotschyana* (n=4) from *C.foetida* (n=5) (*cf.* Sherman, 1946).

Plants containing one extra chromosomes are known as 'trisomics' (2n+1). There are diploids with one chromosome represented three times. Those containing two extra chromosomes are called 'tetrasomomics' (2n+2), and so on. Plants may be deficient of one or two chromosomes. Those with loss of one are known as monosomics (2n-1), and with loss of two as nullisomics (2n-1). The gain of chromosome is equally feasible at any ploidy level. Loss of a chromosome at the diploid level is usually lethal. It can be tolerated more readily at the higher levels of ploidy because of the buffering effect of the multiple genomes. Trisomics are known in *Nicotiana* (Smith, 1943), *Lycopersicon* (Rick and Barton, 1954),*Zea* (Rhoades and McClintock, 1935) and *Antirrhinum* (Rudorf-Lauritzen, 1958). *Datura*, which has n=12, has 12 possible trisomics. These were all produced experimentally and identified (*cf.* Blakeslee, 1922, 1930, 1934). Mosonsomics may arise in obligate apomicts as *Taraxacum* species in which plants are triploid (3x-24) and derived from probably sexual diploid ancestors, and may give rise to seed with 23 chromosomes by the loss of one chromosome of one of the sets.

(ii) *B–Chromosomes*: They occur in variable numbers in some members of a population and which may be absent in other individuals without noticeable effects. They are generally small and heterochromatic. They do not pair with the regular '*A-chromosomes*' at meiosis (*cf.* Muntzing, 1959). However, they pair with each other. Their effect on the plant is not specific as with A-chromosomes, although the effects are various. Their genetic effects may be little or none. In plants, they were detected first in *Zea mays* (Longley, 1927; Randolph, 1928, 1941). They may be of recent derivation or they may have arisen originally from *A-chromosomes*. They may represent centric fragments of telocentric arms of A-chromosomes as a result fo misdivision (Darlington, 1956). This is the case reported in *Oenothera* (Cleland, 1951) and *Caltha* (Reese, 1954). In *Clarkia*, the B-chromosomes represent duplication of part of the normal genome complement (Lewis, 1953). Their adaptive significance has been reported in *Phleum phleoides* (Bosemark, 1956b), *Secale cereale* (Muntzing, 1957, 1959), in *Festuca pratensis* (Bosemark, 1956b), in *Centaurea scabiosa*, (Frost, 1958a), etc. In *Dipcadi* (Liliaceae), some correlation between the edaphic conditions and number of *B-chromosomes* is inferred (Naik, 1974). Nonetheless, they are usually of limited taxonomic significance, even more so because they often vary during the ontogeny of the plants containing them. They have been documented for at least 250 angiosperm families (Stebbins, 1971). Diploids of *Ranunculus ficaria* (Ranunculaceae) have no *B-chromosomes* but are present in the *tetraploids*. They occur in both diploids and tetraploids of *Festuca ovina*. B- chromosomes do occur in bryophytes and are very small (*m-chromosmes*) and are of highly diagnostic value.

Dnyansagar and Dhanraj (1972) made a detailed study of *B-chromosomes* during meiosis in *Solanum viarum*. He recorded them in 20.64% plants and found behaviour varied from plant to plant and even cell to cell. They were visible from *diakinesis* onwards. Their number varied from one to two in PMC's and size varied from short to medium sized. The lagging *B-chromosomes* were excluded from the daughter nuclei and formed micronuclei. The sterility was high in plants with *B-chromosomes*. All these types of behaviour are indicative of little taxonomic significance.

(iii) Chromosome Size and Morphology: Apart from chromosome number, form and relative size of different chromosomes of the same set, number and size of satellites and secondary constrictions, absolute size of chromosomes, euchromatin and heterochromatin differ in karyotypes of various plant groups (Stebbins, 1950). These features have served as good taxonomic characters and have bearing on phylogeny and classification. These are discussed at length here.

Delanunay (1926) formulated for the first time the concept of karyotype based on the features noted above. Karyotype, in his opinion, is a group of species resembling each other in the number, size and form of their chromosomes. Levitsky (1931) revised this concept as the phenotypic appearance of the somatic chromosomes in contrast to their genic contents. The *karyotypes* are represented in the form of '*karyogram*' or '*idiogram*'.

The absolute size of the chromosomes of a karyotype is fairly constant for a species and is genotypically controlled (Thomas, 1936).Size is not indicative of genic content e.g. in the Droseraceae where the total volume ratio of the chromosomes of *Drosera*: *Drosophylum* is 1:1000; in the Liliaceae, *Trillium* has chromosomes 100 times larger than those of the related genus *Medeola*. Even species within a genus may vary in the same ratio as in *Euphorbia* (Darlington, 1956). The length of chromosome generally ranges from 0.5 to 30 (Warmke, 1941). It is characteristic of some plant groups. Many evolutionary trends in chromosome size have been described within families and lower taxa. For instance, large chromosomes are found in the genus *Paeonia,* a genus formerly included in the Ranunculaceae. In *Dianthus* (Caryophyllaceae) and *Chrysanthemum* (Asteraceae), the polyploids have smaller chromosomes than diploids, this is thought that reduction in size is an adaptation to increase in number (Darlington, 1956). Annuals have smaller chromosomes than perennials e.g. in *Crepis* (Babcock and Stebbins, 1942). In the Gramineae, genera adjudged as the most advanced have larger chromosomes than those considered less evolved (Stebbins, 1956). Davis and Heywood (1963) warned that misuse of what is an extremely valuable phyletic tool is a danger to be guarded against.

The size and shape of the chromosomes relative to each other within the same complement are also the most obvious characteristics of *karyotypes*. There are distinctions between chromosomes based on the position of *centromere* : (i) centromere near the middle (*metacentrics, V-shaped*), (ii) centromere slightly away from middle (submetacentrics, L-shaped), (iii) centromere near one end (*acrocentrics, J-shaped*) and (iv) centromere with a termind position (telocentrics, *rod or i-shaped*). Another useful aspect of chromosomal gross morphology is the position of secondary constrictions, which delimit the occurrence of satellites. The presence of *satellites* has important taxonomic value. All these aspects of chromosome, together with chromosome size and number have been found extremely useful at all levels of the taxonomic hierarchy. For example, in *Helleborus* (Ranunculaceae) the chromosomes are mostly V-shaped. They gradually differ in size and shape as pointed out earlier. V-type is thought symmetrical and least specialised and others as asymmetrical and more advanced. In general, these changes in symmetry and shape are due to various structural alterations like *translocation, inversion, duplication* and *deletion*. When possible to trace the ways by which they have arisen in a particular *karyotype*, it gives valuable information on the origin of the group concerned. Jones (1970) based on his studies on the Commelinaceae stated that probably, it is only the adherence to the asymmertic law that would force us to read the evolutionary direction in only one way. In Commelinaceae, Jones (*loc.cit.*) observed the reverse trend of specialisation from *telocentric* to *metacentric* chromosomes.

The importance of chromosome size and *karyotype* symmetry is best known in *Yucca-Agave* complex. *Yucca* and others (with superior ovary) and *Agave* and relatives (with inferior overy) are generally referred to the Liliaceae and Amaryllidaceae respectively. They find composite treatment as under Agavaceae in some taxonomic treatments (*cf.* Cronquist, 1968, 1981; Hutchinson, 1973). Their placement under one taxonomic entity is strongly supported by evidence. They are characterised by a very distinctive bimodal karyotype consisting of 5L+25S chromosomes. The *karyotype*, in this case, is a more precise manifestion of the genome than chromosome number alone, and hence its applicability is greater when investigating genome homologies. Babcock (1947) has well illustrated that in *Crepis,* where the use of chromosome morphology together with number has solved many taxonomic problems at the generic, sectional and specific levels. This also elucidated how many species are evolutionarily interrelated. The genera such as *Aconitum* and *Delphinium* of the Ranunculaceae afford another example with more asymmetrical karyotypes and too associated with more specialised morphological features. Variation in karyotypes can also occur within species, as shown in variations in satellite chromosomes in *Elymus striatulus*, a graminaceous member (Heneen and Runemark, 1972). This obviously emphasises the need for extensive studies before arriving at certain final conclusions.

(iv) Chromosome Behaviour at *Meiosis*: A study of the way the chromosomes behave at meiosis can reveal valuable information to make distinction between taxa. Two major aspects of interest in this regard are: (i) synapsis during the meiotic prophase and (ii) *chiasma* frequency. The regularity of pairing largely determine the fertility of a plant. It also enables for chromosome comparison of the degree of homology between genomes.

Traditionally we believe that the pairing of chromosomes is due to homology. The data on chromosome pairing in hybrids provide evidence of degree of phylogenetic affinity between the parental taxa (*cf.* de Wet Harlan, 1972). It is, therefore, of much use in evolutionary taxonomy. Gerber (1958) employed *chiasma* frequencies as a means of grouping species within the genus *Collinsia* (Scrophulariaceae). Darlington (1937) gave a classification of organisms according to the bahaviour of *chiasmata* at *metaphase*. Meiotic abnormalities can be easily detected and interpreted. Such behaviour may give some clues for affinity. Great weight is often placed on the degree of chromosome homology in hybrids as an estimate of the degree to which the parental species are related. Greater degrees of genomic non-homology result in non-pairing at meiosis (*asynapsis*), or loose pairing without the formation of chiasmata so that the bivalents fall apart before *metaphase* (*desynapsis*). Degrees of chiasma frequency and terminalisation can be used to suggest relationships (*cf.* Patil and Deodikar, 1972). In extreme cases, the whole genome fails to pair, in other situations varying proportions of the chromosomes form bivalents. The degree of pairing is broadly proportional to the level of homology of the genomes.

However, genomic homology is not the only factor which determines synapsis. Solbrig (1968) listed 15 genes attesting the regularity of meiosis, several of which are concerned with pairing itself. The most conclusive evidence is presented by Riley and Chapman (1958) who found that in hexaploid wheat there is a gene which suppresses multivalent formation. The traditional aforesaid methods of genome analysis have, therefore, fallen out of favour with many workers, especially as genes suppressing multivalent formation have been shown widespread.

(v) Chromosome Banding: Of late, studies on chromosomal banding patterns have revealed structural heterogeneity and shed light on functional properties of chromosomes. Ability to distinguish morphologically between chromosomes has been enhanced by new staining techniques using Giesma and Fluorochrome dyes. Under certain controlled conditions, chromosomes are stained in a consistent banding pattern instead of with a uniform intensity. These extra morphological criteria provide distinctions between chromosomes which were earlier indistinguishable or scarcely distinguishable. Each stain and technique give a different pattern of banding. In case of plants, *C-banding*, *G-banding*, *Q-banding* and *Hy-banding* are being employed. Banding is known to be associated with regions of heterochromatin in the chromosomes. *Euchromatin* and *heterochromatin* differ in the timing of their cycle of condensation and have been found of some use taxonomically. The differential banding patterns enable (a) identification of homologous pairs of chromosomes which are normally indistinguishable by conventional methods, (b) easy separation of different karyotypes and (c) grouping of similar karyotypes.

These techniques are applied and found useful for taxonomic purpose. For example, Greilhuber and Speta (1978) made quantitative analysis of C-banded karyotypes in systematics of cultivated species of *Scilla* (Liliaceae). Schwizer and Ehrendorfer (1976) analysed Giesma-banded karyotypes in systematic and evolution of *Anacyclus* (Asteraccae). Schwarzacher and Schweizen (1982) studied *karyotype* and heterchromatin in *Cephalanthera* (Orchidaceae) using Geisma, *C-banding* and Fluorescent counterstaining techniques.Siljak-Yakovlev and Cartier (1982), in *Crepis praemorsa* pointed to the fact that ssp. *dinarica,* is separable clearly from other species and deserves the status of a distinct species. Another technique newly applied to plants is that of *in situ* hybridisation of satellite DNA from one taxon to another using radioactively labelled DNA of one species and auto-radioagraphy. These give additional distinguishing markers to the chromosomes and allow more detailed comparisons between complements of different taxa. All these new techniques promise to provide important new insights into plant variation and evolution with inevitable far-reaching taxonomic consequences.

(VII) Chemistry

Chemosystematics (Chemical Systematics, Chemical Plant Taxonomy or Plant Chemotaxonomy) is the study of chemical variation in various plants and their relationships. It is one of rapidly expanding and more fashionable areas of plant taxonomy. The principles, procedures and results of investigations into chemical variation of plant groups are applied mainly for two purposes: (i) to provide taxonomic features which may help improve existing plant classification and (ii) to add to the knowledge of phylogenetic or evolutionary relationships. Various opinions have been expressed by the experts in this field about the role of plant chemistry in taxonomy. Davis and Heywood (1963) opined that chemical characters may have particularly a high taxonomic value when they can be shown to be stable, unambiguous and not easily, if at all, changeable. Stuessy (2002) expressed a view that certain kinds of chemical data are likely to be of great value in determining phylogeny, whereas others are of equally great value in understanding predator-prey relationships. Chemical characters have continued to prove valuable in solving different kinds of taxonomic problems, and it is now clear that chemistry is here to stay as major source of comparative data for understanding relationships. Stace (1989) hold a view that chemotaxonomy has arisen as a major science in only the past 30 or 50 years and, perhaps because of that, has two common fallacies associated with it. These are that *chemotaxonomy* is a recent innovation and that it in some way deals with more fundamental information than does say, morphology or cytology, so that the data which it produces are more important in classification. Swain (1963) stated that chemical plant taxonomy then, although a very convenient designation for the activities which it encompasses, must not be taken to indicate that it is an attempt to classify plant solely on the basis of their chemical constituents. In fact, its application has up to now usually added confirmatory evidence to agreed characters. The use of chemical criteria, however, adds a powerful weapon to the armoury of the taxonomist. However, as contributory discipline in taxonomic and evolutionary studies, *chemotaxonomy* has gained an important status in recent years. I wish to attract attention of the readers to the general statement made by Pirie (1955). He stated that 'many different types of information can be, and should be, woven into the classificatory scheme, and there is no obvious advantage in attaching overriding importance to any one types.'

The vast explosion in the exploration of the distribution of natural products over the last few decades is a result in part of a growing interest on the part of botanists in the chemistry of plants. There are reasons for the recent rapid growth of chemotaxonomy. First, the development of many new techniques, notably various forms of chromatography, electrophoresis, spectroscopy, etc. which have rendered the analysis of plant products simpler and quicker. Second, the realisation that behind the universal occurrence in plants of many vital biochemical pathways there is an enormous variation between taxa in many other less vital pathways. Third, the current belief that evidence from as many sources as possible should be employed in plant classification. The extent to which chemical data are used in classification varies enormously. The expertise, facilities and man-hours are necessary in order to carry out adequate investigations on a range of plant materials. The use of chemical information is risky without full investigation. There is also problem that there are very few workers who are really highly skilled in both chemistry and plant taxonomy. Almost all work is carried out by people trained in one field or the other, with obvious disadvantages. Chemical features evolve at different rates and are not necessarily synchronised with other characters. They can, however, suggest the direction of change and avoid possible errors of interpretation, but they are only one facet in a complete mosaic.

Indeed, the science of chemotaxonomy has ancient and diverse origin. The accumulation of information on the chemical content of a very wide range of plants are primarily due to herbalists and pharmacists. Actually, the early history of botany is largely a history of the uses of plants in medicine; the root-gathers and herbalists of the past began, many centuries ago, to group plants having similar virtues or medicinal properties. However, this grouping of plants came to have a modern look not until the end of the 17th century. Even before the use of drugs, ancient man sought plants as food. He must have realised that similar species often possessed common properties. He discovered the high protein content of seeds of the grasses and legumes. A second ancient origin has been through morphology and anatomy. For instance, different kinds of crystal can be considered as either morphological or chemical, and colour can be thought as either morphological or chemical. It is now a well known fact that colour is based upon any one or a combination of molecules. Similar is also the cases of various crystals. The taste and smell of plants are another source of origin of chemotaxonomy. These help detect and are useful as indicative of their food, drug or cosmetic value by mankind. In recent years, much attention has been focused on the co-evolution of plants and animals in their use for food. Pathogenic fungi e.g. rust show affinity for related host-plants and have been used to investigate the taxonomic problems. These crude origins, however, have their limitations and can hardly be thought part of the modern science of chemotaxonomy.

The connection between the chemistry and taxonomy of plants has a long history. Amongst the modern pioneers, who made a serious use of chemistry in plant taxonomy is Helen C. de S. Abbott (1886). She studied saponins in plants and made several useful general assertions about the role of chemical information played in evolutionary studies. The early work on starches by Reichert (1916, 1919), on terpenoids of *Eucalyptus* by Baker and Smith (1920), and on serology and plant relationships by Mez and co-workers, (1922) must be mentioned. McNair (1934, 1935, 1945) studied taxonomic and ecological role of alkaloids, cyanogenetic and sulphur compounds. The direct and more general application of plant chemistry in solving taxonomic problems mostly appeared after mid-1950s. Hegnauer (1954, 1958) used the term '*chemotaxonomy*'. Bate-Smith (1958) employed phenolics in classification. A wide variety of literature on this subject matter have been published and it is impossible to review them here. It, therefore, appears plausible to switch on direct on type and use of chemical data in plant taxonomy in the following.

In practice, some sorts of molecules are more valuable. Broadly, three categories of compounds can be recognised *viz.*, (i) Primary metabolites, (ii) Secondary metabolites and (iii) *Semantides*.

Primary metabolites are related with vital metabolic pathways. They occur universally or at least in a very wide range of plants, e.g. citric acid, aconitic acid, or sugars, figured in the photosynthetic carbon cycle, etc. Their presence or absence is of less taxonomic significance. Secondary metabolites (secondary plant products) perform non-vital (or not universally vital) functions. They are not widespread in plants and hence render them more valuable as taxonomic information. These include phenolics, alkaloids, glucosinolates, terpenoids, carbohydrates, oils and waxes, etc. Their value in taxonomy is judged simply by their distribution and correlation with other characters. Generally, function of a chemical constituent is not weighted/ emphasized as a character for taxonomic purposes. Their functions have a little relevance to taxonomists. Semantides are the information-carriers. DNA is a primary semantide, RNA a secondary semantide, while proteins are tertiary semantides. The sequences of nucleotides and amino-acids in these substances provide all the taxonomic information necessary for classification. In practice, there are difficulties in gathering of the sequence data. A convenient and biologically meaningful distinction exits between micromolecules and macromolecules. The former are low molecular weight compounds and include secondary plant products / metabolites. The macromolecular are high molecular weight compounds and include protein and nucleic acids (DNA and RNA) found in the nucleus, mitochondria and chloroplasts.

Chemical characters of taxonomic importance are considered, for convenience in presentation as under: (I) Directly visible constituents (raphides, crystals, starch grains, etc.), (II) Primary constituents (chlorophyll, *semantides* and polysaccharides),(III) Secondary constituents (alkaloids, flavonoids, terpenoids, etc.), (IV) Biosynthetic pathways.

1. Directly Visible Constiutuents: In plants, very few chemicals are directly visible. They are seen in sections of plant material. Their distribution is well known and are constantly present or absent in some plant families or groups. Their characters have attracted attention of taxonomists.

(i) ***Starch Grains*:** Starch develop as grains in the plastids. The grains may occur singly or as compound structure. They differ in shape and size considerably. They also exhibit layering and accordingly are said concentric and eccentric.

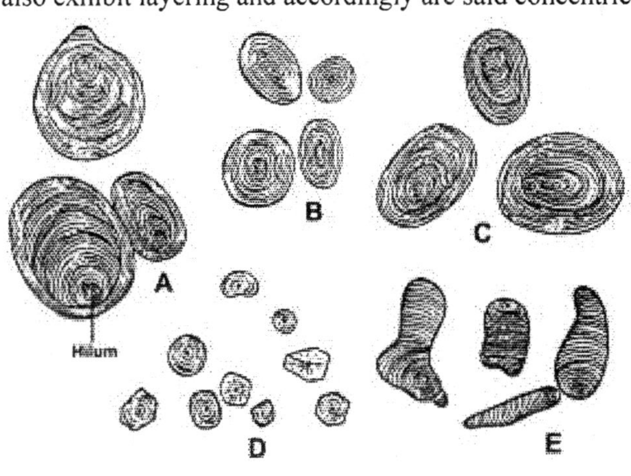

Plate-VII: Starch Grains (Different Shapes)

Reichert (1913) recorded their differentiation and specificity of occurrence in relation to genera and lower taxa in about 350 species. He showed that nearly-related plants often have starch-grains which are very much alike in appearance. The spectra of starches may indicate, in his opinion, nearness or remoteness of relationship. The spectra from starches of nearly related plants are very similar. Thus the types of starch grains sometimes yield useful taxonomic information. Reichert (*loc.cit.*) listed starch grains using terms e.g. potato type, legume type, wheat type, and so on. Tateoka (1962) studied the form of starch grains as an aid to the classification of the members of Gramineae alongwith other features. For instance, typical genera of the tribe Hordeae have compound grains, while *Nardus, Lolium* and *Parapholis* have simple grains which are not encountered in that tribe. Thus distinctness is shown by characters of grains which can be employed in taxonomic consideration.

(ii) Raphides: These are slender, acicular crystals arranged parallel to each other in tight bundles and occurring in special raphide-sacs.

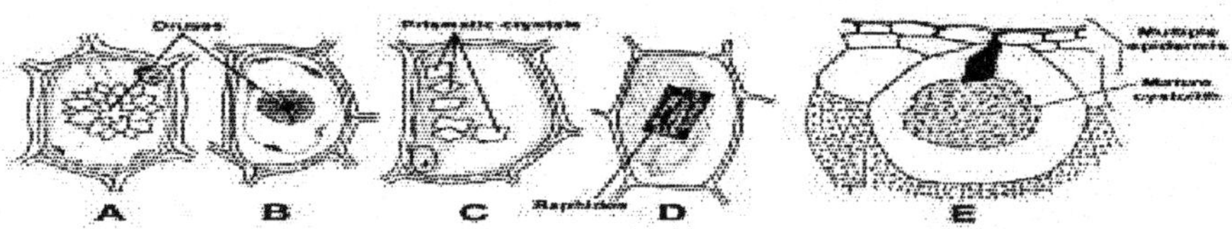

Plate-VIII: Types of Crystals And Cystolith.

They are observable in suitable section of plant material and are made of calcium oxalate. Gibbs (1958) opined that the ability to store oxalate in this form appears to be restricted to certain groups of plants and may, therefore, have some value as evidence of relationship. Metcalfe and Chalk (1950) listed their occurrence in about 55 dicotyledonous families. They further listed their presence in the woods of 11 dicotyledonous families. Gibbs (1958b) noted their absence in presumably primitive monocotyledonous orders *viz.*, Alismatales, Hydrocharitales Scheuzeriales, Potamogetonales, Triuridales and Najadales. Gulliver (1866) made a careful study of raphides. He noted raphides in three families *viz.*, Balsaminaceae, Onagraceae and Rubiaceae of British Dicotyledons. Today, we at least agree with him that they occur in Rubiaceae but are restricted mostly to the subfamily Rubioideae. The genus *Trapa* is included in the Onagraceae. (*cf.* Bentham and Hooker, 1862-1883). However, some taxonomists referred it to a separate family of its own-the Trapaceae (Hutchinson, 1973; Engler and Diels, 1936). However, raphides are lacking in *Trapa* and present in other onagraceous taxa. This is suggestive of its separateness from the family Onagraceae as evidenced by embryology (*cf.* Just, 1946).

Raphides occur in several familes of the Centrospermae (*sensu* Engler) and also in the Cactaceae. It is also supported to be allied by the other evidence

(Maheshwari, 1945; Martin, 1946). Further, *Hydranagea* is sometimes included in the Saxifragraceae (*cf.* Bentham and Hooker, 1962-1883; Engler and Diels, 1936). It lacks raphides, wheareas other taxa of Saxifragaceae do show their presence. This warrants its distinctiveness from the Saxifragaceae in a separate taxonomic entity. Tomlinson (1962a) grouped eight scitaminean families into four natural groups on the basis of raphide occurrence as: (i) Heliconiacceae, Musaceae and Strelitziaceae having raphides associated with symmetrical guard cells, (ii) Costaceae, Marantaceae and Zingiberaceae lacking raphides but associated with asymmetrical guard cells, (iii) Cannaceae devoid of raphides but associated with symmetrical guard cells, and (iv) Lowiaceae are raphide-bearers and associated with asymmetrical guard cells. In his opinion, taxa lacking raphides are more advanced.

Raphides are not reported from the Poales, Cyperales Restionales, Hydatellales, Eriocaulales, and Juncales. They are widespread in the Dioscoreales, Taccales and Asparagales (Stenar, 1949) and also in Orchidales, Haemodorales Pontederiales, Philydrales and Bromeliales. They are rare in the Liliales. They are reported in the Zingiberales especially in the taxa with six or five anthers and not in taxa with single or half anther (Tomlinson, 1969). Raphides occur in most members of Arecales and Pandanales and in all Cyclanthales (Dahlgren and Clifford, 1982).

Calcium oxalate may be accumulated in other crystal forms as: (i) simple short crystals, (ii) simple elongate prismatic crystals (*pseudo-raphides*) and (iii) crystal *druses*. These are also worth notable in various groups of angiosperms.

(iii) Silica Bodies: Hydrated silica is deposited in plant cells either as distinct bodies, which are comparatively large and few or solitary or may be small and in great number (*silica sand*). Silica body containing cells are generally epidermal or belong to the mesophyll. Such epidermal cells are different from other epidermal cells and may be termed '*stegmata*' or '*short cells*'. They have typical shape, characteristic of a plant group (family, order). In the monocotylendons, they are concentrated in the Poales, Plate, Cyperales, Restionales (partly), Arecales, Zingiberales and Bromeliales. In case of grasses, the shape of the silica bodies (in surface view) is taxonomically important at subfamilial level e.g. (i) Panicoideae have cross and dumb-bell shaped or nodular, (ii) Eragrostoideae with saddle-shaped (rarely like Panicoideae), (iii) Pooideae with elongate parallel to the long axis of the leaf and have sinuate or crenate edges (Clifford and Watson, 1978). In the Rapateaceae, silica bodies are spheroidal with rough or spiny surface. They are not mentioned by Tomlinson (1969) in the Eriocaulaceae, but reported in *Paupalanthus* species by Hegnauer (1963). The Commelinales are mostly deviod of silica bodies, except *Callisia*, *Coleotrype*, *Forrestia*, *Gibasis*, *Hadrodemes* and *Tripogandra.* In the Zingiberales, *silica bodies* occur in all families. They are characteristic in shape e.g. trough-like in Musaceae; rectangular in Heliconiaceae and hat-shaped in Lowiaceae; stellate or druse-like in Strelitziaceae, Zingiberaceae, Cannaceae and Costaceae, irregular-spherical in some Zingiberaceae and variously shaped in Marantaceae.

Metcalfe (1971) observed silica bodies almost universally in the Cyperales, except few genera. They are restricted to the epidermal cells. They form cones with their bases resting on the inner walls of the epidermal cells with their apices directed outwards. They vary in number from one to numerous. They may be separate or fused, equal or unequal in size. In some taxa, the shapes are characteristic e.g. wedge-shaped in *Scirpodendron* and *Thoracostachyum*, bridge-shaped in *Mapania*, etc. Tomlinson (1961) noted small, solitary, either spherical, conical or hat-shaped *silica bodies* in the hypodermal cells in the Arecaceae (Palms). Solereder and Meyer (1930) reported striking patterns in subfamilies of the Orchidaceae.

(iv) ***Gypsum***: Crystals of gypsum ($CaSo_4$. $2H_2O$) occur in some plants. They have been noted of taxonomical significance. For example, Brunswick (1920) reported their occurrence in every member of the Tamaricaceae but did not find them in the nearly- related Frankeniaceae and Fouquieriaceae. The latter, instead have crystals of calcium oxalate, Gypsum is also noticed in some Capparidaceae, Compositae, Polygonaceae and Loasaceae.

(v) ***Lapachol***: It is a yellowish powder found characteristically in the cells of the woods of a number of Bignoniaceae. It is 2-hydroxy-3(3-methyl-2-butenyl)-1,4-naphthoquinone. It is not reported from allied families. It is, however, observed in *Avicennia*, a mangrove taxa, of the Verbenaceae. Its occurrence in this genus also warrants its distinctiveness from the family Verbenaceae.

(vi) ***Cystoliths***: These occur in the form silicified bodies with a cellulose skeleton. They are usually encrusted of carbonate of lime (Solereder, 1908; Metcalfe and Chalk, 1950). They are of different sizes and shapes. They occur in different associations and even exhibit different colours. Mostly they are distributed in both axis and leaf, especially in the epidermal and subepidermal layers. They are found in all tissue except wood. They have been reported commonly in the Acanthaceae and species of *Ficus*. The Acanthaceae are the subject matter of many investigations, only few of these with taxonomic significance are discussed here.

Solereder (1909) reviewed cystoliths in Acanthaceae. He noted various features of cystoliths e.g. their presence or absence, occurrence as single, double or in chain; shapes like elongated, spindle and rounded; apices blunt or pointed at either one or both, etc. These features are used to key out the various groups of genera. Metcalfe and Chalk (1950) also summarized and reviewed more or less similar features of *cystoliths* in the family. They thought their nature and distribution valuable for the recognition of genera and species.

Shashi Kumar and Paliwal (1975) in their foliar antomical study in Acanthaceae recorded complete absence of *cystolithis* in the tribes Thunbergieae and Nelsonieae, except *Thunbergia laevis*. Soleredor (*op.cit.*) and Metcalfe and Chalk (*op.cit.*) have also recorded their absence in these tribes. Selvaraj and Subramanian (1983) in their epidermal studies of the family noted cystoliths of different shapes such as globular, cylindrical, conical and bilobed. They, however, noted cystoliths present in *Thunbergia erecta* and *T. grandiflora*. Ahmad (1976) noticed simple, spindle, sickle or cigar-shaped cystoliths, either single or double, with end pointed and another broad in some species of *Hygrophila* and *Dyschoriste* (Acanthaceae). He (1975, 1979) also utilised features of cystolith alnogwith other epidermal features in the taxonomy of some species of *Lepidagathis* and *Dyschoriste* (Acanthaceae). It appears from the foregoing data that *cystoliths* can be employed after careful investigations at and below generic levels. Patil and Patil (2011) reassessed occurrence and taxonomic significance comparatively in the members of Acanthaceae. These should be, in their opinion, used in combination with other endomorphic/ exomorphic features for better considerations.

2. Primary Constituents

(A) Chlorophyll: It is one of the primary constituents. Edwards (1976) while re-evaluating the kingdoms of organisms proposed a new system of seven kingdoms, subordinate to the super kingdoms *viz.*, Procaryota and Eucaryota. He used the distribution of chlorophyll 'a' as major taxonomic parameter. He placed all photosynthetic plants with only chlorophyll 'a' under Cyanochlorobionta of Procaryota. Photosynthetic plants under Eucaryota are recongnised in three kingdoms: (i) Erythrobionta without accessory chlorophyll or occasionally with chlorophyll 'a', (ii) Chlorobionta with accessory chlorophy'b' and (iii) Ochrobionta with chlorophyll 'c'. As all vascular plants have similar chlorophylls, carotenoids, etc. They have been assigned to a single division Tracheophyta. He opined that chlorophyll 'a' is a primitive character state as it is found in all photosynthetic plants. This is also indicative of a fact that all these plants have a common origin. The other types of chlorophylls *viz.*, b, c and d are of rather confined occurrence and therefore are probably good phylogenetic markers (Schwartz and Dayhoff, 1978).

The chlorophylls other than 'a' and other pigment groups (various carotenes and xanthophylls) differ characteristically between the eleven divisions of Algae. Euglenophyta and Chlorophyta have similar set of pigments, which is also observed in all higher plants. This is suggestive that green algae and higher plants must have a common ancestry, different from that of other algal groups. Different types of pigments, about which we are quite aware, are nowadays one of the major characters employed in algal systematics.

(B) *Semantides*: These are, in modern times, popular sources of taxonomic information since they provide either the primary genetic information itself (DNA), or the secondary (RNA) or tertiary (protein) derivatives of it. Of course, their potentiality of the huge amount of phylogenetic information comprising the base sequences of cellular nucleic acids has been recently recongnised. Recent advances in DNA technology have created a wealth of new opportunities for taxonomy, and now it is possible to sequence any form of DNA. The term 'molecular taxonomy' or 'molecular systematics' is used to mean the use of DNA, RNA and protein to infer elationships among plant groups.

(i) Nucleic Acids: Usually the total amount of DNA is highly constant and hence useful taxonomically at different levels. Particularly in angiosperms, this amount varies at the diploid level from 2×10^{8}-10^{11} base pairs, or 0.5 to 254 Spg giving considerable scope for variation.

Nowadays, rather than total DNA being investigated, specific fractions are targeted. Good results have been obtained e.g. single-copy DNA in *Atriplex* (Belford and Thompson, 1981), various defined repeated sequences in *Secale* (Jones and Flavell, 1982), ribonsomal- RNA genes in legumes (Jorgensen *et al.*., 1982) and chloroplast-DNA in potato (Hosaka, 1986), etc. Such data is valuable at all taxonomic levels. At the lowest levels, individual genotypes can be characterised with application especially in cultivar delimitation and population biology.

The nuclear genome also contains nonsense sequences of varied length which are repeated many times, forming the so-called *satellite* DNA. For example, studies on 13 species of *Cucurbita* revealed that all the species possess mainband and satellite DNA whose densities are constant throughout the genus.

All ribosomes are composed of a larger and a smaller subunit, each in turn comprising a RNA molecule (*r* RNA), the size of which vary between plants, and a number of protein molecules. *r* RNA has made impact on molecular taxonomy. For example, Pollard (1964) noted specificity of ribosomal RNA in cabbage, cauliflower, celery, corn and parsnip, He showed that 28 S and 18 S ribosomal RNA from these taxa have distinct base compositions, which are characteristic of the species.

Apart from the nuclear DNA, plants have the circular double-stranded mitochondrial DNA (mt DNA) and Chloroplast DNA (cp DNA). The latter type of DNA are relatively conservative as compared to the former one and therefore of greater taxonomic significance at the specific, generic and familial levels. Comparison of nuclear genome with cp DNA offers a potentially powerful tool to reveal past hybridisation events or introgression. Also, analysis of restriction fragment of cp DNA help distinguish between single and multiple origins of hybrid taxa e.g. presence of two quite different chloroplast genomes in the hybrid species *Aegilops triuncalis* indicates that it has arisen through hybridisation events, one with *A. caudata* as the female parent and one with *A. umbellulata*.

Of late, other recent techniques are being used e.g; (i) The RAPD (Random Amplification of Polymorphic DNA) allows to determine relationships between closely related species. It helps assess kinship relationships to the production of species-specific probes to detect hybridisation and gene flow between two samples. (ii) RFLP (Restriction Fragments Length Polymorphism). technique is significant in the fact that in many plants the distribution sites in the genome where a particular enzyme can cut is relatively fixed, generally showing variations between species or genera. (iii) DNA Hybridisation technique is effective to measure homology of the DNA and RNA in various groups of organisms and has been found to be useful for taxonomic research. For example, based on DNA hybridization studies, the three cereals, *viz.*, *Secale*, *Hordeum* and *Triticum*.Wheat (75%) is more closely related to rye (100% than is barely (57%). (iv) DNA- RNA Hybridisation technique involve DNA of one organism and a fraction of the RNA of another organism. The quantity of association occurring between them is thought as a basis of similarity. For example, the Caryophyllaceae (which contains anthocyanins only) is related to the betalain-producing families but the relationship is not so close as the latter are to each other.

(ii) Proteins: The importance of proteins in chemotaxonomy lies in certain facts. They are large, complex molecules showing little qualitative variation with changing environmental factors. They are universally distributed in appreciable amount. They are easier to extract and many simple, rapid methods are developed for their analysis. Proteins have been employed in taxonomic research in various ways.

(A) Protein Banding Patterns : Electrophoresis is being used extensively in protein investigations. Soluble proteins migrate within an electric gradient at a rate that depends on their net electric charge and on their molecular size and shape. The rate of migration of each protein is constant under identical conditions and therefore can be employed as a reliable character for the detection of homologous proteins. Separation of proteins can be carried out either by paper or gel electrophoresis. The taxonomic applications of the results of electrophoresis have been used mainly at and below the generic level. Much work has been directed at food reserve proteins e.g. (i) phylogenetic affinities in Triticinae, (ii) close relationship between *Vicia* and *Lathyrus*, (iii) interspecific variations among eight species of *Cassia*, etc.

(B) Amino-Acid Sequencing: Amino-acid sequencing of homologous proteins and comparison from different taxa is more modern development and powerful tool for evolutionary and taxonomic investigations. It is possible to break off the amino-acids from the polypeptide chain and then they are sequenced separately. They can be separated by chromatography and /or electrophoresis (*fingerprinting*) and proved taxonomically valuable. Amino-acid sequencing investigates the variation in the precise sequence of amino-acids in a single homologous protein throughout a range of organisms. The degree of similarity in the amino acid sequence is presumably proportional to the degree of genetic relationship. For example, (i) phylogeny of various taxa like Ranunculaceae, Magnoliaceae, Malvaceae, Polygonaceae and Solanaceae have been elucidated.

(C) Serological Study: It is concerned with the nature and interactions of antigenic material and antibodies. The substance capable of stimulating the formation of an antibody is called 'antigen' and the highly specific protein molecule produced by plasma cells in the immune system in response to the antigen is called 'antibody'. Proteins widely used as antigen in *serotaxonomy* are those, which carry useful taxonomic information. Both structural and reserve proteins can be used for this purpose.

Serology has proved a useful taxonomic tool at all levels-below the species and above the family. For instance, (i) Hawkes and Lester (1968) elucidated the ancestry and close relatives of the cultivated potato, (ii) Jenson, (1968, 1974) studied Ranunculaceae at tribal and generic levels which supported relationships based on classical evidence, (iii) At lower levels, Smith (1972, 1983) studied the relationships in brome-grasses (*Bromus*) etc.

3. Secondary Constituents: Different classes/ types of secondary chemical constituents are potentially useful in plant taxonomy. However, some of these have been selected by virtue of their more common presence in various plant groups and their more definite distribution pattern indicating possible use for phylogenetic and taxonomic considerations. Secondary constituents generally lack nitrogen and do not take direct part in plant metabolism. They include constituents like phenolic compounds, polyphenolic compounds, alkaloids, terpenoids, iridoid compounds, pigments, etc.

Flavonoids: The phenolic molecules fall into a general class known as flavonoids. They are further classified into flavones, flavonones, flavonols, isoflavones, isoflavonoids, anthocyanidins, etc. They are found in a variety of types and patterns of side groups which characterise the individual compounds. They are widespread, some are rare, and the pattern of combination of occurrence have on many occasions proved valuable as taxonomic evidence in the flowering plants at all levels from order downwards. Flavonoids can be used as taxonomic markers because they are structurally variable, chemically stable, widespread throughout the plant kingdom and identifiable easily and rapidly.

Flavonoids have been most frequently used compounds in plant taxonomic studies. The many positive correlations that they display stimulated phytochemical taxonomist to consider them also as evolutionary markers. Harborne (1977) and Gornall *et al.* (1979) evaluated information on flavonoids in relation to phylogeny of angiosperms. Swain (1975) suggested earlier that during the course of evolution flavonoids increased in their structural complexity from the primitive glycolflavones in *Chara* to the proanthocyanidins in ferns, gymnosperms and putative angiosperm progenitors. Certain phylogenetic inferences based on flavonoids have been made. Giannasi and Lans Chuang (1976) employed flavonoid data in the systematics of the genus *Perideridia* of the Umbelliferae. They recognised three discrete groups of species: (i) those producing flavonols only, (ii) those with flavonols and a few flavones, and (iii) those producing flavones predominantly. They attempted at correlating these three discrete groups and respective geographical distribution in America. They found the first group centered in California, whereas the other two groups in the Pacific North West and eastward to mid-western USA. Hegnauer (1971) conducted an extensive survey of flavoniods in the families Umbelliferae and Araliaceae. They inferred that these two families are intimately related. They also concluded that these two families may be regarded as constituting the order Umbeliales, which in turn, may have derived from a Rutalean stock. Giannasi (1978a) estimated generic affinities in Ulmaceae, whereas Crawford and Evans (1978) studied relationships amongst the species of *Chenopodium.* Dahlgren and Clifford (1982) reviewed status of investigations on flavonoid distribution in the monocotyledons. They inferred that flavonol, especially myricetin is a useful taxonomic marker at tribal level within the Zingiberaceae. They observed tricin, one of the flavones, almost universal in the Poaceae. Similarly, glycoflavones are much more common in tropical and subtropical Orchidaceae. They are restricted to the Restionaceae of South Africa. Interestingly, a rare compound 'luteolin 5-methyl ether' is unique to two families *viz.*, the Cyperaceae and the Juncaceae. This links these two families together which for many other reasons are regarded as closely allied.

***Anthocyanins*:** Harborne (1963) reviewed distribution of anthocyanins in higher plants. The unusual chemical structures provide the best kind of taxonomic marker. Earlier studies in his opinion, suggested general systematic points of interest such as: (i) Pelargonidin and delphinidin were found to occur with greater frequency in the more advanced plants, replacing cyanidin, which was considered to be the most primitive anthocyanin. (ii) It was discovered that normal anthocyanins were replaced by nitrogenous pigment in the flower of eight families, all of which belong to the order Centropermae, (a point dilated separately) (iii) Pigments which differed slightly. Harbone (*op.cit.*) collated all the data and considered them in plant relationship.

The *anthocyanins*, which are all based on the flavylium skeleton (I), are almost confined to the higher plants. They are present with much greater frequency in Angiosperms. They have been, however, recorded in some Gymnosperms and rarely otherwise. Anthocyanins are responsible for nearly all the red, pink, mauve and blue colours in higher plants. They are, however, replaced by another kind pigment *viz.*, betacyanins in eight families of the order Centrospermae. Harborne (*op.cit.*) reported seven different anthocyanidins known in nature. Three anthocyanidins, *viz.*, pelaragonidin, cyanidin and delphinidin occur for more frequently than any other types. Of these three, cyanidin is the most common. Delphinidin is next in order of frequency, followed by pelarhgonidin. Delphinidin is absent noticeably in Rosaceae, Papaveraceae and Orchidaceae. It is particularly abundant in Boraginaceae, Campanulaceae, Polemoniaceae and Hydrophyllaceae. There is a strong correlation between the presence of delphinidin and blue flower colour.

Anthocyanins lacking a 3-hydroxyl group are very rare and found in Bignoniaceae, Gesneriaceae, Sterculiaceae and Theaceae. These occur in unrelated plants. However, it is interesting that Bignoniaceae and Gesneriaceae, which are closely allied, contain them.

Two methyl ethers of cyanidin (peonidin and rosinidin) and four of delphinidin (petunidin, malvidin, hirsutidin and capensindin) are known. Only malvidin, petunidin and peonidin are all common.

Twenty three classes of anthocyanidin glycoside are known. Harborne indicated their taxonomic distribution. It is suggested that each genus has a characteristic glycosidic pattern and species which do not conform to the genetic pattern are exceptional in other respects. Most genera that have been examined have only one glycosidic type.

Many *anthocyanins* occur in nature acylated with organic acids. They can be readily recognised by spectral and chromatographic means. They have been identified with certainty in ten families *viz.*, Cruciferae, Vitaceae, Violaceae, Melastomaceae, Polemoniaceae, Labiatae, Solanaceae, Commelinaceae, Liliaceae and Iridaceae. When present in a particular genus they are present with a certain regularity.

Harborne (*op.cit.*) remarked that there must be a considerable number of new anthocyanin structures to be discovered. As stated earlier, studies on anthocyanins provide the best kind taxonomic marker. Further studies will reveal their more taxonomic significance.

Betacyanins: These replace anthocyanins in higher plants especially in the families of the order Centrospermae. The characteristic pigments of this order are betacyanins, formerly called 'nitrogenous anthocyanins'. Betanin, the pigment of *Beta vulgaris*, is now known as to have a structure identical with (II). It is quite different from the flavylium pigments and presumably derived biogenetically from two molecules of 3, 4-dihydrocy-phenylalanine. Betacyanins differ conspicuously from anthocyanins in that heterocylic nitrogen-containing aromatic rings are present. However, they carry out the same functions as anthocyanins. Closely related to betacyanins, which are yellow pigments apparently with similar functions to various yellow or cream flavonids, are known as anthoxathins. Betacyanins and betaxanthins are collectively also called 'betalains'.

From the taxonomic point of view, the position of the Caryophyllaceae and Molluginaceae which lack betacyanins is of great interest. Normal anthocyanins have been found in all the genera of these families. The chemical data therefore suggests their removal from this order. It is worth adding that the ability of members of the Cactaceae to synthesise betacyanins helped to secure them a place in the Centrospermac. Cactaceae is usually referred to another taxonomic entity in most systems of plant classification.

The genus *Gisekia* traditionally assigned to Molluginaceae, has been found anomalous within this family because it has betalains instead of anthocyanins in other members of the Molluginaceae. Recently, Takhtajan (1980) referred the genus to Phytolaccaceae, where it appears more at home than any other family of Centrospermae.

The genus *Dysphania* is assigned to different families such as Caryophyllaceae, Chenopodiaceae and Illecebraceae by different systamatists. It is found to possess betalains and also sieve-tube plastids characteristic to Centrospermae (Mabry and Behnke, 1976). These authors strongly supported its placement to Chenopodiaceae.

Cronquist (1965) indicated defining the Centrospermae by the presence /absence of betalains in the flowers and excluded the Caryophyllaceae and Illecebraceae and included the Cactaceae and Didiereaceae. He remarked that this sort of treatment would require the investigator to ignore the evidence available from other disciplines. He also indicated that 'one-character taxonomy, like a one-mouse experiment is always suspect'. Mabry (1966) indicated that not just a single chemical characteristic was evaluated in arriving at such a suggested taxonomic change. He indicated two factors must be considered as they are of particular taxonomic significance: (1) the totally different structures of the two classes of pigments (i.e. anthocyanin and betacyanin) indicate different biosynthetic pathways, and (2) the betacyanin distribution is limited to only a few taxa. This clearly explains the pit-fall of evaluating all 'one-character', bits of information the same in taxonomic investigations.

Leuco- anthocyanins: This term was applied to colourles compounds which may become visible when boiled with mineral acid, giving brilliant red pigments having the characteristics of anthocyanidins. These are present in all vascular plant groups. In Angiosperms, their distribution is irregular (Bate-smith, 1954). Bate-Smith and Lerner (1954), Bate-Smith and Metcalfe (1957) and Bate-Smith and Riberrall-Gayon (1959) confirmed the correlation between the presence of *leucu-anthocyanins* in petals and woodiness. As a rule, the herbaceous members of generally woody families possess *leuco-anthocyanins*, while whole families, especially of the Gamopetalae, whether woody or not, are devoid of them. The tendency for leuco-anthocyanins to be associated with woody habit also occurs within woody and herbaceous families e.g. Leguminosae. Bate-Smith found that 0.1% of the members of woody families contained leuco-antholyanins, while only 15% of the members of herbaceous families possessed them. Bate-Smith noted them in the gamopetalous families *viz.*, Apocynaceae, Asclepiadaceae, Rubiaceae and Caprifoliaceae exclusively, while they are totally wanting in the woody members of the families such as Oleaceae, Verbenaceae, Scrophulariaceae and Bignoniaceae. Interestingly, the herbaceous Primulaceae and Plumbaginaceae are rich in leuco-anthocyanins. The subfamilies *viz*., Mimosoideae and Caesalpinoideae nearly all contain leuco-anthocyanins, whereas the subfamily Papilionideae, which is predominantly herbaceous but also has arboreal members, has taxa with and without leuco-anthocyanins. Dormer (1946) divided the Papilionoideae into two groups *viz.*, leaves pulvinate and epulvinate. This is interestingly correlated with the presence or absence of leuco-anthocyanins (*cf.* Dormer, 1946; Bate-Smith, 1958).

Alkaloids: The beginning of alkaloid chemistry go back when Serturner (1817) recognised that morphine was the principle responsible for the effect of opium. (*cf.* Hegnauer, 1963): Its nitrogen content was then brought to light. Alkaloids often have considerable physiological effects on both man and animals. They have been used for centuries as medicines, poisons and magical potions. They are easily identifiable by their physical and chemical properties. Many authors complained of their precise definition. Attempts, however, have been to define what is meant by the term alkaloid. Winterstain and Trier (1910), for the first time, defined them in the broadest sense. Their definition corresponds to present –day conceptions. They defined alkaloids as basic, nitrogen containing compounds that are distributed in both the vegetable and animal kingdoms. They further defined the 'alkaloid proper' as compounds containing heterocyclic nitrogen, having a more or less distinctly basic character, and a complex molecular structure. Some consider that their definition should take into account the chemical, botanical and pharmacological properties. The above definition makes no sharp distinction between alkaloids and many other nitrogenous compounds in plants e.g. colchicine is counted as an alkaloid because of its pharmacological activity and contains no heterocyclic nitrogen. Winterstain and Trier (*loc.cit.*) also clarified later that the alkaloids can be thought of as by-products of protein metabolism, and the main precurssors being proline, lysine and few others. For the purpose of chemical plant taxonomy, a definition is provided by Hegnauer (1963) as follows :

'Alkaloids are more or less toxic substances which act primarily on the central nervous systems. They have a basic character, contain heterocyclic nitrogen, and are synthesised in plants from amino acids or their immediate derivatives. In most cases, they are of limited distribution in the plant kingdom'.

This definition excludes a number of nitrogen-containing plant constituents which are normally thought as alkaloids.

Alkaloids can be categorised as: (i) true alkaloids having a nitrogen-containing heterocyclic nucleus derived from a biogenetic amine, (ii) protoalkaloids-derived from amino acids but lacking any heterocyclic ring, and (iii) pseudoalkaloids-which are biogenetically unrelated to amino acids and derived from terpenes, sterols, aliphatic acids, nicotinic acids or purines. Alkaloids thus comprise a structurally diverse group of compounds. However, much greater efforts are always needed prior to their application about their chemistry, biogenesis and distribution to benefit chemotaxonomy.

Alkaloids are widely distributed in plants and some plant families are alkaloids-rich e.g. Ranunculaceae, Annonaceae, Menispermaceae, Papaveraceae, Berberidaceae, Leguminosae, Solanaceae, etc. Alkaloids characterising species of a particular taxon are frequently of the same chemical or biogenic group e.g. the members of the Papaveraceae synthesising isoquinoline alkaloids always include protopine, while Leguminosae have lupin alkaloids and Solanaceae tropane derivatives. However, Compositae and Gramineae produce many different types of alkaloids. In *Rauwolfia* (Apocynaceae) about so different alkaloids are known to occur. (*cf.* Hegnauer, 1966). Some specialised e.g. morphine from *Papaver somniferum* (Papaveraceae), coniine (few Umbelliferae), strychnine species of *Strychnos* (Loganiaceae), etc. It should be also noted that plants from different provinces (geographical areas) of the same species may differ considerably in their alkaloid complexes e.g. *Acontium napellus* (Ranunculaceae) variation in aconitine and other alkaloids is detected between plants from the Tyrbl. Auvergne and Haute-Marne (*cf.* Dillemann, 1960). Such differences may be racial or environmental in nature. Within genera containing alkaloids, species which are alkaloids-free are rare; small quantities or only traces may be found, but total absence is unusual. Inspite of certain limitations, alkaloids are very useful for taxonomic purposes as illustrated in the following.

Distribution of alkaloid has proved useful in the taxonomy of some petaloid monocotyledons. For instance, (i) Bentham and Hooker (1862-1883) recognised the families Liliaceae and Amaryllidaceae based on ovary superior and inferior respectively. However, Hutchinson (1959, 1973) emphasised umbellate and non-umbellate inflorescence in this allianee. He included taxa with umbellate inflorescences, *viz.* subfamily Amaryllidoideae in the Amaryllidaceae proper and shifted the non-umbellate taxon *viz*; subfamily Agavoideae (*sensu* Bentham and Hooker, *loc.cit.*) in a separate family, the Agavaceae alongwith the Dracaenoideae. This treatment appeared satisfactory since the Agavoideae are alkaloid-free. On the other hand, the Amaryllidoideae invariably contain alkaloids. (ii) Hutchinson (*loc.cit.*) also included the majority of genera of the Alloideae, which are referred under the Liliaceae by Krause (1930) and Wettstein (1935) and Bentham and Hooker, (*loc.cit.*), in the Amaryllidaceae because they also possess umbellate inflorescences. The members of the Alloideae contain no alkaloids. Moreover, they contain unusual sulphar-containing amino acids (e.g. alliin) not encountered in the Amaryllidaceae proper.The rearrangements made by Hutchinson is not at all satisfactory. It would be better to leave the Alloideae in the Liliaceae, or if this is not desirable on other grounds, to elevate them to family rank even under a separate rank, the Alliales as defined by Treub (1975).

Occurrence of alkaloid has taxonomic implications in rutaceous taxa (Price, 1963). The Rutaceae are regarded a phylogenetically complex family. Willis (1960) stated that the groups of which the Rutaceae is made up differ among themselves to extent that some have been regarded as independent families. According to Price, (*loc.cit.*) the picture presented by the alkaloids suggests that the major subfamilies of the Rutaceae represent a highly homogeneous group. This is so because the furoquinolines are found throughout four subfamilies *viz.*, Rutoideae, Flindersioideae, Toddalioideae and Aurantiloideae of the five subfamilies. Also quinolines are isolated from these four families. The benzylisoquinolines are a particularly widespread group of alkaloid in general. This group is frequently encountered in the Rutaceae but lack in the morphologically related to the Rutaceae e.g. Meliaceae, Burseraceae, Simarubaceae, Zygophyllaceae and Cneoraceae. Further support for the view that the Rutaceae are a distinct and homogeneous group is provided by its essential oils and coumarine. The essential oils are present at least in four subfamilies *viz.*, Auratioideae, Rutoideae, Toddalioideae and Flindersioideae. Coumarines also occur in these subfamilies. The coumarines on the contrary, are not reported in the Meliaceae, Burseraceae, Simarubaceae, Zygophyllaceae and Cneoraceae. The relationship of the subfamily Findersioideae to thè remainder of the Rutaceae has been the subject of some difference of opinion. Hutchinson (1973) placed it in the Rutaceae. Dadswell (1938), Dadswell and Ellis (1939) considered it not as a typical either the Rutaceae or the Meliaceae and favored a separate family, the Flindersiaceae. The Flindersioideae contains two genera *viz.*, *Flindersia* and *Chloroxylon.* Both these genera contain furoquinolines which are not reported outside the Rutaceae. Coumarines are also found in *Chloroxylon* and *Flindersiea*, which are absent in the Meliaceae. Price (*loc.cit.*) is inclined to regard the Flindersioideae homogenous with other rutaceous subfamilies and distinct from the Meliaceae.

Wettstein's order Polycarpicae (1935) contains total 25 families. Of these, thirteen families are mainly monotypic or oligotypic of uncertain position. Alkaloids are isolated frequently from the rest twelve families. The majority of these alkaloids are benzyltetrahydroisoquinolines. Two groups of families are recognisable: (i) Annonaceae, Aristolochiaceae, Berberidaceae, Magnoliaceae, Menispermaceae and Ranunculaceae with both compounds *viz.*, magnoflorine and berberine, (ii) Hernandiaceae, Lauraceae, Monimiaceae and Nymphaeaceae with only berberine. Several taxonomists (Cronquist, 1968a, Kubitzki, 1969, Takhtajan 1969, Thorne 1968, 1976, Dahlrgren 1975), based on the ability to synthesise aforesaid alkaloids, made certain rearrangements in their treatments especially of the families with apocarpous gynoecia.

The complex indole alkaloids are known from the Loganiaceae, Apocynaceae and Rubiaceae. This indicates that these families may be related as suggested by Hallier

(1912) and Hutchinson (1959). Hutchinson (*loc.cit.*) derived the Rubiaceae from the Loganiaceae (*cf.* Hegnuer, 1963).

Terpenoids: This is another class of secondary volatile compounds used extensively in plant systematics. Mostly terpenoids are polymerised isoprene residues. They are derived from the geranyl pyrophosphate pethusous and eventually lead to steroids. They are almost universal in distribution. The primitive human societies identified various groups of plant by their characteristic smell due to various types of terpenes in them, for example, mints (Lamiaceae), umbellifers, grasses like vetiver (*Vetiveria zizanoides*), *Cymbopogon* species, *Citrus* species (Rutaceae), tribe Tagitinae (Asteraceae). Probably, the relationship between plant constituents and classification started with terpenoids.

Terpenoids are classified based upon the number of isoprene units present in their molecule: (i) Hemiterpenes (single unit e.g. tiglic acid), (ii) Monoterpens (two units e.g. menthol), (iii) Sesquiterpenes (three units e.g. farnesol), (iv) Diterpenes (four units e.g. phytol), (v) Teriterpenes (six units e.g. squalene), (vi) Tetraterpenes (eight nits e.g. carotenoid) and (vii) Polyterpenes (e.g. rubber). Various groups of terpenes have been studied in relation to plant taxonomy.

Some specific terpenes are noted in certain families e.g. cucurbitacins in Cucurbitaceae, asperulloside in Rubiaceae and sesquiterpene lactones in Asteraceae. The monoterpernoids, also called 'essential oils' from their 'essence' or 'odours' are found commonly, as stated earlier, in some plant families like Lamiaceae, Rutaceae, Apiaceae and some gymnosperms. For example, (i) in *Juniperus* (Cupressaceae) the limits of intra-individual, ontogenetic, seasonal and population variation is studied (Adams 1975b, 1979; Adams and Hagerman 1976, 1977; Adams and Powell 1976). (ii) Mirov (1961) found the turpentine composition of each valid species of *Pinus* was specific and little modified by the growing season or transplanting. The two closely allied Mediterranean species *P. halepensis* and *P. brutia* can be distinguished. Two terpenes, *viz.* -pinene and carene are isolated from the latter. They are but absent in the former. (iii) Baker and Smith (1920) found the oil composition for a given species of *Eucalyptus* (Myrtaceae) was more or less constant and therefore regarded the oils as having equal weight to other morphological characters used in taxonomy of the genus. These authors correlated groups of species containing different combinations of oils with bark structure and leaf venation. For example, (a) obtuse 'feather' venation indicated low oil yield, (b) more acute venation indicated a high oil yield, (c) 'Butterfly' venation indicated a very high oil yield, etc. (iv) Irving (1968) found that there are three *sympatric* taxa of *Hedeoma* (Lamiaceae) coexisting without any sign of natural hybridization and used their terpenoids in their identification. (v) Herz (1973) suggested that sesquiterpenes characteristic of Compositae might serve as good taxonomic markers in the family. (v) Emboden and Lewis (1967) indicated that a survey of monoterpenes in 19 species of *Salvia* (Lamiaceae) aided greatly in their identification. They also noticed that the study of terpene composition was as valid as morphological characteristics in the analysis of introgression within this group.

Iridoids: The monoterpenoid cyclopentanoid laetones, also know as iridoids, attractal the attention of taxonomists in recent times. It is also consideral as evolutionary marker. Bate-Smith and Swain (1966) indicated their possible significance in taxonomy. Based on the iridoid distribution, several re-arrangements or alternations have been made in various plant groups. The iridoid compounds are mainly restricted to plant groups with unitegmic ovules. This, according to some, suggests that they developed along an evolutionary line where the ovules were about to evolve from bitegmic to unitegmic condition. Jensen *et al.*. (1975) opined that the families Daphniphyllaceae and Altingiaceae (Hamamelidales) are perhaps remnants of primitive iridoid-containing plant groups bearing bitegmic ovules. Jensen *et al.* (*loc.cit.*) brought out taxonomic importance of iridoids in angiosperms. They inferred that the group '*Sympetalae*', as conceived by Takhtajan (1969) is an unnatural assemblage. It is but polyphyletic as divulged by chemical evidence. The orders Dipsacales, Gentianales and Scrophulariales (exclusive of Solanales) are characterised by the presence of iridoids, whereas other orders are devoid of them. The *monogeneric* family Fouquieriaceae is quite distinctive from the related orders like Ericales, Cornales and Solanales. It is also included by some under Tamaricales. In the light of iridoid distribution it should be referred as an order of its own (Dahlgren *et al.* 1976). Kubitski (1969) attempted to link the Rosalean and Guttiferalian complexes based on their distribution. Cronquist (1968) opined that these two groups sprung from Magnoliidae, which, however, is totally wanting them.

Saponins: These comprise a heterogenous groups of glycosides with terpenoid aglyca. They are chracterised by the ability to foam in water solutions. They include two principal groups: (i) steroid saponins, (ii) triterpense saponins. The former have aglyca with furostanol and spirostanol structures, whereas the latter are derivatives of oleanans and related compounds. Also, the former are most widely distributed in the monocotyledons, whereas the latter are the commonest type in the dicotyledons. Apparently, these two types of saponins have a tendency to be mutually exclusive although this is by no means absolute. Their distribution is of taxonomic and phylogenetic significance. The steroid saponins occur rather generally in the Dioscoreales, Asparagales and Liliales. Perhaps they are absent in the alkaloid-rich Amaryllidaceae and probably also in the Colchicaceae. Steroid saponins are generally poisonous (blood haremolysing) and thus seem to serve mainly as a means of protecting the plants. This may explain their probable absence in families having rich amounts of poisonous alkaloids e.g. Amaryllidaceae and Colchicaceae. Steroid saponirs are, however, widely scattered in the dicotyledoons e.g. in disciflorous taxa (Zygophyllaceae, Anacardiaceae, Celastraceae), gamopetalous taxa (Asclepiadaceae, Apocynaceae, Lamiaceae, Solanaceae, Scrophulariaceae, etc.) apetalous taxa (Moraceae, Euphorbiaceae, etc.). Further studies on their distribution will reveal their probable taxonomic significance.

Tannins: These are identifiable by their characteristic appearance in microscopic sections of plant organs when fixed in FAA (formalin- acetic- alcohol) and stained with iron salts. They are significant from phylogenetic point of view. Bate-Smith and Metcalfe (1957) surveyed extensively the angiospermic families for tannins and as well as for leuco-anthocyanins. They found about 83% of tanniniferous taxa conciding with the presence of leuco-anthocyanins in their leaves. Most of the tanniferous families are woody or probably allied to woody families, and most non-tanniniferous families are herbaceous. Some families are highly tanniniferous, some contain little or none. The occurrences of tannins thus have also taxonomic value. The taxonomic importances of ellagitannins have been well studied by Harborne (1973), Bate-Smith (1973) and Sporne (1975a). In their extensive survey of flowering plants they revealed that this is exclusive to dicotyledons and indicate fundamental splitting between the Magnoliidae and the Hamamelidae-Dillenidae. The former lack ellagitannins, whereas the latter do contain them. Probably, this in turn, also points to the origin of monocotyledons from a magnolian stock.

Cyanogenetic Compounds: These are usually found in the plants as glycosides. They are poisonous compounds produced by the plants when their cells are injured. Plants containing them are called cyanophoric plants. Hegnauer (1977) defined the term 'cyanogenesis' as the ability of certain plants to release hydrocyanic acid (prussic acid) after injury of cells. Such plants may contain one to several cyanogenetic glycosides. The activity is located in the leaves and seeds. Cyanogenetic glycosides may be transported from one tissue to another and thought to bė stored in the vacuole. The most frequently occurring cyanogenetic glycosides fall under five groups: (i) prunasin group, (ii) taxiphyllin group, (iii) linamarin group, (iv) gynocardin group and (v) cyanolipid group. The type of biogenesis of cyanophoric compounds differs in the major groups (i) pteridophytes have phenylalanine as precursor, (ii) gymnosperms having tyrosine as precursor and (iii) angiosperms have precursors such as valine, leucine, cyclopentenyl glycine, phenyalanine and tyrosine. The cyanogenetic compounds are derived from different precursors and therefore vary biogenetically. They are widely distributed amongst the various trachaeophyes (angiosperms, gymnosperms, pteridophytes) and even in fungi (Paris, 1963). *Cyanogenesis* therefore appears to have considerable significance in plant taxonomy and phylogeny.

In the monocotyledons, cyanogenetic compounds are known from various orders but show no definite pattern. Certain families are particularly rich e.g. Poaceae, and Arecaceae. They appear absent in certain large orders such as Ericaulales, Bromeliales, Orchidales and nearly so the Liliales. However, more investigations in these are still awaited. Their richness in *Scheuchzeria, Triglochin* and *Lilaea* of the Zosterales can be consideral as indicative of their close relationship. A cerṭain tendencies for concentration of cyanogenetic compounds in the Poaceae, Arecaceae, Flagellariaceae and partly in Cyperaceae is perhaps discernible. This is also the case of Ariflorae.

Cyanogenetic compounds are formed along the tyrosine pathway in Magnoliaceae, Calycanthaceae, Trochodendraceae, Ranunculaceae, Berberidaceae, Papaveraceae, Fumariaceae, etc. The monocotyledons evolved from ancestors resembling the present-day Magnoliidae appear well supported by the study of the distribution of cyanogenetic compounds in the tracheophytes. In both, the monocotyledons and

Magnoliidae, cyanogenesis proceeds in the same way that is with tyrosine as the precursor (*cf.* Hegnauer, 1973; Valen Van., 1978a).

Cyanogenetic compounds are also of taxonomic significance at subfamilial levels (Hegnauer, 1977). For instance, the rosaceous taxa with a base chromomosome number nine have either prunasin or amygdaline. Linamarin is characteristic to the tribes Trifolieae and Phaseoleae of Fabaceae and the tribe Calenduleae of Asteraceae (Hegnauer, 1977). The subfamilies of Poaceae also show characteristic occurrence e.g. dhurrin in Andropogonoideae, triglochin in Festucoideae and Eragrostoideae, and taxiphyllin in Bambusoideae (Van Valen, 1978b).

The cyanogenesis in nearly all monocotyledons proceeds along the tyrosine pathway, except *Chlorophytum capense* and *C. comosum* (Liliàceae). These species, however, proceed with the phenylalanine pathway (Valen Van F., 1978b). In certain cases, cyanogenesis is quite variable. For example, the genus *Glyceria* comprises cyanogenic, facultatively cyanogenic and non-cyanogenic species. Investigations on cyanogenesis in 12 species of the genus *Lotus* (Fabaceae) from Israel revealed extreme *polymorphism*. Some species are cyanogenic and others acyanogenic. Few species contained both cyanogenic and acyanogenic populations (*cf.* Band *et al.* 1981). Paris (1963) is of opinion that the distribution of particular compound is perhaps of taxonomic interest e.g. macrozamin (XVIIa) found in species of the genus *Macrozamia* (gymnosperm).

Oils, Fats and Waxes: Oils and fats are distinguishable by their physical state at normal temperature. The former are liquid, whereas the latter are solid. Lipids are the esters of fatty acids with glycerol, and made up of carbon, hydrogen, a carbohydrate group, phosphorus or some other group, in addition to carbon, hydrogen and oxygen. These are called complex (conjugated) lipids. The waxes are esters of long-chain alcohols with long-chain fatty acids and may contain free alcohols, free fatty acids, aldehydes, ketones or hydrocarbons.

Fatty acids specific to particular plant groups can be considered for taxonomic purposes, for instance, erucic acid is found in Cruciferae, petroselinic acid in Apiaceae, ximenyric acid in Olacaceae and Santalaceae. Some species of Flacourtiaceae characteristically possess chaulmoogric acid (*cf.* Shorland, 1963). Proportions of fatty acids in lipids are constant in both genera and species in Palmae (Hilditch, 1952). The seed fats of Lauraceae and Palmae are rich in lauric acid (Shorland, *loc. cit.*). Several workers have used leaf waxes taxonomically. The waxes of two subgenera of *Eucalyptus* differ in their melting point ranges (*cf.* Eglinton and Hamilton, 1963). Kreger (1958) attempted a taxonomic approach based on the nature of wax constituents as deducted from the X-ray constituents as deducted from the X-ray powder photographs of unfractionated waxes. He summarised his results as: (i) closely allied species generally produce waxes of corresponding or identical composition, (ii) plant species from different families may have cuticle waxes of almost the same composition, (iii) cuticle waxes which consist predominantly of primary alcohols appears to be restricted to monocotyledons, (iv) there are certain indications that morphological features common to plants of different families are often accompanied by a similarity in the constitution of the cuticle wax. Full understandings of the biogenetic pathways involved will be essential for the satisfactory use of leaf wax constituents in plant taxonomy.

4. Biosynthetic Pathways: Physiolgy of a plant just as morphology of a plant, is largely determined by genetic factors, which presumably act through enzyme systems controlling chemical syntheses. Physiology is concerned with the biosynthesis of relatively simple secondary constituents such as terpenes, coumarins, alkaloids, flavanoids, etc. Each stage of such a synthesis requires an enzyme system which is genetically controlled and which can be modified or lost as the result of genetic mutation. Studies of the results of such mutations in term of alterations of the structures of secondary constituents could therefore give information on evolutionary sequences and to this extent it could be very useful aid to taxonomy. Chemists attempt to define the nature of the biosynthetic alternations and assess their trivial or fundamental nature. However, botanists will attempt to define how far he wishes to use such information in classification. In chemotaxonomic studies, the absence or presence of a specific chemical substance or a set of chemical constituents are taken for granted in deducing the relationships. Chances of convergent evolution are generally overlooked. The same compound in different plant groups might come as result of different process and from different precursors.

The variations in the photosynthetic process among green plants are widely acknowledged. Three variation *viz*., *C_3 type* (Calvin cycle), *C_4 type* (C_4 – dicarboxylic cycle) and *CAM* (Crassulacean Acid Metabolism). Majority of flowering plants and also the lower plant groups have C_3 type of photosynthetic process. C_3 photosynthesis evolve in response to changes in carbon and oxygen concentrations in the atmosphere in past (Smith, 1976). Instead of the usual phosphoglyceric acid, the initial product of Calvin-Benson cycle (C_3), the C_4 plants fix carbon initially into 4-carbon organic acid, oxaloacetic acid, and then malic or aspartic acid (Webster *et al*., 1975). Such plants have been called C_4 plants by Haton and Slack (1970). The chloroplasts have a slightly modified granar and wall structure and are present in a highly specialised chlorenchymatous leaf bundle sheath. These anatomical specializations are collectively called 'Kranz Syndrome'. The plant species in which they are observed are called 'Kranz species' (Tregunna *et al.* 1970; Smith and Brown, 1973). Smith and Brown (*loc.cit.*) recorded Kranz syndrome in the three tropical subfamilies *viz*., Panicoideae, Eragrostoideae and Arundinoideae.

C_4 photosynthesis is noted in seven out of eleven families of the Centrospermae. However, the basal group Phytolaccaceae shows C_3 photosynthesis (Marbry, 1977). CAM has been recorded in the subgenus *Euphorbia* of the genus *Euphorbia*. C_4 photosynthesis is but observed in the subgenus *Chamaesyce* of the same genus. Webster *et al.* (1975) therefore opined that these subgenera have totally independent origin. These authors pointed out the necessity to raise these subgenera to the level of distinct genera. All these cases are really physiological data of sorts involving biosynthetic pathways of basic metabolites or process rather than structures. Taxonomic and phylogenetic information can be obtained with the revelation of closely interwoven biosynthetic routes.

(VIII) Reproductive Biology

Reproductive biology often provides significant taxonomic information. In fact, the very cornerstone of modern taxonomy, the 'biological species' concept is centred round breeding behaviour of plants. Plants, as we know, constantly interact with abiotic and biotic elements of their ambience and do not remain isolated. Their strategies of life history permits them to adapt to their surrounding and survive. Some of the aspects of reproductive biology are useful for comparative taxonomic purposes. The data like (1) Phenology (2) UV patterns (3) Pollinators (4) Floral nectaries (5) Dispersal agents and more importantly (6) Breeding system are directly helpful. These are focussed here.

Before switching over directly to the aforesaid reproductive strategies, it appears imperative to review briefly the history of reproductive biology in plant taxonomy. In Arabian countries, the male flowersof date palm are shaken over the female flowers on the occasion of a ceremony. Theophrastus (30-287 B.C.) referred this fact in his 'Enquiry into plants'. This is also represented in the ancient pictures of date palm workers in Egypt or Mesopotamia. It is, however, Camerarius (1694) who for the first time proved sex in plants experimentally. He noted that a female mulberry tree, near which no male plants were growing, formed only abortive seeds (ovules). He experimentally confirmed this fact in *Ricinus communis* (Euphorbiaceae), *Zea mays* (Poaceae) *and Mercurialis annua* (Euphorbiaceae). Later, the role of insect in flower poliination was studied by Kolreuter (1761-1882), Sprengel (1793) gave detailed account of the floral adaptations of more than 500 species in his 'Das emdeclite Geheimniss der Nature in Bau und in der Befruchtung.' This understanding was further strengthened by Darwin (1809-1882) in his books (1) On The Various Contrivances By Which Orchids Are Fertilized By Insects (1862) and (2) The Effect Of Cross And Selffertilization In The Vegetable Kingdom (1876), Muller (1879, 1881, 1883) also published his in depth observations on pollinating insects from central Europe. Other important studies on this line are available through the works of Knuth (1906-1909), Frisch (1914), and Clements and Long (1923). More recent studies are known through the contributions of Barker (1983) and Ducker and Knox (1985) and few others.

(i) ***Phenology***: When two closely related taxa grow in different periods, whether they are sympatric or allopatric, then there is a likelihood that they are good species (*sensu* biological species concept) since they will not interbreed. Many times, there is an overlap period that must be assessed critically in reproductive terms, e.g. the possible production of hybrids, for taxonomic consideration. The comparative flowering behaviour of certain flowers, particularly those with complex morphologies is also an important aspect in such consideration, e.g. *Anthurium* (Araceae) with the characteristic spadix of unisexual flowers enclosed by the spathe. Similary, protogyny, which promotes out-crossing, can occur in certain flowers. The different patterns of stamen emergence relate to differences in pollen presentation.Such phenological data from plants are useful for taxonomic purposes.

(ii) Ultraviolet Patterns: Patterns of ultraviolet absorbance and reflectance from reproductive parts of flower, particularly corolla, are important in phenomenon of pollination. They are useful as comparative information in taxonomy. These patterns are decided by cellular structure of the epidermis and various chemical contents e.g. carotenoids and flavonoids (Harborne and Smith, 1978). These data, however, are better used amongst the closely related taxa, and not across broad group of plants.

(iii) Pollinators: Mainly, there are two pollination factors *viz.*, (1) Abiotic and (2) Biotic. Usually structural modifications go hand in hand in such factors. Water and air are the principal agents of the former. Water is a less common vector. The flowers, involved in water pollination, have reduced perianth and generally exhibit unisexuality. Pollination is along the water surface e.g. *Vallisneria* (Hydrocharitaceae) or even through the water e.g. *Amphibolis antarctica* (Cymodoceaceae).Wind pollination is common in the angiosperms. Even entire families e.g. Betulaceae, Ulmaceae, Poaceae, Cyperaceae, etc. are anemophilous. Entire subclass e.g. Hamamelidae have wind pollinated flowers. Flowers, in case of wind pollination, are usually small, unisexual, with no or only inconspicuous perianth segments.

Various agents are responsible for biotic pollination e.g. (1) insects (*entomophily*), (2) birds (*ornithophily*), (3) and mammals. Insects are large class of pollination agents. They seek pollen and or nectar as food sources. Flowers adapt to specific vectors, for example, with bee flowers pale- coloured or white with tubular corollas and fragrances, fly flowers dull-coloured (rarely purple) and sometimes foul- smelling, beetle flowers with strong fragrances and food tissues, etc. Flower orientation is related functionally to vector behaviour. Bird pollinated flowers generally have red or bright orange corollas with long tubes. Humingbirds and sunbirds are important as pollinators. Bat pollination (*chiropterophily*) is also observed especially in tropical regions. These flower open at night, have strong scents, often greenish or white in colour and with stamens or styles exerted to reveal.All these pollinating mechanism require long – term and careful investigations. The understanding about the various modifications/ adaptations can be very helpful taxonomically.

(iv) Nectar: Different types of floral nectaries exist in angiosperms e.g. hypogynous, epigynous, sepaline, petaline (or tepaline), staminal and ovarian. The ovarian nectaries can be septal as in several petaloid monocotydonous taxa or in the form of epigynous glands of some scitaminean taxa (Pai, 1966; Pai and Tilak, 1965; Patil, 1985; Fahn, 1953). The opening of septal nectaries also vary. They open at the base, at the middle or above the middle of the style. All these types of nectaries and their variations are obviously adaptation for pollination. These can be conventially employed in taxonomic considerations.

Nectaries secrete nectar and the composition of nectars are also useful comparative data for evaluating relationship. Sugars and amino acids are formed in the nectaries. Their variations, especially quantitative ones, can aid in taxonomy.All such studies need detailed long term observations before arriving at certain generalisations.

(v) **Dispersal system:** Dispersal systems of propagules or diaspores play important role in the reproductive process. Vectors of different types help more them away from the parent plants. Some seeds are dispersed by mechanical means of plant itself e.g. jaculators in some Acanthaceae and explosive fruits in Balsaminaceae. Others are distributed by animals, birds, mammals, reptiles, fish and ants. They are carried sometimes accidentally or deliberately by animals. Depending upon the type of agents, the propagules have developed. In general, fruits or seeds are brightly coloured, fleshy, slimy in nature or contain sugar or food contents useful for them. Structures such as wings, pronges, parachutes, hairs (comose), etc. are developed as adaptations. Light seeds e.g. in Orchidaceae, very thin membranous seeds as in *Oroxylum indicum* (Bignoniaceae) or circular flat are also in response to easier dispersal. All these modifications, adaptations or special structures are taxonomically useful in angiosperms.

(vi) **Breeding Systems:** Breeding system of plants should be investigated during a taxonomic study. It is important taxonomically since (1) the extent of interbreeding largely defines the pattern of variation and hence the delimitations of taxa, (2) a knowledge of the breeding system frequently helps to understand taxonomic complexity, although often it does not solve the problems associated with it,(3) and a study of the breeding system is often essential in revealing evolutionary pathways. Neverthless, it is debatable to which extent the breeding system determines variation patterns. First the proportion of outbreeding or inbreeding within any specified species is traditional thought to determine the variation within and between populations. In outbreeding species each population is variable but similar to nearby populations, on account of gene exchange which usually effected between such populations. However, inbreeding species obviously try to exist as comparatively uniform populations. Secondly, the extent to which a taxon interbreeds with different taxon is usually thought a measure of the phenotypic discretness and therefore also validity of that taxon. Lewis (1967) states that 'rate of gene exchange has no taxonomic significance whatever'. He further argues that taxonomic species are defined on phenetic discontinuities, not on interfertility. However, the degree of phenetic discontinuity is itself, at least partly, determined by the ability or not to interbreed, and on the length of time during which barriers have existed. It is clear that a detailed consideration of the breeding system is essential for a complete taxonomic understanding of any plant group. An understanding phenetic complexity may permit one to interpret patterns of comparative data among taxa. Inbreeding and apopmixis particularly need documentation since these often lead to pockets of variation in separate populations.

A taxonomically ideal species is the one which possess no taxonomic problems. It is always recognizable as a distinct entity and does not merge with other species. However, there are some taxa which pose intrinsic taxonomic problems and do not correspond with the above situation, instead either (1) they hybrid with other taxa so that the genetical limits are wider than the morphological ones or (2) they exhibit breeding barriers between different elements of a morphologically recognizable taxon. These two categories are dilated a little bit in the following:

(a) **Hybridising species:** A taxonomic hybrid, as here to be considered, is the product of breeding between distinct taxa. It can be interspecific or intergeneric hybrid. Generally taxonomists do not speak of hybrids between taxa within one species. Often few criteria are employed to assess the likelihood that a plant is hybrid. These criteria are: (1) phenetic intermediary between the putative parents, (2) reduced fertility, (3) F_2 generation, (4) distributional eviolence (5) artificial resynthesis. There is no invariable formula or set of rules for the identification of hybrids. Each case has to be judged on its own merits, using as many of these criteria as possible.

(b) ***Semi-cryptic species*:** It is a taxonomic problem associated with hybridization. There are certain species which do not differ morphologically but do so cytologically. Being semi-cryptic in nature, taxonomic problem/ argument arises with respect to the correct status of the ultimate taxa. The species recognised by some taxonomist become relegated to subspecies even varieties by others. Such taxa appear to be based on different genetic situations: (1) Outbreeders internal barriers: Many species or species complexes known in which there are internal breeding barriers. In most instances, the underlying cause is chromosomal, usually a difference in chromosome number, but sometimes in their structure, (2) Inbreeders : Inbreeding species tend to generate small intra but relatively wide interpopulational variation, (3) Apomictic taxa: It is the habitual reproduction by non-sexual means.

In a nutshell, data from reproductive biology have not been employed widely as a source of comparative information in taxonomy in past. They are important for clarifying the biology of the characters and states of the features possessed by the plants. They are helpful in understanding of the function of structural characters used by taxonomists. They help understand patterns of partitioning of characters state variation within and between populations.

(IX) Genetics and Cytogenetics

Genetics is the science of heredity and variation, and Cytogenetics is the study of chromosomes and their genetic implication. These two *viz.*, genetics and cytogenetics are considered together as a source of comparative data for taxonomic purposes. The promise of evaluating relationships of plants directly by comparing genetic resembalances and differences and also their organization into chromosomes have always attracted attention of the taxonomists. Understanding what the chromosomes contain genetically and how they behave during meiosis is a deeper level of insight. The landmark studies by Clausen, Keck and Hiesey (1940, 1945, 1948) laid the foundation for the work on their line on the nature of plant species. The plant taxonomists have great attraction of genetic and cytogenetic data because these probe the real hereditary bases of evolutionary divergence and hence offer the possibility of a more refined yardstick for classification

The genetic similarities and differences among taxa can be determined in two ways: (1) crossing studies to determine the genetic basis of selected characters. (2) isozymes analyses to yield an estimate of genetic distances between taxa.

(i) Genetic Data :

(a) Genetic analysis: Very limited investigations have been done on the genetic basis of features in a purely taxonomic context. Following are classical examples. Rollins(1958) studied pubescence on the fruits of *Dithyrea wislizenii* and *D. grifithsii* (Crucifereae) on genetic basis. The former was regarded as typically pubescent and the latter glabrous. Rollins performed the necessary artificial crossing to obtain F1 ratios that showed convincingly that a single gene was responsible. Rollin concluded that it is safe to reject the phenotypic characteristic of glabrous siliques as having no significance for taxonomic purposes. The variation in fruit pubescence encountered within populations of these taxa made one suspect the taxonomic validity of the featurte, but the genetic studies confirmed it.

Recent study by Jackson and Dimas (1981) revealed that one of the distinguishing features separating *Haplopappus phyllocephalus* group (Asteraceae) from sect. Isocoma of the same genus, the presence vs. absence of ray florets, was controlled by a single gene. This information and other considerations, led to the transfer of the *H. phyllocephalus* group from sect. Bhepharodon to sect. Isocoma. For taxonomic decisions, the important point is not to know only the genetic basis of a taxonomically useful character, but more significantly to understand its consistency within and between groups and its correlation with other features.

(b) *Isozyme* Analysis: This analysis offers a more rapid means of estimating genetic distance between taxa. The enzymes are extracted from plants, spotted on gels, and separately electrophoretically yielding bands of proteins. These reflect isozyme phenotype and are interpreted in genetic terms. The frequencies of their occurrence within taxa are calculated. The degrees of genetic polymorphism including number of alleles per locus, as well as the proportion of the loci which are heterozygous can be calculated. The genetic distance between taxa can also be determined by various statistics. Additional crossing studies can be done to determine the actual genetic basis for each of the isozymes separated and used. Such information is now available through the works of Hunziker (1968), Gottlieb (1977), Crawford, Stuessy and Silva, (1982).

(ii) **Cytogenetic Data:** The fundamental idea is to find naturally or to produce artificially, intertaxon hybrids and analyse the resulting progeny in terms of a spectrum of interrelationship ranging from F_1 seeds to advanced hybrid generations.It is assumed that more robust and fertile the hybrids, the closer will be the genetic relationship between the taxa involved. Frequently in closely related flowering plants, intertaxon hybrids are obtained. But they should attain maturity so that they will be helpful to evaluate relationship. Just to obtain seed set between the two parents does not suffice the purpose of investigation. They should germinate, survive as seedlings and test their fertility upon maturation. Their fertility can be tested quickly by the pollen viability in the F_1 hybrids using stains e.g. lactophenol cotton blue, tetrazolium dyes, etc. The degree of seed set in the F_1 can also be measured. The degree of bivalent and multivalent formation, meiotic irregularities, abnormalities such as bridges and lagging chromosomes, etc. help assess relationship amongst the taxa under consideration. These are indicators of genetic relationship.

Data obtained from genetics and cytogentics is mostly applicable at the lower levels of the taxonomic hierarchy. Artificial intergeneric crosses in some families e.g. Gramineae and Orchidaceae (Knobloch, 1972) are on record, but thre is a very low percentage of successful possibilities within the genus, at the specific, series, sectional, and subgeneric levels.However, artificial crossibilities commonly achieved. The most useful cytogenetic data are obtained and applied. There is no recorded successful interfamilial cross. Isozymes are most useful at the specific and infraspecific levels. Isozyme data can be compared at higher level in the hierarchy.

(X) Plant Geography

There are certain aspects of plant geography which have much relevance to plant taxonomy. Each taxon shows a certain pattern of distribution which is one aspect of its definition. When evolution is to be thought of, the coincidence or not of areas occupied by related taxa has a bearing on the classification of the group. Also, geographical patterns have practical significance since plants are gathered from distinct areas, catalogued in herbaria and described in Floras which are based on these same areas. Taxonomists are thus often experts on regions as on taxa. Candolle (1855) clearly set out views on 'botanical geography' from the taxonomic perspective. In taxonomic work, it was thus expected to find mention of geography and habitat and so it has been to the present time.

(i) **Geographical Distribution:** Geographical differentiation exists between taxa at each level of taxonomic hierachy and in all degrees of spatial separation. Since different taxa may have different powers of migration, effective spatial separation differs greatly. Taxa occupying mutually exclusive geographical areas are called '*allopatric*', while those occupying similar or overlapping areas are called '*sympatric*' or 'partially sympatric'. Sympatric taxa which are closely related, often exhibit different types of genetic, ecological and structural differentiation from those shown by closely related allopatric taxa (*cf.*

Ehrendorfer, 1968), because allopatry is itself an important isolating mechanism, whereas sympatric taxa can rarely reply on spatial separation.

Geographical distribution may differ at the species level e.g. agricultural weeds, certain aquatic species, etc. are cosmopolitan. On the contrary, there are certain species which are known from a single population. In case of weeds and certain water plants the wider-than-usual area distribution may be noted. They may exhibit relatively recent and affective methods of dispersal. Some taxa reproduce sexually rarely. In such cases, a very slow rate of evolution into distinct taxa in separate areas may be indicated. Distribution patterns, at the genus and family level, are obviously wider, in the area occupied. Many sizeable families are restricted to certain areas, say to the Americas or to Australia, e.g. *Eucalyptus* (about 500 species) occurs only in Australia, certain geographical patterns are found to recur consistently when the distribution of a large number of taxa is analysed. These patterns of taxa which exhibit them are called 'floristic elements' and their demonstration may be a valuable method of floristie analysis. A knowledge of the floristic elements, of a recently glaciated region helps understand the pathways of post-glacial migration into it and may also aid in taxonomic decisions in cases of doulet. In certain instances, the distinction between geographical distribution and ecological distribution pose difficulty, for example, taxa restricted to alpine or coastal areas, and thus separated from lowland or inland taxa, can be equally thought in either category.

(ii) ***Disjunction***: Some taxa have distribution patterns which are interrupted by considerable areas from which the taxon is absent. Such taxa said to exhibit disjunct distribution. These might have originated in various ways. The disjunction has often arisen by the long-range dispersal of the taxon from one area to another. The presence of plants e.g. coconut on a great number of far-distant volcanic Pacific islands is an obvious, case. In other cases, disjunctions may represent the relics of earlier continuous, wide distribution patterns, the intervening areas having been depopulated. It is possible that long-range dispersal could have been effected via continuous or 'stepping-stone' bridges of suitable habitats linked the two areas. The connection between Britain and France up to only about 7500 years ago forms a classic example. Migration of species in past between certain European or African mountain species were not restricted to the high mountains (*cf.* Morton, 1972). Alternatively, taxa in two separate areas might have arrived there from their area from which they are now absent. It is also possible that the taxon would have originated independently in its separate areas of distribution, thus indicating evolutionary parallelism or convergence e.g. certain amphidiploids have arisen polytopically (Stace, 1975). The two separated elements of a disjunct taxon have often been found to differ slightly e.g. *Empetrum rubrum* in North and South America, *Sphaerocarpos texanus* in Europe and America, which might lead to their recognition as distinct taxa. The wild African bananas are presently separated from *Musa* into the genus *Ensete*. Similarly, the New world and the Old World members of the Gesneriaceae are now largely kept under distinct subfamilies.

(iii) ***Vicariance***: When two similar taxa occupy separate geographical or ecological areas, then they are said to be 'vicariants' or 'vicariads', and the phenomenon as '*vicariance*' or *vicarism*'. The difference between vicariance and disjunction is one of degree only. For instance, the genus *Cedrus* is a disjunct taxon, occurring in four separate areas: (i) Atlas Mountains of Morocco and Algeria, (ii) Cyprus, (iii) Lebanon, Syria and South-eastern Turkey, and (iv) Wester Himalayas. It provides an example of vicariance at the species level because four species occupy four separate areas viz., *C. atlantica, C. brevifolia, C libani* and *C. deodara* respectively. Vicariants may also originate in different ways: (i) by a taxon migrating to a new area and evolving there into a different taxon, (ii) by a formerly continuously distributed taxon becoming separated into different areas and there undergoing divergent evolution, (iii) by parallel evolution of two taxa from a common ancestor into separate areas, and (iv) by convergent evolution of two taxa from separate ancestors.

It is now certain that disjunction and vicariance are always different. Geographic disjunctions do not arise overnight, but gradually. There is no reason to believe that plants did not migrate to fresh areas at about the same time that these areas were becoming isolated. Rose (1972) believed that some of the rare plants of south-eastern England arrived there from continental Europe by vicariance events and others by dispersal events.

(iv) ***Endemism***: Endemics are the taxa which occur only in single confined geographical area, and the phenomenon is called '*endemism*'. The concept of endemism is usually applied only when there is a considerable restriction in the area of distribution. Endemism is basically two types *viz*., (i) *neoendemism*, where a taxon is evolutionarily young and has not yet been able to spread to other areas, and (ii) *palaeoendemism*, where a taxon is now restricted but once enjoyed a far wider distribution. For example, (i) *Senecio cambrensis* is a neoendemic. It is a new amphidiploid which arose in North Wales this century and is still endemic to Britain, (ii) *Crepis fuliginosa*, which arose by chromosomal rearrangements in the Balkan peninsula from its more widespread ancestor *C. neglecta*. Similarly, *Ginkgo biloba* is a palaeoendemic. It is now restricted to a small part of China but widespread as a fossil in the North Temperate Zone. Richardson (1978) recongnised another intermediate category and termed it as '*holoendemics*'. *Holoendemics*, in his opinion, are plants which are not of recent origin but have retained a narrow distribution.

Based on cytotaxonomic criteria, Favarger and Contandriopoulis (1961) recongnised three different types of neoendermics: (i) Schizoendemics, derived from or having given rise to a more widespread taxon with the same chromosome number, (ii) *Patroendermics*, confined diploids which have given rise to more widespread *polyploids*, and (iii) *Apoendemics*, confined polyploids which have arisen from more widespread diploids.

Usually, *neoendemics* outnumber *palaeoendemics* in the flora. Therefore, the percentage of endemic species in a flora is generally proportional to the degree of isolation of the region involved. The degree of isolation may be measured by (i) the distance from other similar areas or by (ii) the length of time that the taxa has been isolated. Often either of these factors are applicable. Larger areas generally have a greater percentage of endemics than smaller areas. The taxonomic treatment of a flora obviously affects the statistics concerning its endemism.

(v) **Centres of Diversity:** It is be generally observed that there are one or more areas with a marked concentration of species within a genus. Such an area is termed a 'centre of genetic diversity' for that genus. Often there is a single such area for the genus or even other greater / lesser taxon concerned. There are progressively fewer species observed as the distance from the centre of diversity is increased. Vavilov (1951) during 1920 to 1930 made detailed study on this line for crop plants like legumes and cereals. This data was then updated by Zeven and Zhukosky (1975). They defined 12 such centre. All of them lie in tropical, subtropical or warm- temperate areas. Virtually they have provided all thc world's important crop plant. It is also now concluded that genetic diversity of non-crop species also coincides with those of the crop species. Vavilov (*loc. cit.*) also opined that the centres of diversity were also the centres of origin of the taxa concerned. The centres of diversity today, however, hold the main remnants of the ancestral genetic material of the taxa concerned. The older the taxon, the less likely the coincidence of the centres of origin and diversity.

A *monophyletic* taxon has a single centre of origin by its definition. Investigations on the characteristics of the taxon in its centre of diversity in relation to its characteristics can help deduce the major patterns of evolution and pathways of migration. The adaptations to the different environments during its migration are thus revealed. Such an 'outward' pattern of evolutionary migration is referred as 'adaptive radiation'. Cytologic study is important especially in temperate taxa. Plants near the centre of diversity are diploid, while the percentage of *polyploids* is towards the periphery of the range. Polyploidy is obviously an important aid for adaptive radiation in many cases e.g. *Paradisea liliastrum* has diploids in the Pyrenees but polyploid in the Alps where it migrated. Sometimes there appear to be more than one centre of diversity for a taxon. This indicates its polyphyletic nature. However, often all but (or even all) of the centres are 'secondary centres of genetic diversity'. Thorough biosystematic investigation can help to decide their primary or secondary centres of diversity. Decrease in variability (the number of biotypes) can be helpful to uncover pathways of migration. For example Bocher (1972) observed *Ranunculus glacialis,* an artic-alpine species, for richer in biotypes in the Alps than in the arctic.

(XI) Plant Ecology

Ecological data deal with the plant environment interactions and not with the features of plants, whether macro - or micro - structures. These interactions are indeed manifestations of all the features of the plant with all the aspects of the environment. These interactions are expressed in two basic ways, biotically and abiotically. The former considers part of the environment including herbivores, competitors and even pollinators, whereas the latter includes the abiotic part of environment such as soils, temperatures, moistures, etc. The role of pollinators is, however, explained under the role of reproductive biology. Ecological studies permit for a manipulation of plants into various environments to determine the environmental *vs*. genetic components of observed patterns of variation. These may be greatly helpful to reach to certain taxonomic generalizations and decisions. It is obvious that habitat and environmental factors play an essential role from evolutionary point of view and evolution, as we know, is the basic process which directs taxonomic differentiation.

1. Useful Ecological Data: The plant environment interactions are to be explained under (A) Abiotic and (B) Biotic factors.

(A) Abiotic factors: These include distribution, vegetation zones, local environmental factors, geology and such other abiotic factors. Distribution has been already included under the role of plant geography in this book.

- **(i) Vegetation Zones:** Vegetation zones on a broad scale have been now made available for the different continents of the world and various parts of it. These are based on elevation, latitude and precipitation. This is, however, only moderately useful since many closely related taxa occur in the same broad vegetational zones. The dominant species in certain local environment hence need to be documented. A clear view of classifications of plant communities in a particular area is often needed.
- **(ii) Local Environmental Factors:** Both abiotic and biotic factors should be considered in the local environmental factors. The latter is considered in the local environmental factors. The latter is attempted later. One of the most important of the abiotic factors is soil. It is a well known fact now that soil differences affect distribution of plants. Sandy soils, lime stone derived soils and many after variations often can be factors responsible for producing various distributional patterns. Soils play a significant role in speciation. In some cases, plants have apparantly evolved tolerances to high levels of inorganic ions in soils.
- **(iii) Geology:** Geology of bedrock can also affect plant distributions. This also clearly relates to the development to separation of taxa, but patterns of underlying bedrocks do affect. e.g. *Carduus nutans* (Asteraceae) in northwest Ohio correlates positively with limestone bedrock below the soil surface. Two closely related species *viz.,Delphinium* trealease and *D. alabamicum* were confined to dolomite and limestone outcrops respectively.

(iv) **Other Abiotic Factors:** These include rainfall, humidity, soil moisture, seasonal and diurnal temperatures, available light and elevation. Elevation is related to oxygen, UV levels and temperature. More local data are essential for seeking meaningful correlations with distribution of specific and infraspecific taxa. Characteristic past climates (palaeoclimates)can also be useful data for the taxonomists. They can be useful to understand the overall ecology of a plant group and its relatives and also to interpret present day distributional patterns.

(B) Biotic Factors: These factors are equally important for determination of plant distributions. The associated plant species in the local community are very valuable as comparative data. These are generally cited as general plant communities described by dominant taxa. The more complete the data on the associated species, the better understanding of the environment. The insects, birds, and mammal predators would be important to know. Pollinators and dispersals agents play a significant role. Their role is explained under reproductive biology. Mutualistic associations also play a social role.

2. Ecological Variations: Each individual is referred now by its *genotype* and *phenotype*. The former describes its innate genetical constitution, whereas the latter refers to the sum total of attributes, which a manifestation with the environment. The capacity of phenotype to respond environmental influences is itself under genetic control (*cf. e.g.* Bradshaw, 1965), and there is some evidence that environmental modifications suggest that it is not random or biologically insignificant. It is frequently adaptive (Cook, 1968, Heslop-Harrison, 1964) e.g. Turesson (1922) brought out two species of *Lysimachia* (Primulaceae) responding differently to height and intensity. One species shows phenotypic response, while another shows genotypic differentiation. It means both altered genotype and phenotype may occur mixed in a single population (e.g. Nelson, 1965). The balance of the two mechanisms may vary between species and between populations (Marshall and Jain, 1968; Jain, 1969).

3. Ecotypes and Taxonomy: When adaptive variants or *ecotypes* show correlated differences susceptible to morphological treatment, they may be considered eligible for taxonomic consideration following normal taxonomic principles (Davis and Heywood, 1963). Often *ecotypes* are found to coincide with recognisd taxa. Bocher (1943) found that in *Plantago* species the units of orthodox taxonomy correlated very well to certain genetical races (Ecotypes) or to ranges or continuous character gradients. Daubenmire (1959) encouraged recognition of ecotypes taxonomically for ecological purposes. In his opinion, it is clearly useful for variants to have names so that they can be easily referred to.

Many Pflanzenreich monographs, and in floras at the beginning of 20th century mentioned the subdivision of the species all the way down to the 'forma' level. It received limited support on the ground that the lower units are artificial, not repeatable, are descriptions of indivisual specimens, or simply too small to occupy the serious attentions of the taxonomist. Many practicing taxonomists find the need for occasional use of all the infraspecific units for dealing with the variation in well studied group.

This above basic consideration led to the introduction of the '*deme*' terminology. It is believed necessary to keep separate general purpose orthodox classifications which are morphologically expressible from those concerned with micro evolutionary units. The system was originally proposed by Gilmour and Gregor (1939) and developed fully by Gilmour and Heslop-Harrison (1954). The subject matter has been already attempted at proper place in this book.

4. Phenotypic Plasticity: *Phenotype* is a manifestation of *genotype* and environment together. It is implied that the phenotype is not merely a manifestation of genotype. Environmental factors play a part in modifying the latter to produce the former. The ability to express a genotype as different phenotypes according to external conditions is referred to as 'phenotypic plasicity'. One may call it 'plastic responses'. Different phenotypes which are merely the product of differing environments are also called '*ecotypes*' or '*ecades*'. It is important for taxonomists to know plastic nature of plants and to decide whether or not the different ecophenes should be taxonomically named.

To give rise to extreme ecophenes in a wide range of species, certain environments are well known, e.g. adaptations to wet/ dry conditions, shady/ sunny habitats, alpine/ lowland conditions, etc. Certain species or genera produce a wide range of ecophenes. The taxonomists have to care for such taxa. Phenotypic plasticity should not be underestimated in taxonomic and evolutionary work in plants. Use of it for taxonomic delimitation may give rise to a multitude of 'paper' species and (such taxa are later relegated as mere synonyms) varieties without taxonomic value. In *Epilobium* e.g. sunny plants have small thick leaves, much anthocyanins, many hairs and a short structure, whereas shade plants have the opposite characters. In Epilobium, the type of indumentum is the most important diagnostic character. It is hence essential to examine the quality of the hairs and not their quantity, when making determinations. In many cases, dry conditions promote a dwarf habit. Plastic responses do not affect exomorphic features. They do affect chemistry and anatomy of plant species. Stace (1965) noted many cases of plasticity in leaf epidermal characters. Certain genera of flowering plants adopt different phenotypes according to the time of year at which they germinate and flower. These are called seasonal varieties and this phenomenon is often referred to as 'psuedoseasonal polymorphism'. It is mostly observed in montane plants which may be subjected to different length of season and temperature regimes according to the altitude at which they grow

(XII) Palaeobotany

The science of palaeobotany plays a potentially important role in phylogeny and taxonomy of plants. Plants parts e.g. root, stem, leaf, flower, fruit, pollen, etc. are preserved under conditions which allow preservation of antigenic materials or even DNA. DNA can be isolated and emoployed in taxonomic considerations. Thus palaeobotany is a window into the past and provides historical record of plant life. Study of fossils is helpful in the phylogeny particularly of angiosperms and reveals information on the origin and diversification. The evolutionary history of present day plants is often furnished by fossil records, besides palaeoclimates. Limitations of fossils records are but obvious.

Information can be obtained from macrofossils as well as from microfossils. The plant parts are available in the form of fossils like petrifaction, compression, impression, etc. Preservation of parts depends on their nature. Roots and stem have elaborate vascular system and hence preserved better than other parts. Pollen grains are well preserved because of resistant wall layers. While assessing synthetically the role of other disciplines in taxonomy, the evidence from palaeobotany is always sought. This is so because any theory of evolution is incomplete without palaeobotanical evidence. The palaeobotanical studies have proved helpful in various ways. These are summed up in the following:

Fossils provide details of the structure or form, besides estimates of the ages of the plant group under considerations. They afford additional data which can be scored for character states. These aid in a phylogenetic analysis. The evolution of the flora of the past can be determined on the basis paleaobotanical evidence. The fossil records also help divulge and explain the ecological conditions of the past. Microfossils (fossils of pollens and spores) can aid to restructure past environmènts. The assemblage of pollen in the sediment of a particular area is representative of the vegetation of that area. Pollens accumulate continuously year after year. They can help determine when and how rapid changes occurred. Some lakes have annual layers that make it possible to reconstruct short term climate variability over hundreds or thousands of years. Fossil DNA is one of the best tools of information of the past plant groups. DNA sequence information from fossils can also help to resolve many phylogenetic questions related to both recently and not so recently demised plant species. The PCR (Polymerase Chain Reaction) helps to make gene sequences available from at least some recently fossilized plants. PCR allows to amplify many million fold whatever small amounts of DNA is present, which can be further sequenced as if it were new. It is possible to obtain phylogenetically useful data.

6

Numerical Taxonomy

Taxonomy has no sources of its own. It is amoeba-like engulfing sources from all sides. It draws information from various disciplines and principles. One such aspect of study is *Adansonian Principles* (Adanson, 1763) and the classification based on these postulates is called '*Adansonian Classifications*'. These are established first on construction of classification selecting each attribute of equal weight and secondly, taxa are based on correlations between the attributes selected. Since 1960s Adansonian Principles were welcomed to such an extent that culminated into new method in taxonomy. Adanson (1763) was the first to give equal weightage for all characters to be employed in classification. Numerical taxonomy may be defined as numerical evaluation of the apparent (phenetic) similarity among the groups of organisms and ordering them into higher ranking taxa (Heywood, 1967b). It is empirical and aimed at developing objective methods. Obviously, they can be verified through repetitive experiments for assessing taxonomic relationships and raising taxa in hierarchical system of taxonomy. Moreover, this method affords clues in phylogenetic analysis to assess evolutionary rates accurately better. It is based on two foundations of taxonomic groups and then their discrimination. Being a distinct discipline of study, numerical taxonomy has its own method and terminology.

1. Sokal and Sneath (1963): They defined '*Numerical Taxonomy*' as the numerical evolution of affinity or similarity between taxonomic units and the ordering of these units into taxa on the basis of their affinities.
2. Heywood (1967b): 'Numerical Taxonomy' is defined as a numerical evolution of similarity between groups of organisms and the ordering of these groups into higher-ranking taxa on the basis of these similarities.

Different names, have been coined for '*Numerical Taxonomy*' (Sneath and Sokal, 1973) such as: (i) *Mathematical Taxonomy* (Jordine and Sibson, 1971), (ii) *Multivariate Morphometrics* (Blackith and Reyment, 1971), (iii) Taxometrics (Mayr, 1966), (iv) Taximetrics (Rogers, 1963), (v) *Numerical Systematics* (Wagner, 1961). However, the name by Sneath and Sokal (*loc.cit.*) has gained wide currency .

Principles

Sneath and Sokal (1973) put forth some principles of numerical taxonomy as;

(i) The greater the content of information in the taxa of a classification and the more characters on which it is based, the better a given classification will be.

(ii) A priori, every character is of equal weight in creating natural taxa.

(iii) Overall similarity between any two entities is a function of their individual similarities in each of the many characters in which they are being compared.

(iv) Distinct taxa can be recognized because correlations of character differs in the groups of organisms under study.

(v) Phylogenetic inferences can be made from the taxonomic structures of a group and from character correlations, given certain assumptions about evolutionary pathways and mechanisms.

(vi) Taxonomy is viewed and practiced as an empirical science.

(vii) Classifications are based on phenetic similarity.

Operational Steps

Two major bases have been recommended to design natural system of classification *viz.*, (i) Phenetics and (ii) Cladistics (Phylogenetics).

(i) ***Phenetics*:** Phenetic classification is the similarity (relationship) of individual under consideration based on a set of phenotypic characters (characters related to the appearances). It is based on overall affinity as deciphered by using as many characters as available. It is does not include any relationship through ancestry. The dendrograms, with taxa and groups of taxa, as they are connected at different levels according to their overall similarity, do not reflect evolutionary relationships accurately. These dendograms are referred as 'phenerograms'.

(ii) Cladistics: It is expressed in terms of correlation among individuals with regard to their evolutionary history. The relationship means common ancestry. It is aimed at producing objective and repeatable branching diagrams (evolutionary trees) that depict hypothetical evolutionary histories. The evolutionary trees with anastomosing branching leading to different taxa are referred as 'cladograms'.

The following successive steps are used in the construction of taxonomic groups:

1. **Operational Taxonomic Units (OTUs):** It is the basic unit in numerical taxonomy. It can be an individual, species, genus, family, order or class. They are termed as OTUs since they are not always comparable to formal taxonomic units. Comparison of OTUs of equal rank is always made in numerical taxonomy. Adequate representation of different polymorphic forms is essential in case of OTUs above the level of individual e.g. (a) when genera are compared, they should be represented by different species. (b) when families are compared, they should be represented by different genera, and so on.

2. **Unit Characters:** The characters used in numerical taxonomy are known s 'Unit Characters'. Unit character is defined as a taxonomic character of two or more states, which within the study at hand can not be subdivided logically, except subdivision brought about by changes in the method of coding (Sokal and Sneath, 1963). The phenotypic characters are exclusively employed as unit characters e.g. presence or absence of a spine on a leaf.

(a) Types of Unit Characters: There are two types of characters *viz.*, (i) Binary, (ii) Multistate.

(i) Binary Characters: The characters which exist in two states are called binary characters e.g. presence or absence of spines. They can be represented by the simplest form of coding, where characters are divided into + or – or as 1 and 0. The positive characters, in such cases, are recorded as + or 1 and likewise negative characters are recorded as – or 0. When the organ possessing a given character is missing in an organism, the character is scored NC (means 'no comparison').

(ii) Multistate Characters: The characters which express in more than two states are called multistate characters. The multistate characters may be: (a) qualitative and (b) quantitative. The former contain forms and each form is ranked on equal footing e.g. colour of flower, which can be in any number of states such as yellow, red, purple, white, etc. The latter represent measures of the size on a continuous scale e.g. leaf length, which can be 2, 3, 4 or more in cm.

(b) Admissible Unit Characters: These are featured out as: (i) They can not be subdivided, (ii) They must exhibit variation between the taxa under comparison, (iii) They should be inherent in the organism, (iv) They should not be susceptible to environmental changes or affected by experimental conditions. (v) It must be of some diagnostic significance.

(c) Inadmissible Unit Characters: In the application of numerical taxonomy, the proper selection of unit characters is a critical point as in other disciplines of taxonomy. There are certain characters that can not be employed in numerical taxonomy (Sokal and Sneath, 1963). These are: (i) Attributes that are not a reflection of the genotypes of the organisms themselves. (ii) Any property that is a logical consequence of another whether wholly or partly. (iii) Characters that do not differ with the entire sample of organisms.

(d) Proper Selection of Unit Characters: There are some suggestions for proper selection of unit characters such as: (i) They should come from all parts of the organism, (ii) They should belong to all stages of the life cycle of an organism, (iii) Variable characters within the group should be used, (iv) The characters related to morphology, ecology, distribution and physiology of the organisms should be given proper attention.

3. Estimation of Resemblances: Among any two operational taxonomic units, the resemblance is estimated in terms of similarity (percentage of characters in which they agree) or in terms of dissimilarity (percentage of characters in which they do not agree). The dissimilarity is particularly useful in the construction of taxonomic models or maps in numerical taxonomy because of the taxonomic distance between the positions of the operational taxonomic units in the phenetic appearance.

Phenetic resemblance between taxonomic groups is generally estimated by three methods: (i) coefficient of association, (ii) coefficients of correlation, and (iii) measurement of taxonomic distance between OTUs.

(i) Coefficients of association are measured by the following formula:

$$S = \frac{Ns}{Ns + Nd}$$

Where S is the numerical index, Ns=the number of positive features shared by any two OTUs; and Nd=the number of positive features in one OTU and number of negative features in the other OTU.

(ii) Coefficients of correlation (r) are measured by the following formula:

$$rjk = \frac{\Sigma^n\left[(Xij - \bar{X}j)(Xik - \bar{X}k)\right]}{\sqrt{\left\{\left[\Sigma^n (Xij - \bar{X}j)^2\right]\left[\Sigma^n (Xik - \bar{X}k)^2\right]\right\}}}$$

Where j and k stand for two units under comparison; Xij stands for the value of the character i in unit j; Xik stands for the value of the character i in unit k; $\bar{X}_j$ and $\bar{X}_k$ stand for the mean for all the characters in units j and k; n stands for number of characters.

(iii) Taxonomic distance (d) is measured by the following formula:

$$djk = \sqrt{\left\{\left(\Sigma^n (Xij - Xik)\right)\frac{2}{n}\right\}}$$

Where Xij is the character state of unit j for the character i; Xik is the character state of unit k, symbol stands for the sum over n characters; the value of the taxonomic distance (d) is the distance in a phenetic space divided by $\sqrt{n}$.

4. **Cluster Analysis:** Cluster analysis (clustering) is a type of multivariate statistical analysis. It is used to group organisms into separate clusters based on their statistical behavior. The main purpose of clustering is to decide similarities between organisms, and then group similar organisms together to assist in understanding relationships that might exist among them. Different OUTs are grouped together on the basis of degree of similarity. These groups of OTUs are called 'clusters'.

   (i) **Cluster analysis characteristics:** Cluster analysis is based on a mathematical formulation of a measure of similarity. There are several characteristics that can distinguish different approaches to cluster analysis. (i) Numerical, statistical, and conceptual clusters. (ii) Agglomerative vs. divisive – Agglomerative methods start with individual taxa and seek to connect them into pairs, etc. in such a way that similarities between pairs or group members are maximized at each level. In contrast, divisive clustering splits are sought in the group of taxa that will in some way maximize the collective phenotype disparity between the two groups formed by the split. (iii) Overlapping vs. disjoint cluster. (iv) Incremental vs. non-incremental. (v) Flat vs. hierarchical representatives.

(ii) Clustering methods: Clustering can be achieved in two ways: (a) Monothetic system: This system employs the attributes one at a time. The monothetic method obviously leads to artificial clustering. (b) Considering all their attributes simultaneously, this method gives a natural grouping.

One cluster or a group is separated from the other by a dividing line. This line indicates a distinct gap between the two. There are often few discontinuities in case of intraspecific clusters.

Phenetic clustering methods: Phenetic clustering methods have largely been overtaken by cladistic methods in order to relate data to evolutionary history, and they are now rarely employed in taxonomy as they often lead to substantially different clusters when applied to real data. The most widely used phenetic clustering methods in taxonomy are the following:

(a) Nearest neighbour clustering methods: It is alternatively referred as single linkage clustering. The phenograms are constructed by joining OTUs and groups on the basis of their most similar members (i.e., the shortest distance). Whether an OTU will join an existing cluster depends on its maximum similarity (or minimum distance from) any member of that cluster. The phenograrn is initiated by joining the two (or more) most similar OTUs.

(b) Furthest neighbour clustering method (complete linkage): This method tends to emphasize differences between clusters. Construction of furthest neighbour phenogram is essentially similar to that for nearest neighbour ones, with the only difference that the OTUs are joined at a level corresponding to the greatest distance between their component members instead of the shortest distance as in case of the latter.

(c) Unweighted pair-group method using arithmetic averages (UPGMA): An OTU is joined to existing clusters on the basis of their average (mean) distance to the members of that cluster. The mean distance between two clusters of OTUs is calculated by adding together the distances between all possible pair-wise combination of OTUs in the two clusters and dividing by the number of combinations. The tree construction method is similar to nearest neighbour clustering. It is not so subjected to the effects of aberrant OTUs since this method uses mean distances. It is hence preferred over nearest neighbour methods.

(d) Weighted pair-group method using arithmetic averages (WPGMA): This is a set of methods rather than a single one. They are basically similar to UPGMA except that, they weight distances such that there is a tendency to emphasize separations between large clusters. They are subjected to the effects of single or small groups of outliers and therefore preferred over nearest neighbour methods. The distances are usually weighted by some function of the number of OTUs in each cluster.

(e) **Centroid clustering:** Although basic aim is similar to UPGMA and WPGMA, it requires the actual taxa versus character raw data or more precisely normalized data, rather than inter-taxon distances. This method makes use of the Euclidian distances within the character hyperspace between the centroids of clusters of OTUs. Euclidian distance is equivalent to the real physical distance by which the taxa would be separated if they were plotted in a space in which each axis represents one normalized variable and each axis is perpendicular to all others.

(f) **Differential shading of the similarity matrix:** This is a method to describe structure in matrices of similarity coefficients, in which similarity coefficients are grouped into five to ten evenly spaced classes, each of which are represented by different degrees of shading in the squares of half matrix. A darker shade is usually used for the highest value. The lightest shade is used for the lowest value. The half matrix can be observed as a pattern to different shades, limited by a diagonal of squares with the darker shade.

Applications

Taxonomists employed numerical methods since its beginning and are successful in the study of similarities and differences. Work on this line is extended on animals and micro-organisms, apart from higher vascular plants. For example: (i) *Oryza* (Morishima and Oka, 1960), *Solanum* (Soria and Heiser, 1961), *Sarcostemma* (Johnson and Holm, 1968), *Apocynum, Crotalaria, Cucurbita, Chenopodium, Oenothera, Salix, Zinnia*, etc. besides cultivars of wheat, barley and maize (Sneath and Sokal, 1973) and *Cassia* (Mondal *et al.*, 2000).

Advantages

Sneath and Sokal (1973) earmarked certain advantages of numerical taxonomy as:

(a) Numerical taxonomy has power to integrate data from a variety of sources e.g. morphology, chemistry, physiology, amino acid sequences of proteins, afflictions between DNA strands, etc. This is difficult to do by conventional taxonomy.

(b) Greater efficiency is promoted through the automation of large portions of the taxonomic process. Much taxonomic work thus can be done by less highly skilled workers.

(c) Data coded numerically may be used easily for the creation descriptions, catalogues, maps, keys and such other documents.

(d) The numerical methods are more sensitive in delimiting taxa and therefore they provide better keys and classification systems in comparison to the conventional methods of taxonomy.

(e) Since more and better-described characters are used in numerical taxonomy, the quality of conventional taxonomy is thereby improved.

(f) A fundamental advantage of numerical taxonomy has been the re-examination of the principles of taxonomy and purposes of classification.

(g) The existing biological concepts have been reinterpreted in the light of numerical taxonomy. The method is thus coming into its own as a best tool in biological research.

Demerits

Numerical taxonomy is not free from demerits. It has proved to be disadvantageous on the following viewpoints:

(a) The numerical methods are useful only in phenetic classifications and not in phylogenetic (Davis and Heywood, 1963).

(b) The specific limits bound by numerical methods may not be agreeable to the proponents of 'biological species' concept (Davis and Heywood, 1963).

(c) If characters chosen for comparison are inadequate, the statistical methods may give less satisfactory solution. Character selection is thus the weak link in this approach (Davis and Heywood, 1963).

(d) Different taxometric procedures may yield different results. A major difficulty is to choose a procedure for the purpose and the number of characters (from 40-100) needed in order to obtain satisfactory results by these mechanical aids. It is essential to ascertain whether a large number of characters would really give satisfactory results than those using a smaller number (Stearn, 1968).

7

Molecular Taxonomy (SYSTEMATICS)

Since few decades, genetic data sometimes contradicted earlier ideas about taxonomic integrity or distinctions which were based especially on morphological descriptions. This trend led taxonomists to reject given names of subspecies, recognize new ones or elevate some to the species rank and so on. Of late, the study of identification and classification of plants has become much more precise due to refined techniques using its genetic material DNA. These are particularly important if interpretations of data from disciplines such as morphology, embryology, anatomy and palynology conflict. DNA help resolve such conflict and lead to a clearer definition of relationship among plant species. This obviously provides a better understanding of the evolution of plant structures and breeding systems. Analysis of gene sequences and molecular markers can be used to answer to specific biological questions to reconstruct the evolutionary history of plants. It also provides the phylogenetic data that underpin molecular systematics.

Molecular Approaches

Molecular taxonomy generally assumes that a classification of species, genus, family, order and so forth should reflect the evolutionary history of the group. Groups must be established on the basis of shared, derived characters. A shared derived character is a character which first appeared in an ancestral speceis. It is exhibited by its descendents. Alternatively, a character must be homologous in all of the species placed in a group and also the result of a common ancestor based on molecular information.

Although study of taxonomy began with morphological information, presently it is being revolutionized by molecular approaches. The mitochondrial and nuclear DNA have proved beneficial to arrive at inferences about the relationships of different taxa ranging from the species to recognization at higher levels in plant kingdom. The application of nucleic acid data to problems in taxonomy is the most exciting developments in the past decade. The term molecular systematics is employed to mean the use of DNA, RNA and protein to infer relationships among plant groups. The proponents of molecular taxonomy thought that molecular data were more likely to reflect the true phylogeny than morphological one ostensibly since they reflected gene level changes that were considered to be less subject to convergence and parallelism than morphological markers. This assurance now appears debatable and molecular data are, in fact, subject to most of the same problems that morphological. The major difference is that there are many more molecular characters available, and their interpretation is generally easier. An adenine is always an adenine whereas compound leaves, for example fan form in quite different ways in different plants. The molecular data are, of late, widely used for generating phylogenetic deductions.

A number of organelles are the source of molecular markers as single copy, middle repetitive, high copy number sequences and specific genes of specific storage proteins. The most significant molecular marker sources are chloroplast, mitochondrion and nuclear genes, in plant kingdom. The chloroplast DNA is used as marker in many phylogenetic studies of plants. Analysis of restriction site variation is the method of choice remains a good way to evaluate relationships among species. Restriction site variation often becomes hard to interpret for taxa that are more distantly related. Systematics have been involved in a community wide effort to generate a large database of sequences of the chloroplast gene, rbcL. This gene encodes the large subunit of the photosynthetic enzyme ribulose-1, 5-bisphosphate carboxylase/oxygenase (RuBisCO). It is the major carbon acceptor in all photosynthetic eukaryotes and cyanobacteria. The secondary structure of the gene is known and amino acids can be assigned to particular structural components of the gene. This gene was chosen because it is almost universal among plants, except the parasites. It is fairly long (1428 bp) and occur in many copies in the cell. One of the limitations of using rbcL as a marker is its slow rate of change. It is basically a conservative molecule that is highly constrained at the amino acid level. It is hence particularly useful for inferring relationships within or between closely related genera. Instead, other chloroplast genes have been used for such purposes, notably those for subunit F of NADP dehydrogenase (ndhF, in the small single copy region), those for the and subunits of RNA polymerase II [rpoA and rpoC2, in the large single copy region), and a maturase gene in the intron separating the coding region of tmK (matK also known as ORF) for open reading frame, K]. Since these are all part of the same non recombining genome as rbcL, they all track the same maternal history. The other gene encoding the subunit of ATP synthase. Atp is also used to address the same problems as rbcL. It seems to evolve at about the same rate. It provides additional phylogenetically informative characters.

The only nuclear genes are ribosomal genes with a high copy number for easy study. They are arranged in tandem arrays of several hundred to several thousand copies. A smaller (5.8S) gene separates the small subunit (18S) and large subunit (26S) genes, and the whole set of genes is transcribed as a single unit. There are short transcribed spacers (ITS) between the three genes. Each set of three genes is separated from the following set by a large spacer referred as intergenic spacer (IGS), extragenic spacer (EGS) or non-transcribed spacer (NTS). The last one is a misnomer, in that some of the sequence immediately upstream of the 18S gene and down stream of the 26S is actually transcribed are called external transcribed spacers (ETS). The middle portion of the spacer is not transcribed. It is made up of variable numbers of short repeated sequences (100-300 bp each). The 5S rRNA genes are in tandem arrays of several thousand copies. They are separated by nontranscribed spacers. Such highly repetitive sequences undergo homogenization processes known as concerted evolution. When a mutation occurs in one copy of a sequence, it is usually corrected to match the other copies. Sometimes, the nonmutated copies are corrected to match mutated one, so that nucleotide changes propagate throughout the array. This means that many copies of the sequence are generally more similar to one another than they are in other species. Within species, variation does occur since concerted evolution is slower than the mutation rate. Thus, highly repetitive sequences can be employed to assess variation within and among populations of the same species. Ribosomal genes were studied initially by restriction site mapping of the intergenic region. This region is quite variable even within populations. Its major role has, therefore, been in studies of closely related plants, where it has been useful in determining population structure and in assessing patterns of hybridization. All rDNA and chloroplast DNA *genotypes* found in asexual plants were also found in sexual plants, but in different combinations. This finding is evidence that asexual plants are produced frequently by hybridization between sexual ones. Methods for direct sequencing of RNA were developed earlier than rapid DNA sequencing methods and PCR. This led to some early hope that the 5S rRNA gene sequences would be the key to evolutionary history. Researchers soon recognized that the 5S rRNA genes were too short (120bp) and too conservative to illuminate relationships. The nontranscribed spacer region has also been used with some success in closely related groups. Never the less, sequences of the 18S and 265 genes are more promising that are large about 1800-3300 bp, respectively. They have some regions which are highly conserved and quite variable. These help to distinguish phylogenetic groups. It also provides tools for understanding gene flow and population level variation. Other nuclear genes are those of phytochromes for phylogeny reconstruction. The small heat shock proteins and glutamine synthetase. Considerable data must be acquired to determine the taxonomic level at which the genes vary before such genes can be used reliably for phylogenetic studies.

Molecular Methods

Various techniques for analysis of DNA, RNA and protein can be applied for the classification. These techniques can be divided into (a) that involve molecular marker methods and (b) that based upon the comparison of gene sequences at specific loci The type of molecular method used to measure genetic distances in plants will differ depending upon the magnitude of the genetic differences being assessed. Techniques such as RAPD analysis may be useful for .distinguishing different genotypes within a plant cultivar. The sequence analysis of the ribosomal genes may allow species or higher taxonomic level. The three most common markers are protein based, hybridization based and PCR based markers. Each marker has its own advantages and disadvantages over the other in different levels of the taxonomical hierarchy as individual, variety, species, genus, family, etc.

(i) **PCR based Molecular Markers:** With the advent of PCR, techniques became available which overcome many of the limitations of other marker methods. The discovery of DNA amplification using polymerase chain reaction (PCR) is an important milestone in life science research. In PCR, the target DNA is amplified many million fold using two oligonucleotide primers of about 20 nucleotides in length that specifically hybridize to opposite DNA strands flanking a region to be amplified. The reaction is carried out with a thermostable polymerase in the presence of deoxynucleotides and primers in a thermocycler. The reaction mixture is alternately heated to high temperatures and cooled several times. A typical cycle would involve denaturation of DNA and separation of the strands by heating, annealing of the primers by cooling following extension of the primers by DNA polymerase by addition of nucleotides to them. The polymerase chain reaction is thus a cyclic process with each cycle, the number of DNA targets doubles. This method provides new ways of generating molecular markers for plant genome analysis.

(ii) **Arbitrary Primed PCR (AP-PCR):** Arbitrary Primed PCR is as same as that of RAPD. It but requires high primer concentration in the PCR cycles and the amplification is in three parts each with its stringency and concentration of constituents. The primers of variable length are often designed for other purposes.

(iii) **Restriction Fragment Length *Polymorphism* (RFLP):** The detection of polymorphic locus characterized by a number of variable length restriction fragments is termed as restriction fragment length polymorphism (RFLP). It was initially used to fingerprint plant genomes. The first step in the analysis is to derive a set of clones that can be used to identify RFLPs. Genomic clones which represent sequences at random are a poor choice as hybridization probes. Many of the clones will contain repeated sequences and hybridization with those clones containing repeated sequences generate many hybridization bands which are genetically difficult to analyze. The two primary sources of these clones for RFLP mapping of plants are cDNA clones and PstI-derived genomic clones. These two clone sources usually represent expressed genes. They are in low copy number. cDNA clones are DNA copies of expressed genes. PstI clones are based on the suggestion that expressed genes are not methylated. Usually GC and GXC methylation is the most prominent form of methylation in plants. The enzyme Pstl is C-methylation sensitive. The enzyme hence will only cut nonmethylated sites. If a gene is expressed, then its sequence will not be methylated and will be susceptible to PstI digestion. Once a series of clones are derived, DNA will be then digested with restriction endonucleases, which recongnize and act on specific sequences in the DNA. The cleaved DNA is subjected to electrophoresis on agarose gels to separate the fragments as per to their size. The separated fragments are transferred to a membrane support. They are hybridized to labeled DNA probes. As a result of the polymorphic fragments detected by the chosen probe, fingerprints appeared as variant banding patterns on X-ray films. The number of fragments observed and their sizes would depend upon the number of restriction sites. Their positions in the DNA which, in turn, would be characteristic of the genome.

(iv) **Random Amplified Polymorphic DNA (RAPD):** RAPD markers are based on the differential PCR amplification of a sample of DNAs from short oligonucleotide sequences. This molecular marker is based on the PCR amplification of random locations in the genome of the plant. A single oligonucleotide is used to prime the amplification of genomic DNA. Since these primers are 10 nucleotides long, they have the possibility of annealing at a number of locations in the genome. For occurrence of products, the binding must be in inverted repeats of sequences generally 150-4000 base pairs apart. The number of amplification products is directly related to the number and orientation of the sequences which are complementary to the primer in the genome. The first step involves the isolation of total genomic DNA from the individual plants. Then establish the conditions for the amplification of the DNA from specific species of interest. The considerations are: $MgCl_2$, primer and $_dNTP$ concentrations and the quality and concentration of the target DNA. The goals of RAPD experiments are to compare populations at the individual, varietial and species level. This requires well to well and run to run consistency. Thus, one need to be sure that each well of the cycler will produce the same result given the same, target DNA, primer and $_dNTP$ concentration. The PCR amplification steps are 94°C, 1 min for denaturation, 35°C, 30 sec for annealing and 72°C, 2 min for extension. This profile is performed for 45 times and the products are then completed by a single extension for 7 min at 72°C. The amplification products are separated on a 0.8% agarose gel, stained and photographed. Variability is scored as the presence or absence of a specific amplification product. Molecular characterization based on RAPD markers is quickly becoming one of the most popular methods as it is technically simpler than using RFLPs. A number of steps can be done away with such as preparation of genomic or cDNA clones, restriction digestion of DNA, southern blotting or hybridization with radioactively labeled probes. As RAPD analyses need much smaller amounts of DNA and less pure DNA than RFLP analyses, the RAPD method is especially attractive to researchers working with species recalcitrant to nucleic acid extraction e.g. woody plant species.

(v) **Multiple Arbitrary Amplicon Profiling (MAAP):** This technique involves the use of a single arbitrary primers in a PCR reaction. It results generally in amplification of many discrete DNA products. Each product will be derived from a region of the genome when contains two short segments sharing sequence similarity to the primer. They are on opposite strands and sufficiently close together for the amplification to work. These are collectively termed as Multiple Arbitrary Amplicon Profiling (MAAP). The most commonly used is RAPD analysis in which the amplification products are separated on agarose gels in the presence of ethidium bromide and visualized under ultraviolet light. AP-PCR (Arbitrary Primed PCR) and DAF (DNA Amplification Fingerprinting) differ from RAPDs principally in primer length, the stringency conditions and the method of separation and detection of the fragments. Polymorphisms are detected as the presence or absence of bands mainly result from sequence differences in the primer binding sites. These techniques need DNA probes or sequence information for primer design. The procedure involves no blotting or hybridizing steps. The technique requires a thermocycler and agarose gel electrophoresis apparatus. It requires small amounts of DNA (10 ng per reaction), sample throughput can be high and the procedure is automatable. It is absolutely critical to strictly maintain consistent reaction conditions to achieve reproducible profiles. Data quality is limited because MAAP gives dominant marker. It may, sometimes, consist of co-migrating products and band identities are difficult to assign.

(vi) **DNA Amplification Fingerprinting (DAF):** DAF consists of AFLP (Amplified Fragment Length Polymorphism), RAPD (Random Amplified Polymorphic DNA) and uses short 10 base primers to randomly amplify polymorphic DNA sequences (RAPDs). Subsequent to the polymerase chain reaction cycles, the amplified products are separated by electrophoresis either on agarose gels and detected with ethidium bromide or on polyacrylamide gels and detected by silver staining. DAF uses low stringency amplification conditions so that primers can anneal arbitrary at multiple sites on each template strand. A number of mechanisms are responsible for the variation in length or number of amplification products observed in DAF fingerprints.

(vii) **Amplified Fragment Length Polymorphism (AFLP):** It combines restriction digestion and PCR. The first step involves restriction digestion of the genomic DNA with two specific enzymes, one a rare cutter and the other a frequent cutter. Adapters are added to the ends of fragments to provide known sequences for PCR amplification. The adapters are necessary as the restriction site sequence at the end of the fragments is insufficient for primer design. Short stretches of known sequence are added to the fragment ends through a ligase enzyme. If PCR amplification of the restricted fragments was carried out, all the fragments would be amplified which may not be resolvable on a single gel under conventional means. Primers are thus designed so that they incorporate the known adapter sequence plus 1, 2 or 3 additional base pairs (any one out of the four possible: A, G, C or T). PCR amplification will only occur where the primers are able to anneal to fragments that have the adapter sequence plus the complementary base pairs to the additional nucleotides. The additional base pairs are referred to as selective nucleotides. If one selective nucleotide is used, more fragments will be amplified than if two are used, and even fewer fragments will be amplified with three selective nucleotides. Generally, two separate selective rounds of PCR are carried out. In the first round, only one selective nucleotide is used, whereas in the second round the same selective nucleotide plus one or two additional ones are used. This results in between 50-100 fragments being amplified, which can be separated on a polyacrylamide gel by electrophoresis. The amplified products are usually visualised after exposure to X-ray film, where radiolabelled primers are used, but the technique has been adapted to fluorescent, nonradioactive and silver staining procedures. It has been automated. AFLPs provide an effective means of detecting polymorphism in a single assay. It requires more DNA (0.3-1.0 mg per reaction) and are more technically demanding than RAPDs. Their automation and the recent availability of kits means that the technology can be brought in at a higher level. Since, AFLPs are dominant markers, relatedness and variations among the large number of plant species, their origin, domestication and rarity, can be identified using gel scanners. The AFLP markers were used to determine the overall gene diversity among genotypes. The phenotypic information gained from field studies, greenhouse studies and the AFLP marker information on overall gene diversity can be used to find out the origin, its management systems and domestication of cultivated legumes.

(viii) **Sequence Tagged Sites (STS):** Sequence tagged sites (STS) is a popular alternative to RFLP markers as they are PCR-based and also do not require radioactive probing. STSs are developed by first sequencing the ends of a RAPD product or a clone used as the RFLP probe. From the sequence information, oligonucleotide primers of 18-20 nucleotides long are synthesized which are complementary to each end of the RAPD product or clone. These new primers are then used to amplify DNA by PCR. Because the primers are longer than those used for RAPD mapping, the PCR reaction can be run at a higher annealing temperature. This simple change in reaction temperature results in specific and simple amplification pattern. Since they are portable from lab to lab, it is then possible to develop STS markers that can permit the quick location of any gene on a molecular map. It is essential to define from two to three evenly dispersed STS sites for each chromosome of the species with which you are working. Whenever a new gene of interest is identified, the linkage relationship between that gene and each of the STS loci can be established. The new gene should map within 25 cM of-one of the STS loci. When the new gene is located in relationship to an STS, then one can switch over to a RFLP or RAPD map. One can also select probes or primers that will permit to identify markers more closely linked to the new gene.

(ix) **Single Primer Amplification Reactions (SPAR):** Microsatellites are effective in Single Primer Amplification Reactions (SPAR). Oligonucleotide primers of 16-20 bases in length and mimicking the simple di-, tri-, tetra-, and pentanucleotide microsatellite repeats were used in SPAR to amplify DNA of evolutionarily diverse plant species. The results showed that tetranucleotide primers were most effective amplifying the polymorphic patterns. The polymorphisms contained in such patterns were able to identify lines or varieties of plants.

(x) **Sequence Tagged Microsatellites (STMS):** It is also called Simple Sequence Repeats (SSRs). Generally, microsatellites are abundant and scattered throughout eukaryotic genomes. They are an attractive source of primers for PCR based genetic marker systems. Because the flanking sequences at each SSR may be unique, cloned and sequenced, primers to the flanking regions can be designed to define a sequence tagged microsatellites (STMS). The important advantages of STMS are: (a) a single locus which, because of the high mutation rate, is often multi allelic codominant markers (b) can be detected by a PCR assay that can be exchanged between laboratories and their data are highly informative. It is common to distinguish STMS on polyacrylamide sequencing gels where single repeat differences can be resolved and all possible alleles detected. The assay that is relatively quick and throughput, can be increased by selecting a small number of different STMS with alleles of non-overlapping size ranges and multiplexing either the PCR reactions or the products of the separate reactions, so that all the alleles of the different loci can be run in a single lane on the gel. Multiplexed STMS have been also automated. Retrieval of microsatellites are not easy in plants because of their comparatively low abundance compared with animal genomes. STMS often show limited cross transferability within the genus and to species of other genera.

(xi) **Inter Simple Sequence Repeats (ISSR):** It is a modified technique of STMS using anchored microsatellite oligonucleotides as primers which directs amplification of fragments other than the repeat regions itself. It uses oligonucleotides based on a simple sequence repeat (SSR) anchored at their 5' or 3' ends by 2-4 arbitrarily chosen nuclèotides that trigger site specific annealing and are dominant, generally.

(xii) **Cleaved Amplified Polymorphic Sequences (CAPS):** This techniques is also called PCR-RFLP. In this, the PCR primers are constructed for a particular locus. The PCR amplified product is then digested with a restriction enzyme and visualized on an agarose gel using ethidium bromide staining. Polymorphisms are detected by differences in restriction fragment sizes.

(xiii) **Sequence Characterized Amplified Regions (SCAR):** These markers are generated by cloning and sequencing RAPD markers are of particular interest. When the sequence is known, it is possible to design primers which are longer than the usual RAPD markers. They are exactly complementary to the ends of the original RAPD fragment. When those primers are used in a PCR, single loci are identified which corresponds to the original fragment are called sequence characterized amplified regions (SCAR). The results are highly reproducible. The markers are codominant than RAPD and other arbitrary primed methods.

(xiv) **Variable Number of Tandem Repeats (VNTRs):** Hypervariable regions are interspersed within the genomes of higher organisms. These are comprised of tandemly repeated DNA sequences. There are two classes: (i) microsatellites or simple sequence repeats (SSRs), where the basic repeat unit is around 2-8 base pairs in length, and (ii) mini satellites for longer repeat units of around 16-100 base pairs. Hybridization to restricted DNA with micro- or minisatellite probes gives multilocus patterns. They can resolve variation at the levels of populations and individuals. The variation results from changes in number of copies of the basic repeat. It is often referred to as Variable Number of Tandem Repeats (VNTRs). VNTR loci are, in principle, codominant markers. They often behave as dominant markers in RFLP analysis.

Sequencing of DNA

DNA sequencing of genes, parts of genes or noncoding regions is becoming more and more common. It is now widely used in systematics. Sequencing determines the precise order of nucleotides-adenine (A), cytosine (C), guanine (G), or thymine (T) – in a stretch of DNA. The central difficulty of sequencing has always been getting enough DNA to work. This was done initially by cloning genes into bacteria and allowing the bacteria to replicate the genes along with their own genomes. This method is quite slow but is reliable. It avoids some of the possible artifacts of more efficient methods. This rapid method has allowed systematics to study the same region in many species of a particular group. One disadvantage of PCR is that the polymerase itself introduces occasional errors, which could in theory affect an estimate of phylogeny, particularly if the sequences being compared are extremely similar. In practice, this may not be a serious problem, in that the errors are not likely to be biased in favor of any one grouping. One way to reduce potential error in sequencing is to sequence both strands of the molecule The decision of how accurately to sequence depends on the relative costs of an error versus the costs of repeatedly sequencing the same region. Systematists must often make a choice between highly accurate sequences from fewer taxa or less accurate sequences from more taxa. Direct sequencing of the PCR product will not usually reveal minor variants of the sequence if they are present. This is, sometimes, a problem with highly repetitive genes such as ribosomal genes, for which the many copies often are not identical. It also can not distinguish between alleles of the same gene.

DNA *Bar Coding*

This is a taxonomic method which uses a fragment containing the first half of the COl gene of mtDNA to identify it as belonging to a particular species. It is based on a comparatively simple concept: most eukaryote cells contain mitochondria and mtDNA has a relatively fast mutation rate and more differences than the ribosomal genes, which results in significant variance in mtDNA sequences between species and, in principle, a comparatively small variance within species. Molecular bar coding takes into account the isolation of the organism, amplification of the target gene, cloning, sequencing and phylogenetic analysis leading to the assessments of species content, abundance and diversity. However, DNA *bar coding* is only as good as the reference database. It can not be employed to decipher species not already catalogued. *Bar coding* will also be most reliable for identification of potential putative new species, but only for species groups whose genetic diversity has been well surveyed. However, DNA bar coding can not replace the traditional methods of describing species.

Some Applications

(i) Gilmore *et al.* (2007) studied chloroplast and mitochondrial DNA in case of *Cannabis sativa.* They revealed comparative sequencing of *Cannabis* individuals across 12 chloroplast and mitochondrial DNA loci at 7 polymorphic sites including 5 length variable regions and 2 single nucleotide polymorphisms. They developed simple PCR assays to assay these polymorphisms. They obtained organelle DNA haplotypes for 188 *Cannabis* individuals from 76 separate populations, including drug-type, fibre-type and wild populations. They discussed relationship between haplophyte data and taxonomic opinions of *Cannabis* and the implications of organelle DNA haplotyping to forensic investigations of *Cannabis*.

(ii) Spooner *et al.* (2020) explored the phylogenetic utility or mitochondrial DNA sequences in *Daucus*. They compared the results with prior phylogenetic results using the same 36 accessions of *Daucus* (and two additional groups) with plastid DNA sequences and with other nuclear results. They obtained data from 47 of 71 total mitochondrial genes but only 17 of these 47 genes recovered major clades that were common in prior plastid and nuclear studies. The phylogenetic trees of the concatenated data set of 47 genes were moderately resolved, with 100% bootstrap support for most of the external and many of the internal clades, except of *D.carota* and its most closely related species *D.syrticus*. mtDNA is shown to be useful in addressing important evolutionary and phylogenetic questions at higher taxonomic levels.

(iii) Song *et al.* (2019) studied phylogenetic relationships among species of *Sorghum* which was unsolved due to lower information on the traditional DNA markers, which provides a limited resolution for identifying species of *Sorghum*. They sequenced the complete chloroplast genomes of *S.sudanense* and *S.propinquum*. They analysed the published chloroplast genomes of *S.bicolor* and *S.timorense* to retrieve valuable chloroplast molecular resources for *Sorghum*. According to them, a phylogenetic analysis strongly supported that *Sorghum* is a monophyletic group in the tribe Andropogoneae. They further stated that the genomic resources in their study could provide potential molecular markers for phylogeny and species identification in in *Sorghum*.

(iv) Moudi and Go (2016) made comparative study between nuclear ribosomal DNA and chloroplast DNA is molecular systematic study of 04 sections of the genus *Dendrobium* Sw. (Orchidaceae). These four sections were *viz.*, Aporum, Crumenata, Strongyle and Bolbidium. They investigated 34 species of *Dendrobium* using Maximum Parsimony (MP). Three nucleotide sequence data sets from two distinct genomes chloroplast DNA genes (*rbcl* and *matK*) and nuclear ribosomal DNA (ITS) were used to construct cladograms. The results from ITS (Internal Transcribed Space) gene showed that the nuclear genes are reliable marker for the phylogenetic.study of *Dendrobium* compared to chloroplast DNA with low resolution level among sections. They concluded that the 04 sections of the genus *Dendrobium* formed a monophyletic group. The sections Aporum and Strongyle were genetically closely related, whereas sections Crumenata and Bolbidium were nested together in one clade indicating the high genetic similarity. They further decided to have one section named Aporum for them according to ICBN rules.

Phylogenetics and Phylogenomics

The functions of plant taxonomy are: (i) to name, identify and catalogue organisms (taxonomy); (ii) to discover the ancestral relationships among organisms (phylogenetics), and (iii) to organize information about the diversity of organisms into a hierarchical system (classification). Molecular systematics is the utilization of knowledge of genome information, particularly sequence and structure of DNA, RNA molecules and amino acid chains, for addressing questions regarding the phylogeny and taxonomy of plant species. The molecular data are more suitable for phylogenetic investigation than morphological ones: (i) DNA and protein sequences are strictly heritable entities, whereas morphological characters can be influenced by various environmental factors, (ii) the interpretation of molecular characters, such as the assessment of homologies, is generally easier than that of morphological characters. (iii) The molecular characteristics generally evolve much more regularly than morphological ones. They can provide a clearer picture of relationships. (iv) The molecular characters are much more abundant than morphological ones and many can be generated in a relatively short period of time. (v) Various preserved, deformed and partly degraded materials can often be used for molecular investigations. Using standard protocols and commercial kits, sequence information of certain genes or DNA fragments can be obtained from a single plant or even only a part of it. The specialized methods enable the extraction of short DNA fragments from long-time preserved, formalin-fixed and glycerine-embedded specimens. Recent achievements in molecular biology and the wide application of molecular techniques have revolutionized our knowledge in taxonomy and phylogeny of various plant groups.

Species Concepts

Various definitions of species are put forth to date. Species were at first merely taxonomic units, i.e. the named categories to which Linnaeus and other taxonomists of the 18th century assigned organisms, mostly on the basis of external appearance. As per the typographical species concept, the species is thought a community of specimens described by characteristic features of its type specimen. A great deal of evidence leading to the widely used modern concept of species has accumulated now. This species concept was found on to observations: (i) species are composed of populations and (ii) populations are co-specific if they successfully interbreed with each other. This idea was promulgated by Mayr (1942) in the biological species concept. Species are groups of interbreeding natural populations that are reproductively isolated from other such groups. Many more species concepts have been proposed. The most popular one in systematics is the so-called 'Phylogenetic Species Concept'. This concept does not emphasize the present properties of organisms or their hypothetical future, but rather points at their phylogenetic history. However, the applicability of this concept has remained debatable, for it proposes operational criteria of how to decipher species only as phylogenetic taxa, rather than describing the role that species play in the living world.

The biological species concept has its own limitations. Application of the biological species concept is confined to sexual, outcrossing populations over a short period of evolutionary time, excluding parthenogenetic organisms. Moreover, the diagnosis of biological species is seldom done in practice by testing their propensity to interbreed and produce fertile offspring but is often made by observing the difference in morphology. This should not be a contradiction, because phenotypic characters often, although not always, serve as markers for reproductive isolation. Morphological distinctiveness is a good criterion for separating species but not infallible. Sibling species (biological species but morphologically indistinguishable) may remain undetected, even after careful morphological examinations, unless allozymes or other genetic markers investigated. The genetic difference among related species seems to vary substantially but generally increases with the time elapsed since their reproductive isolation. The degree of the genetic distance among populations, estimated from allozyme frequency, nucleotide divergence, amplified fragment length polymorphism (AFLP) random amplified polymorphic DNA (RAPD), microsatellites or other markers, provide evidence to determine whether or not they should be assigned the status of species. These approaches to decipher species are based on indirect inference of the presence or absence of gene flow.

Nearly all information from the genome and proteome at all levels, including the sequence of fragments of DNA, RNA or amino acids, the structure of molecules, the gene arrangement and presence versus absence of proteins or genes, may be applied to molecular taxonomy. Various biochemical and molecular techniques have been introduced for diagnostics, the estimation of genetic diversity of populations and the inference of phylogenetic relationships between taxa.

8

Characters

We generally express relationship based upon a comparison of characters of organisms. We do so by employing words such as resemblance or difference, similarity or dissimilarity, etc. While involving into this activity we select characters. Selection of characters again depends on the availability of characters, their traditional usage and the general convenience. Use of characters, in a nutshell, is 4 - pronged challenge : (i) to understand what characters are (ii) what their role in classification really is (iii) what terms have been applied to them and (iv) how they can be selected for use in making classifications. Hooker (1855) therefore opined on the problem that 'a knowledge of the relative importance of characters can only be acquired by long study; and without a due appreciation of their value no natural group can be defined.' 'Taxonomists strive for assimilation of the 'facies' of an organism at a glance, but in order to interpret to others what he sees, he must mentally dissect the organism and describe it as a series of characters, some of which may have greater significance than others (*cf.* Kendrick 1964).

(I) Definitions of Character

The science of '*Taxonomy*' has no data of its own. It accepts data from various areas of biology. Data in taxonomic terminology are referred to as 'characters.' Classification, one of the functions of taxonomy, is based upon character distribution and their variation patterns. Taxonomists, to start with, conceives the organisms into characters and it is these characters which bestow evidence for classification. It is the type of character that deciphers the very nature of the resulting classificatory system. A clear concept of characters is a pre-requiste for an understanding of taxonomy. Many definitions of character have been proprosed by various workers. Davis and Heywood (1963) define a character as 'any attribute referring to form, structure or behavior which the taxonomists separates from, the whole organism for a particular purpose such as comparison or interpretation.' Jones and Luchsinger (1986) defined similarly. According to them, the characters of an organism are all the features or attributes possessed by the organism that may be compared, measured, counted, described or otherwise assessed. Crowson (1970) suggested simply 'any feature which may be used to distingush one taxon from another'. There are taxon-oriented definitions. It was doubted that how can characters be defined in terms of delimiting taxa when the taxa must be defined before we can know what the characters are? Alternatively, to avoid this problem, taxon –free or more general definitions of characters have been attempted. For example, Micherner and Sokal (1957) defined character as 'any features of one kind of organism that differenatates it from another kind of organism' Kendrick (1964) thought it any attribute referring to form, structure or behavior which can occur in any one organism as one of two more mutually exclusive states. Cain and Harrison (1958) defined it 'anything that can be considered as a variable independent of any other thing considered at the same time'. It is also thought by Ross (1974) as 'some defined attribute of an organism', whereas Blackwelder (1967b) defined chacter as 'the expression of the feature in the individual.'

It appears pertinent to disscuss about *character* and *character state*. These two terms are often employed loosely and interchangeably. Actually, these denote things differently. Character is the particular attribute that the taxonomist consider for classificatory purposes. It is simpily an abstraction. The various patterns of expressions are called *charcters-states*. The same character may have many *character-states* for example (i) flower symmetry is a character, while actinomorphic and zygomorphic are its character-states; leaf arrangement (phyllotaxy) is character, while alternate opposite or whorled are its character-states, (ii) placenation is a character while axile, basal, parietal, marginal, etc. are its character-states. It is also to be remembered here that when a particular *character-state* is exclusive to a particular taxon in a given assemblage it is said to be '*characteristic*' to the group. For example, leguminous fruit is characterstic to the family Leguminosae and gynobasic style is characteristic to the family Lamiaceae, etc. However, this distinction between 'character' and 'character-state' is recommended for use. Mayr (1969c) opined that historically in systematics the term 'character' has been used in this way for centuries, whereas the concept of 'character-state' has been only recently introduced explicitly. Evolutionary cladists consider character as a feature of an organism. According to Wiley (1981a), such features arise in evolution by modification of a previously existing ontogenetic or cytogenetic or molecular sequence. And thus only characters have information of evolutionary value for constructing classifications and there is no more necessity of character-state.

Eldredge and Cracraft (1980) considered these two as relative terms and opined that these should be construed two mean relative levels of similarity within a given hierarchy. The cladists suggest abandonment of character and character-state for 'feature' for both. But this is not recommeded.

Various other general terms e.g. attribute, feature, quality, characteristic, descriptive term, essence, difference, etc. are employed in relation to or in place of 'charcter-state'. These can be studied from philosophical, logical and biological viewpoints. In philosophical context, Griffith (1974a) treats character and attribute as synonyms and defines them as 'an attribute is that in a thing which constitutes part of our cognition of it, and 'a partial conception so for as it is considered as ground of cognition of the whole conception'. An 'essence' is a feature that attempts to reflect the philosophical essence of an object, as used in Plato's essentialist philosophy. In a strict logical sense, 'any quality or attribute is regarded as being either a difference or a property or an accident'. A quality is said to be a 'difference' if it serves to distinguish the class of entities of which it is a quality from other species of the same genus. A quality is said to be 'property' if it is quality necessarily possessed by every member of the class, yet not utilised to distinguish the class from other species of the same genus. A quality is to be an 'accident'if it may indifferently belong or not belong to all or any of the member of the class (Sinclair, 1951). In the biological persective, a 'characterstic' is viewed as a particular character-state occurring exclusively in certain specimens or species. A 'descriptive term' is a descriptor referring to a condition or state of an organsim that has taxonomic importance and it can be equated with characteristic in the sense of Ross (1974). In a nutshell, difference, property, accident and essence have precise. logical definitions. Feature, quality and attribute all refer generally to some aspect of an organism, whereas characteristic and descriptive terms refer to aspects that are regarded as taxonomically significant. One is then led to concur with Davis and Heywood (1963). In their opinion, no precise general answer can be given to the question what is a character ?

(II) Kinds of Character

Character, in a broad perspective, help describe and measure the perceived reality of the world. And hence we are positioned to assess any and all relationships of interest. Use of character to interpret and appreciate reality in the natural world obviously depends upon our skill and ability in employment. Characters are numerous and their distinctions are necessary for due appreciation and proper use. The various qualitying terms for characters available in the literature are here grouped under four categories *viz.*, (i) General characters (ii) Phenetic characters (iii) Phylogenetic characters and (iv) Cladistic characters.

(i) General Characters: Organisation of states, variability of characters and states, specific utility and general validity distinguish the types of general characters. Often, the fundamental distinction is made in the organsiation of states of a character into quantitative or qualitative units. *Quantititative characters* are assessed by counting or measurement and the character states are expressed in numbers. Thus features assessed by size, length, etc. are quantitative, while those relating to shape and form with states given in descriptive words rather than numbers are considered as *qualitative characters*. Examples are leaf shape in terms of ovate, obovate, lanceolate, linear, etc., ovary superior and inferior; inflorescence raceme, spike, cyme, etc. Different numbers or ranges of numbers of parts are commonly used in taxonomic descriptions. Linear dimensions are also employed in such descriptions. *Quantitative characters* tend to be more useful at the lower levels of the taxonomic hierarchy (Davis and Heywood, 1963).

Another viewpoint on the organization of states is to view characters as *microcharacters* and *macrochracters*. These refer to scaling perspective on the actual size of the states and the types of method of obtaining the data (such as SEM and TEM, electrophoresis, etc). The distinction between these two types of characters is made by those who emphasise one type of data as superior to others in revealing relationships. The claim of supremay of microcharacters is totally unfounded (*cf.* Stuessy, 1973; Grashoff and Turner, 1970). As with all characters, sometimes *microcharacters* are useful and sometimes they are not (Stuessy, 1979b).

The variability of characters and their states also determine type of character. Characters which are invariant within a taxon under consideration (occurning in only one state) are obviously of no value in deciphering relationship between subtaxa within the taxon. Only 'variant' (variable) with two or more states are taxonomically useful. A similar distinction exists between '*fixed*' characters and '*plastic*' characters. The former are invariant characters and vice-versa. The latter are also invariant characters but not the reverse. The term 'plastic' refers to environmental modifications of characters which are without genetic basis, whereas 'variant'refers to genetically governed differences.

Some terms are suggestive of utilities of characters. For example, (a) specific, generic and familial characters, etc... These are employed to delimit taxa at different levels in the taxonomic hierarchy. Terms have relative helpful meaning within a particular group. (b) Linnaeus referred to 'essential characters' which would distinguish them from each other within the same order (Cain, 1958). (c) characters employed in identification, characterisation and delimitation of a group are called '*analytic*', whereas those employed in the classification of same group at a higher level in the hierarchy are called '*synthetic*' (Just 1946). According to Davis and Heywood (1963) there is no inherent difference between such characters. The diference is only of their use.(d) diagnostic or key characters are those used in description, delimitation or identification (Jones and Luchsinger, 1986). (e) *descriptive* characters are those which are useful to describe some aspect of an organism or taxon, however, do not help to show difference from close relatives (f) biological characters have some clear function or vital role in the organism, whereas '*fortuitous charactsrs*' have no such function (Wernham 1912, Davis and Heywood 1963). Diels (1924) recognized three subordinate types of biological characters: (i) Functional characters – those intimately connected with some special function but remain unaffected by external conditions, (ii) Epharmononic characters- those are apparently connected with the mode of life of the plant but remaining constant even under varying external conditions (iii) *Adaptive characters* - those varying according to the exeternal characters. Straw (1956) and van der Pijl (1960) conceived functional aspects of characters as more useful especially in natural classification. The exact nature of functional or adaptive significance of many of the characters is not yet fully understood. The herbarium taxonomists do not emphasize functional value of the character. They lay emphasis on diagnostic value of characters. They are concerned more with the constant nature of the characters selected for discriminating groups (*cf.* Heywood, 1959b). Borrill (1961) stated that they can not be used for taxononic alignment, for example, the flowers in Papilionaceae, Polygalceae and Orchidaceae, may be thought as convergent adaptations in response to similar methods of pollination and therefore can not be employed for taxonomic separation. Whether we understand their functional value or not is largely a matter of our knowledge about them. The constant characters are most probably selected by the nature as such (Davis and Heywood 1963). All characters are theoretically of use in classification. However, their value or use varies according to a number of factors and type of taxonomic activity. There is a general tendency to rely on certain kinds of characters.

(ii) Phenetic Characters: Phenetics (often referred synonymous to numerical taxonomy) is one of the major approach to biological classification. It uses overall similarity to assess relationships. Specialized types of phenetic characters have been proposed. It reflects no evolution. Taxa are related or based simply on similarity and difference of character states regardless of the evolutionary content of the characters and states reflected. Unit character is searched out. It is a taxonomic character of two or more states which within the study at hand can not be subdivided logically except for subdivision brought about by changes in the method of coding. The basis for the search for unit character is aimed at avoiding logical circularity and to obtain data in which each datum represents a new item or bit of information (Sneath, 1957; Sokal and Sneath, 1963; Sneath and Sokal, 1973). There are, however, difficulties in their selection. To decide a logically indivisible character is most difficult for most all features can be dissected further to a lower level of hierachial organization down to the atomic level and even beyond. Griffith (1974a) opined that many seemingly simple character statements can be subdivided logically. Crowson (1970) suggested another definition of unit character. According to him, unit character is the amount of phenotypic change which could be brought about by a single continuous bout (turn) of natural selection. There is another objective of phenetics. It is to use unweighted characters. In phenetics, each unit character is given the same or no weight. This is difficult in practice, if not impossible (Stace, 1980). Again in phenetics, non-correlated characters are to be used. Different correlations exist e.g. developmental, evolutionary, random, etc. It is the most important to avoid in phenetics logical correlation, in which what were regarded as independent unit characters actually are consistently correlated and hence providing redundant information.

The choice of characters and states depends to some extent, upon the particular approach to classification used. Some viewpoints on selection of characters can be generally grouped as: (a) logical (b) biological (c) information theoretical and (d) practical.

(a) **Logical Criteria:** A 'posteriori' rather than a 'priori' selection of characters is the most logical consideration (Crowson, 1970). The former applies to deciding which characters are important after having studied the organisms. The latter refers to the selection before the organisms have been examined carefully (Cain, 1958c). If one is interested to propose a new classification for a group of angiosperms with which no individual or personal familiarity exists it could be assumed a priori that type of inflorescence will be significant. This selection force the classification to develop along these lines which might be erroneous. Type of inflorescence could be extremely variable within this group and shows no correlations with other characters or it also could be invariant, so that a posteriori it would not be a suitable taxonomic character. A posteriori selection is a very important aspect of taxonomic characters because this enables to seek logical correlations of character states among taxa. The pheniticists lay emphasis on the concept of logical irreducibility as one of the criteria of selection of characters. Characters, in general, shoud be viewed as fundamental units of information that informs about relationships among organisms.

(b) Biological Criteria: In the selection of characters and states many different biological persectives can be thought over. Importantly, the biology of the characters (and organisms *sensu lato*) be determined (*cf.* Mayr, 1969c). Phenitcists and cladists belive, in recent times, that they can understand relationships of any group with their methods and they need only basic data matrix. The biology of there characters and states must be understood for a much more accurate relationships. Aspects e.g. breeding systems, ecological and distributional information, etc. can be presented in the data matrix and other can not. The facts such as the genetic basis of characters and states, knowledge of definitions of adaptive characters and their value, evolutionary information content and conservatism, etc. pose difficulties in selecting characters and their states.

(c) Information Theoretical Criteria: This emphasises the selection of characters for their maximum predictive quality. Davis and Heywood (1963) state that it is more useful to regard a character as reliable since its use will allow us to make a large number of deductions about the other attributes of the groups it separates off. It is to be noted that recognition of the importance of this aspect of characters is one thing. While finding ways to measure this information content for predictive quality is another. There has been attempts to relate unit characters to basic units of information (*cf.* Sneath and Sokal, 1973). The supremay of evolutionary characters in leading to maximally predictive classifications have been suggested (*cf.* Farris, 1979b; Brooks, 1981). Further studies are, however, needed.

(d) Practical Considerations: This perspective takes into account the practical matters. Theroretically, it is important to be aware of and conversant about ways in which characters are important and can be selected. Ultimately, one has to come down to selection of characters which work (*cf.* Cronquist, 1957, 1964). The practicing taxonomist usually evaluate conservative nature of characters and also looks for correlations with other characters to establish the suites necessary for taxonomic considerations. Pheneticists emphasise conscious selection of characters. This is desirable in theory but it is not possible approach with the majority of taxonomic efforts in which speed of work is valued. A selection of a few constant characters and constructing a classification from them is much more efficient than the carefully reasoning out all the unit characters (which could number many) and developing the classification along these lines. The selection of character states is problematic equally in practice and has still remained unadressed. The selection of qualitative and quantitative character states is an initial and fundamental problem. Crovello (1974) rasied a question. If a character is to be measured what set of values should be used to record the state of the character in a given individual? Further in his opinion, more difficult decisions involve qualitative atributes e.g. leaf shape. Invetigators's procedure would be simply to recognize those patterns as distinct that best serve the purpose of analysis. This he considered unscientific and stated that no formal decision function exists to determine qualitative character states.

(iii) Phylogenetic Characters: Phylogenetic (phyletic or evolutionary) characters are employed primarily in constructing phylogenetic classifications. It is of prime importance to know the distinction between homologous and analogous characters. Although the difference between these two terms appears simple but philosophical problems do exist. Originally, homologue was thought the same organ in different animals under every variety of form and function and analogue as a part or organ in one animal which has the same function as another part or organ in a different animals (*cf.* Cain, 1976). Prior to Darwin theory of evolution, homologous characters were originally understood as simply basic structural differences. During post-Darwinian period, homologous characters were viewed as the structural modifications of the same organ inherited from a common ancestor in response to different selection processes. *Analogous characters* on the other hand, were considered those developed by different organs to the same selection pressure. The detection of *homologous characters* in two groups in done by knowing that they have descended from a comman ancestor. If this were known, there would no nesessity to use these homologues to reconstruct the phylogeny it would be known before time for selection of homologous characters. Philosophically, there are alfernatives to the problem such as by considering other features of the organims for additional signs of similarity (Cain, 1976). The detection of homologous character is a hard task for any phylogenetic reconstruction. Moreover, pitfalls can occur.

Few other sets of terms are also employed to phylogenetic characters. For example, phylogenetic characters and ontogenetic characters, the former are presumed to reflect information relating to phylogeny of the group, while the latter deal with developmental features. These can sometimes suggest phylogenetic patterns, but they may not always do so. A regressive character is one in which loss of organs/features has occurred. This type of characters should be cautiously employed. 'Adaptive' and 'non-adaptive' characters are used usually in the evolutionary context. The former contributes to the fitness of an organism, while the latter does not do so to fitness. Difficulty lies in agreeing on an acceptable definition of 'adaptive'.

(iv) *Cladistic Characters*: It is the cladistic approach to classification which gave rise to cladistic characters. These attempt at determination of branching sequences of evolution and reconstruct classification upon them. The patterns of branching are divulged through analysis in taxa of distributions of character states. The character states are considered to be important evolutionarily and contained within homologous characters. Exclusively derived character states are regarded as significant cladistically. Primitive conditions are thought as misleading and non-informative. Various sets of terms are employed e.g. primitive *vs.* derived character states or as synonyms generalized *vs.* specialized primitive vs. advanced, plesiomorphic *vs.* apomorhic and plesiotypic *vs.* apotypic. Characters which are useful cladistically are called 'compatible characters' (*cf.* Estabrook, 1978) in which the evolutionary directionality of the states within each character is the same. Some authors recognize distinction between true cladistic characters in an idealized sense and those that are defined operationally in the course of actual studies (Estabrook *et al.*, 1975).

(III) *Character Weighting*

Organisms possess numerous characters. All characters are not usually selected in taxonomy and thus character weighting is perforce exercised. In practice, taxonomists are forced, by strictures of time and manpower, to be selective in their use of characters. Selection of characters is generally based on two primary considerations : (i) they should be useful diagnostic characters which do not differ intragroup (ii) such invariant characters are of use in intragroup studies. Therefore, characters are selected which are variable intragroup. We select characters and thereby give a higher rating weight for those selected. It is called 'selection weighting' by Heywood (1967). Obviously, other characters are rejected. This is also called rejection or residual weighting (Stace, 1980). Characters are weighted by taxonomists for the reasons : (i) because they are most easily or readily observed, (ii) beause they exhibit highest correlation with others in a natural group, (iii) beause they are believed to have some known or inferred phylogenetic importance or they are thought to be more conservative characters, (iv) because they were left out characters while constructing a natural group. These are dilated below :

(i) There are enormous number as characters which are easily observed. It is virtually impossible to describe them all. This kind weighting is general and unavoidable. It is the product of the limitations.

(ii) Weighting based on correlation of characters is called 'correlation weighting' or a '*posteriori weighting*' (*cf.* Davis and Heywood, 1963). This is a fundamental process in making a system of classification. It is one of the bases of good or reliable characters. Correlation weighting is the method which theoretically sound. Characters abstracted by employing this method will be the ones that help diagnose the group. Such abstracted markers will probably be the most likely characters to use for keys. The characters showing the highest average correlation with other similar characters have called the '*indicator character*' or 'indicator' (*cf.* Smirnow, 1924, 1938; Terentjev, 1931). Such indicators have the highest taxonomic significance and a set of these constitute taxonomic coefficient. Correlation weighting can be made clear by the species of *Chlorophytum,* a liliaceous genus (Naik and Nirgude, 1981 a,b). The characters such as the basic chromosome number, level of ploidy, duration of the aerial shoot, ecology and secondary thickening on the lateral walls of the metaxylem vessel elements of different species of the genus exhibit close correlation. This set of characters have the highest taxonomic significance. This is further substantiated by the fact that change in any one of there characters concomitantly brings about a corresponding change in the other characters. The diploid species with base number $x = 8$ have metaxylem elements with spiral thickening on their lateral walls, distributed in the semi – arid situations and are with short-lived aerial shoots. However, the diploid and polyploid species with base number $x = 7$ have their metaxylem vessel elements with scalariform or pitted thickening on lateral walls, a comparatively long-lived aerial shoot and found in the humid situations. Leguminosae also offers a best example of correlation weighting. Members of this family has a monocarpellary gynoecium with marginal placentation. The characters such as a predominantly woody habit; alternate often compound stipulate leaves; racemose or racemose-derived disposition of flowers, basically polypetalous corolla and diplostemonous condition show close correlation with each other. The gynoecial character has thus the highest correlation with other characters. This character together with others constitute the taxonomic coefficient.

(iii) Selection or rejection of characters is generally an intuitive process. Characters are selected because (i) they are believed to be of greater taxonomic importance, (ii) which are genetically more stable and conservative, (iii) which are of greater diagnostic value and (iv) thought to be evolutionary markers (*cf.* Hansell and Wing, 1972). Such selection of characters for their supposed significance before one has any evidence to justify it is said to be a 'priori weightng' of characters (*cf.* Heywood, 1967).

(iv) When we select characters, in turn, many other characters are rejected. These are the characters left when we have rejected others.Such a process is called '*rejection weighting* or *residual weighting*.' A taxon is a combination of characters. Characters are generally rejected beause they are not suitable for disagnosis or the construction of keys or for reasons of phylogenetic interpretation.

(v) The cladists are assumed to restrict attention to cases of true homology. To identify synapomorphies, it is essential to study the character polarity (direction) (i.e. to identity which character state in a specific character is plesiomorph and which one is apomorph). The 'polarity' of a character refers to which state is postulated to be 'primitive' or 'plesiomorphic', relative to which the other state(s) would be 'derived' or 'apomorphic'. Several criteria for establishing polarity have been suggested. Once we delineated character states, we can 'polarize the character'. We can polarize character (-states) a priori, before phylogenatic analysis, or a posteriori. This analysis may be carried out within the group (in-group) or outside the group (out-group).

(IV) Presnence or Absence of Characters

Cronquist (1969a) opined that the presence of a structure or a substance is likely to be more important than its absence. He lent support to his opinion by providing a genetic explanation. It is now obvious that individual phenotypic characters are commonly controlled by a complex system of genes, rather than a single gene unaffected by others.

The whole genetic system is necessary for the appearance of a particular characterstic in the phenotype. If two allied taxa or individuals share the same structure then they probably also share the same genetic system controlling its development. The structure is not developed if any one of the essential genes in the system is lost. And thus a plant that is devoid of a particular structure may differ by only a single gene from the one which has it. It is obvious that the absence of a character may reflect the complete absence of a genetic system which would be required to develop it. Cronquist (1*oc. cit.*) concluded that on practical and theoretical basis, the absence of a character is a less reliable guide to taxonomic affinities than its presence.

9

Floristics

Floristics is taxonomic studies of a *flora*, or of a major segment of a flora of a specified area. It may range from a compiled check list of plants of a small politically bounded area to a thorough taxonomic or biosysystematic analysis of the components of the flora of a continent. It is a type of botanical activity that may be conducted for scientific gain which can be limited in scope to meet the potentials and the needs of any serious minded investigator.

Distinctions and nature of *Flora*, *Manual*, *Monograph* and *Revisions* are already covered under another chapter-taxonomic literature. A word about 'compilatory lists' however appears pertinent. These are of three types : (i) those based on exiting accounts collated from the literature and (ii) based on herbarium specimens without the latter being subjected to a critical study by the compiler of the list (iii) a third type is that represented by a combination of the other two stated. These lists meet an interim need or account for the plants of an area until a more authentic list or flora has been producted. While preparing compilatory lists, the compiler has to consider : (a) prepartation of preliminary list form the latest manual of the region. It should be restricted to those taxa ascertainable from the manual to be indigenous to the area selected (b) to survey literature covering the period since the manual of the region was published for reports concerning : (i) new range extensions into or through the area (ii) new stations for plants within the area (iii) new taxa described for the first time and from localites within the area (iv) nomenclatural changes affecting plants of the area (v) taxonomic status of plants within the area.

It is desired to consult existing floristic, revisionary and monographic works for each of the aforesaid aspects. Supplementary material from these sources should be incorporated into the initial preliminary list prepared from the older manual. The material should be systematically arranged e.g. by families. It should be then compared with specimens of the plants concerned. Care should be taken to list all locations within the area where each nomenclatural element has been collected, with collector's name and date of collection. If there are other nomenclatural elements present among the collections that are not noted in the prepared list, they are to be noted. Compilatory list can be published indicating its character.

To have a complete inventory of the plants involvèd, field study is basic. The size of the area is subject to factors such as political boundaries, physiographic provinces or their subdivisions. It is to be noted that greater unity and stronger interrelationships are present among the plants of flora based on a physiographic area than among those based on other types. It is sometimes desirable to prepare flora of an area within a physiographic and / or hydrographic region. For this purpose the square degree may be acceptable than a politically bounded unit. A square degree is an approximate rectangle bounded by one degree of latitude and one degree of longitude.

Field collectors are the principal activity of floristic study. Maximum possible time is to be spent in the field once the limits of the selected area are established. Possibly field collections should be made in all weeks of growing season so that all collected plant specimens have flowers and fruits. Usually, several years are required to such studies. In temperate regions, it is the winter season important from collection point of view. One can make his observations particularly on leaves and fruits, the buds, apex from late-blooms, early blooms and out of - season blooms. This diverse data can be incorportated in the floristic study. If some localities whose plants are not well reprenented in herbaria, one is desired to prepare several sets. Care should be taken about rare plants. They should not be eradicated from the locality. It is always expected to cover all places and habitats in the area under study. It is important to remind one about field collections. All necessary and pertinent data, concerning characters of the specimen that may not be apparent after drying, should be recorded with particular attention. Field collector can follow either of the ways for field notes and even for indetification – (i) one is to affix a consecutively numbered small tag to each specimen as it is collected and to assign the same number to notations made of the time of collection. The another way is to press the specimen as it is collected and write the number and data on the sheet in which it is pressed, treansfering the data to the notebook later. Generally, observations with respect to features that are made during field study are such as texture, herbaceous, arborescent habit, size of plants, habitat, fragrance or odours of different parts, pollinating agents, frequency, soil type, elevation, vernacular names, uses, if any, etc.

Flora contain compiled information after floristic studies in any one form accompanied by collection number, name of localities, local names, etc. It is to be determined whether the plant species collected are indigenous or exotic – the degrees of establishment of exotic elements should be also mentioned e.g. naturalised, introduced or adventive. Such elements generally occur near places like habitation, cemeteries, dumps, railroad yards, highways of urban localities, etc. Their nativitiy status and degree of influence in floristic scenario should be clearly projected.

The specimens should be sorted out according to an accepted system of plant classification and stored. They must be labelled correctly and numbered as well. Usual indentification of specimens should be accomplished by the usual taxonomic procedure or be consulting experts. Generally, both the procedure are followed. Other approaches for indentifications are suggested under a separate chapter of taxonomic literature. Spot determinations is generally done with respect to families and sometimes genera and probable species. These are, however, treated as tentative until verified. Material unknown is worked out to family status by following the treatments in available manuals or such comprehensive works as Bentham and Hooker. Generic determination can be done by such manuals or monographs. It is always desired to consult the latest literature pertaining to each group. If the investigator has his own opinions different from others, he should explain taxonomic or nomencatural departures from the view of experts.

The preliminary determination of the field collectons may be integrated with the herbarium studies, especially when the investigator is affiliated with or situated near a larger herbarium. Herbarium study may be completed in three phases : (i) verification of preliminary determinations made of the field collections, (ii) determination of the collection records of plants of the area as evidenced by material already in the herbarium, and (iii) examination of authentic material for the verification of names applicable to the collections involved. Moreover, it is also essential to examine types, isotyes or other acceptable authentic material for any realiable floristic study. Such examination increases authenticity of the determination made of the specimens.

A third channel of investigation is the existing literature in any floristic study. It is although generally scattered is very essential for completing such studies. One should first learn whether any manual of the region or especially earlier floras were published. These earlier reports must be critically analysed. Statements made in such evidence must be held in a tentative status until each has been challenged and verified. Their verification is completed through examination of the herbarium specimens on which previous record was established. If there is any univerified element, it should be indicated in the new flora and explanatory note should also accompany its entry. Published reports in the literature of occurrence must be substantiated by herbarium specimens or else be viewed with extrme skepticism. This view is also applicable to all reports of range of extension into the area and of new taxa described from collections from within the area.

It is also expected to study other floristic works of the adjacent area or physiographic region. Taxonomic papers which are much valuable and have considered opinion of taxonomic scholars must be scanned. Other several indexs to taxonomic literature should be investigated pertinent to the study.

10

Taxonomic Literature

Plant taxonomy is one of the earliest of all biological disciplines. It is still important and a very relevant science. Moreover, in modern time taxonomy has become increasingly synthetic science. The data from all relevant fields or sources are being amalgamated. It is to be noted that taxonomy is fundamentally a descriptive and highly documented science. The information is really voluminous and today a vast literature exists. Some important aspects of taxonomic literature is mainly brought out in this attempt. There have been sincere attempts by the plant taxonomits throughout the world to reveal and compile their results in the form of certain publications. The pertinent items of taxonomic literature are introduced for the sake of readers.

Taxonomic literature have been published in the form of the earliest classic works, floras, monographs, index, revisions, bibliographies, catalogues, review serials, keys, periodicals, glossaries dictionaries, maps and such other miscellaneous ones. The subject of taxonomy demands an almost encyclopedic knowledge of the world's pertinent literature. In other biological sciences, the current literature outmodes the older writings and is considered to be of only historic or academic interest. This situation does not find parallel in plant taxonomy. The earlier findings in taxonomy are usually evaluated and considered along with those of contempory writings as a concern of a particular taxon. It is for this reason all taxonomists /readers must know the important sources of taxonomic literature as mentioned above.

(I) Floras and Manuals

A *flora* is a systematic work devoted to the plant species of a particular region, whether small or large, and arranged according to an accepted or available system giving ideally keys, descriptions, scientific name, author citation, synonymy, geographical distribution within the area in question, often illustrations, local names, etc. It is apparent that omission of any of these features not only detracts from completeness but lessens the scienetific value of the contribution.

There are certain minor differences between a *flora* and a *manual*. The latter do not necessarily indicate place of deposition of plant specimens (and so their numbers) included. Mainly they lay emphasis on suiable keys and diagnostic descriptions. However, usually these two words are not used in a strict sense but used synonymously. Davis and Heywood (1963) distinguishd flora based on their purpose : **(i)** Research Floras – which can be used in the field, their back and detailed information probably make them most valuable for further revision and indetification in the herbarium. They stand between the shorter flora and the monograph and provide excellent basis for further taxonomic research **(ii)** Concise or Field Floras – which are much shorter than above one, descriptions are usually brief or even wanting. Types and herbarium specimens are not cited. Synonymy is kept at minimum. Detailed information on distribution and ecology are provided e.g. Clapham Tutin & Warburg's flora of the British Isles (Ed. 2, 1962). Fernald's edition of Gray's Manual of Botany (1950), etc. **(iii)** Excursion Flora – in which space is saved by combining keys and decriptions or by omitting or at least abbreviating certain kinds of information such as distributional, ecological or ethological e.g. Matthew (1991) An Excursion Flora of Central Tamilnadu, India.

The following is generally expected in a Flora. Although varying in amount known about the flora of the area concerned and scope of the work :

(i) **Introduction:** This pays attention to clear delimitation of the area covered showing floristics or phytogeographical division (using map generally). Information with respect to topography, climate, geology, vegetational cover, etc. is to be included.

(ii) **Taxonomy:** It mentions an accepted system used in the flora for convention, besides short descriptions of a family followed by keys to the genera. Generic descriptions may be followed by relevant revisions. Keys to the species are important and usually dichotomous keys, whether indented or bracked, are provided. These are separately discussed elsewhere in the book.

(iii) **Citation Synonymy:** It should be as complete as possible

(iv) Other elements such as description of the species, ecological data, geographical distribution, vernacular names, uses and illustrations, if any, generally constitute the flora of a region as stated earlier.

(II) *Monographs* and Revisions

A taxonomic monograph of a group of plants, whether a family, a tribe or a genus is a comprehensive treatise with analysis and synthesis of existing taxonomic knowledge of that taxon, besides the results of bonafide research. It is a complete worldwide account at a given time.

A taxonomic revision is a less ambitious and differs from a monograph primarily in degree of scope and completeness. A revision is generally confined to section of a genus or the elements restricted to a smaller geographical area or a continent. It is usually based on herbarium studies and many revisions make no attempt to review all earlier works on the plant group. Revision is thus provisional and introductory to more thorough treatments.

Ideally, a monograph takes into consideration the information from fields like morphology, cytology, genetics, anatomy, palynology, phytochemistry, ecology, geography and paleobotany of the group under study, a part from taxonomy. It should be thus based on biosystematic studies. A monograph is generally worldwide in its application and scope. It assesses all taxonomic treatments then available for the taxon. It involves an extensive field knowledge or their behavior under cultivation, all herbarium collections of the world, available type specimens and phylogenetic relationships of the taxon. It accounts for full synonyms, complete descriptions, precise designation of types and their deposition, citations of specimens examined, distribution and habitats, etc.

Taxonomic work, whether a monograph or a revision is largely valued on the basis of amount of original critical observation they contain. There should be clarity about the specimens whether they have been confirmed or examined personally or not by the worker. They should take the names of all the taxa in their group into account. They should also comment on doubtful synonyms and lacunae, if any. These works should be constructive and not mere compilations.

(III) *Flora vs. Monographs*

The Flora is superficially a more attractive proposition from utilitarian, economical and political points of view. Greater efforts are extended in the production of floras than of monographs (*cf.* Stace, 1989). Floras have, however, disadvantages compared with Monographs. Monographs include the total variation of the taxa covered, whereas floras simply encounter variation often completely arbitrarily defined (politically defined) region. It is for this reason floras are usually taxonomically less sound than monographs. The floras are also less thorough as they concentrate primarily on identification of the species present in the region. Monographs do aid in indentification but they are primarily a systematic synthesis of the available data on the taxon under study. Infrapseific variation and biosystematic information are included in monographs which are totally wanting in floras. Floras are aimed at providing a standard point of reference rather than presenting the results of new taxonomic research. Thus they are conservative in nature. On the contray, monographs are far more innovative. They represent real taxonomic advances. Monographs contain more widely applicable and useful data than the floras.

(IV) General Taxonomic Indexes

Indexes help locate quickly the source of original publication of a name, to learn if a particular name has been applied to a plant tribe, a subfamily or an order. There are the volumes compiled and still being complied throughout the world. For example, **(i)** Index Kewensis Plantarum Phanerogamarum 2 vols, 15 suppl Oxford, 1893-1970. It is essential for taxonomic study especially of flowering plants. Moreover, it is the most comprehensive index of the botanical names **(ii)** Gray Herbarium Card Index, Cambridge, Mass. It is a card index accounting for all new names and new combinations applied in any category to the flowering plants and pteridophytes. It accounts of Latin names given to vascular plants since 1885.

(V) Bibliographies, Catalogues and Review Serials

Bibliographies accounts for all books and scientific literature published during a given period on the subjects of the particular bibliography. Many of these include as list of abbreviations used for the series whose papers are cited. Such a list is helpful in locating a serial in a library card catalogue as it includes the full title of the periodical. For example, **(i)** Bay, J.C. Bibliographies of Botany. In, Progressus rei botanca, 3:331-45.6 1910. **(ii)** Bibliogr. Agr. Bibliography of Agriculture (V.S. Dept. of Agr. Library). U.S. Dept. of Agr. Misc. Publi. 337. 1942. **(iii)** Jackson.B.D. Guide To The Literature Of Botany, London 1881.

Catalogues account for the books of special libraries rich in botanical titles. They are more important in taxonomic studies. Data with reference to the full name of a particular author, the unabridged and exact title of a work and time of publication are usually available from such catalogues. For example, **(i)** Catalogue Of The Library Of The British Museum (Natural History) 5 Vols. London, 1903-1915. Supplement 3 vols. 1922 – 1940. **(ii)** Catalugue Of The Library Of The Royal Botanic Garderns, Kew (up to 1898), etc.

Review serials are periodicals issued at regular intervals. They provide **(i)** a bibliography of current literature of a particular subject, **(ii)** an abstract of papers or books in special fields, **(iii)** reviews of titles of current literature, **(iv)** any combination of these functions, for examples **(i)** Bullentin of the Torrey Botanical Club, **(ii)** Lingan Science Journal, **(iii)** Biological Abstracts, **(iv)** Botanical Abstracts, etc.

(VI) Perodicals

A periodical is generally published at regular intervals (half yearly, quarterly, bimonthly, etc.). Each is called a number or fascicle. These numbers / fascicles of calender year collectively comprise a volume of a particular periodical. Scientific periodicals are the results of taxonomic studies carried but by various herbaria, institutions or societies or associations. These are authentic and learned treatises on concerned plant taxa. They are useful in taxonomic resserch. They are generally sponsored by a scientific organsation, society, instituition, university or museum, etc. For example, **(i)** Adansonia, France **(ii)** American Journal of Botany, USA **(iii)** Annals of Missouri Botanical Garden, USA **(iv)** Botanical Review, USA **(v)** Bulletin of Torrey Botanical Club, USA **(vi)** Canadian Journal of Botany, Cananda **(vii)** Journal of Indian Botanical Society, India **(viii)** Kew Bulletin, England **(ix)** Taxon, the Netherlands **(x)** Journal of South African Botany, S. Africa, etc.

(VII) Glossaries and Dictionaries

An alphabetical enumeration of terms together with an explanation of their meanings is called glossary, whereas a dictonary is one of words pertinent to its subject matter. Several comprehensive all - inclusive glossaries are published as separate works of definite character. For example, **(i)** Gilbert – Carter, H. (1950) Glossary Of The British Flora, Cambridge, England **(ii)** Gray A (1887) Gray's Lessons In Botany, New York, USA.

Reliable source for the explanation of technical botanical terms is an unabridged dictionary. Dictionaries are often encyclopedic in scope. Most botanical dictionaries are of plant names, for biographical data of person for whom plants have been named

For example, **(i)** Baillon, H. (1876-1892) Dictionnaire botanique, Paris, **(ii)** Gerth Van Wijk (1911-1916) A Dictionary Of Plant Names, The Hague.

(VIII) Cultivated and Economic plants

Taxonomic literature related to identification of cultivated and economic plants is published usually as separate books dealing exclusively to a special subject e.g. manuals, encyclopedias, monographs of horticultural subjects, etc. These are also the sources of competent taxonomic treatements of cultivated plant, for example, **(i)** Bailey, L.H. (1914-1917) The Standard Encyclopedia of Horticulature, New York, USA. **(ii)** Bailey, L.H. (1949) Manual of Cultivated Plants, **(iii)** Grey, C.H. (1937-1938) Hardy Bulbs, London, UK, etc.

11

Keys

The indentification of an unknown plant is its determination as being indentified with or similar to another and already known plant. Plant idenfication is accomplished by means of one or more of different methods or combination of methods. The different methods of indentification include (i) Expert determination – the experts use treatments of the plant group in question (e.g. monographs, revisions, floras, manuals, etc.). Experts are associated with different herbaria, botanic gardens, universities, etc. (ii) Recongnition – it approaches expert determination in reliability and is based on extensive past experience of the indentifier with the plant group in question (iii) Comparison – it takes in consideration the comparsion of an unknown with named specimens, illusrrations, photogrpahs, etc. (iv) Keys - it is a commonly employed device useful in identifying an unknown. The prepartation and use of keys are attempted in the following.

(I) *Key*

A key is an artificial analytical device or arrangement whereby a choice is provided between two contradictory propositions resulting in the acceptance of one and the rejection of the other (*cf.* Lawrence 1951).

Keys are useful in identification of an unknown. They do not offer descriptions of the plants concerned but state only the essential diagnostic characters by means of which the taxa can be indentified. The most conspicuous and clear – cut characters are employed in prearing keys. Such characters, many a times, are taxonomically the most important. Therefore, the sequence of taxa in a key is often quite artificial hence they are termed frequently as '*artificial keys*'. The aim of a key is ease and certainty of identification. For convenice, naturalness is often sacrificed (*cf.* Davis and Heywood, 1963). Keys are prepared in such a way that any one possessing working knowledge of morphological terms can be able to use them profitably (Naik 1984). However, the construction of keys is an intellectual challenge for the taxonomist /author of the group.

(II) Synopsis

A synopsis (or *synoptic key*) is defined by Hitchcock (1925) as 'segregation of groups of the same category into successively smaller divisions according to characters common to its members.'

Synopsis has the format of a key. It does not help identify individual species. It aids in distinguishing plant groups particularly above the species levels. It reflects natural relationships. Following forms an example of synoptic keys to orders (as adopted from Lawrence, 1951):

1. Perianth wanting or reduced to bristles, bracts, scales(petaloid in few advanced genera of Helobiae).
2. Fls. variously arranged but not in axils of dry chaffy bracts.
3. Perianth represented by bristles or chaffy scales; leaves stiffly long and sword-shaped ....13. *Pandanales*
3. Perianth usually of fleshy or herbaceous bracts; lvs.various or wanting fr. usually drupaceous or baccate.
4. Seeds without endosperm; pollen grains in triads; fr. 1- seeded ....14. *Helobiae*
4. Seeds with endosperm; pollen grains in diads or tetrads; fr. more than 1- seeded ....18. *Spathiflorae*
2. Fls in axils of dry, chaffy bracts ....15. *Glumiflorae*
1. Perianth generally present and in two series, the inner or both usually more or less petaloid (except in the order Synanthae).
5. Plants typically woody; leaves generally decompound or palmately lobed, cleft or divided.
6. Carpels 3 in each ovary, each cell 1-ovuled ....16. *Principes*
6. Carpels 2-4, each cell many – ovuled ....17. *Synanthae*
5. Plants herbaceous, or if woody the leaves never compound nor flabellate.
7. Seeds not minute, endosperm present.
8. Endoseprm mealy ....19. *Farinosae*
8. Endosperm fleshy, horny, or cartilaginous.
9. Functional stamens 3-6, seeds rarely arillate ....20. *Liliiflorae*
9. Functional stamens 1; seeds mostly arillate ....21. *Scitamineae*
7. Seeds very minute, lacking endosperm ....22. *Microspermae*

(III) Kinds of Keys

Artificial keys are of two main types *viz.,* (I) Single – access or sequential keys (II) multi-access keys.

1. Single-Access or Sequential Keys: These are written in the form of 'dichotomous keys.' A dichotomous key consists of a series of couplets (mutually exclusive pairs of statement), each statement (lead) of a pair leading on to a further couplet. A decision to follow one lead or the other has to be taken. In such a way the number of possible taxa with which the unknown specimen can be identified is successively reduced until there is only one possibility.

The *dichotomous key* is currently conventional and most acceptable type. It can be written in two different formats *viz.,* (i) indented (yoked), (ii) bracketed. This are illustrated in the following;

(i) Indented Type:

(1) *Cyrtococcum* Stapf

1. Pedicels short, rarely as long as spikelets:
2. Leaves 2.5-6.0 cm long; spikelets hispidulous; branches of panicle glabrous .... *C. trigonum*
2. Leaves more than 6 cm long; spikelets glabrous;branches of panicle often pilose...................... *C. oxyphyllum*
1. Pedicels longer than spikelets :
3. Spikelets 0.15-0.2 cm long, glumes; not cfusidate...*C. deceanense*
3. Spikelets 0.15 cm long or less, if longer then glumes cuspidate :
4. Panicles lax.. *C. accrescens*
4. Panicles contracted... *C. patens*

(2) *Pancratium* L.

1. Bulbs without a neck.. *P. triflorum*
1. Bulbs with a neck :
2. Perianth –tube short, 20.-2.5 cm long........... *P. sanctae-mariae*
2. Perianth – tube much longer :
3. Perianth-tube 12.5-15.0 cm long....................... *P. longiflorum*
3. Perianth – tube 7.5-12.5 cm long :
4. Scapes stout, c10 cm long .... *P. donaldii*
4. Scapes slender, 15-30 cm long............................... *P. parvum*

(ii) Bracketed Type :

***Terminalia* L. nom.cons.**

1. Fruits without wings..2
1. Fruits with 5 equal wings 4
2. Leaves alternate and clustered at end of twigs; spikes simple 3
2. Leaves alternate or subopposite, not clustered; spikes panicled ...... *T. chebula*
3. Drupes ellipsoid, compressed, with fibrous mesocarp .... *T. catappa*
3. Drupes globose or subglobose, not compressed; .... *T. bellirica* mesocarp not fibrous
4. Bark pale grey, smooth; leaves glabrous... *T.arjuna*
4. Bark dark, rough; leaves usually hairy or tomentose beneath...............*T. alata*

There are essential differences in these two formats of dichotomous keys. The two lead of each couplet appear together in the bracket key. All the possibilities arising from the first lead are dealt with before the second lead mentioned in case of the indented key. This type has advangtage of giving a visual presentation of the group so that the user can readily obtain a picture of the taxon, but when the key is long these is a much wastage of page-space and the user often has to turn pages to find the two halves of a couplet. The theory of constructing and using dichotomous keys is commonly available in many texts, it is therefore avoided here.

2. Multi-access Keys: These are produced on pages in a book, but on cards of suitable size with names of all the taxa (all families, genera or species for which the key is proposed) printed on each one of them. There are two principal types of punched card key (i) Edge-punched key (ii) Body – punched key (Polyclave).

(i) Edge – punched Key: There is one card for each combination of attributes each attribute is then represented by one of the holes punched around the perimeter of the card. If a taxon possesses a particular attribute that position is clipped out to form an open notch instead of a circular hole. All the cards are stacked up with their corresponding holes aligned and an attribute possesd by the specimen to be identified is chosen. A thin rod or needle is then passed through the proper position and the rod/needle is horizontally lifted and gently shaken. All those cards (i.e. taxa) possessing that attribute will fall away from the stack leaving the cards without that attribute on the rod/needle. The latter are put aside, and the others are collected. The process is again followed using further characters, until only one card falls out of the stack.

(ii) Body –punched Key (Polyclave): The holes are punched in rows in the main body of the card. Each card represents an attribute. Each taxon occupies standard position on the card. If the taxon possesses that attribute its position is punched out. To identify a specimen some of its attributes are listed and the appropriate cards are selected. These are aligned and taxa which possess all the attributes being tested will show a hole right through the stack when held up to the light. Cards (attributes) are added to the stack until only one whole (taxon) remain.

These types of keys are fascinating to handle. They enthuse and interest the learners in the identification exercise. However, they are costlier affairs. Moreover, the printed matter on the cards is blurred due to constant handling. They have thus short life. It is, therefore, dichotomous keys are preferred in many floras, manuals, monographs or revisions.

12

Plant Nomenclature

(I) Introduction

Any object that is acknowledged to mankind must possess a name as it may not be convenient or possible to describe it in order to communicate ideas in its reference. As all known objects are given names, plants are also given names or should be given names. Actually, to name an object is humen instinct. The art of naming objects, in fact, a science of application of names is nomenclature. When it is in reference to plants, it is called plant or botanical nomeclature. This science of naming plants is governed by international rules approved by scientific community called ICN (International Code of Botanical Nomenclature). After Melbourne Code (2011), this Code is now being called ICN (International Code of Nomenclature). The science of naming plants based on international rules proposed by botanists to ensure a stable uniform and universal system is called Botanical Nomenclature.

Plant names serve mainly two purposes: **(i)** as an aid to communication **(ii)** to indicate relationship. The use of a name avoids the use of a descriptive phrase every time we wish to refer to an object or group of objects. A name therefore stands, as it were, in place of the object itself and acts as a means of reference it. These (names) conventional symbols or ciphers in biology serve as a means of reference and as a key to the enormous amount of literature published on them.

A name must stand for the same object or group of objects for all who use that name. It must be unambiguous. We must not only avoid the giving of the same name to different things, we must also avoid giving different names to one and the same thing. Otherwise, we will lead to confusion. A name is a convention that facilitates talking and writing about an object. A name is thus an aid to the flow of information.

(II) *Common Names*

Plants are referred by common, vernacular or local names in a particular region or language or dialect. They differ from region to region, language to language and even from dialect to dialect. In a multilingual country, like India, the number of local names is exceedingly large for each plant species. They are, thought by the scientific community, neither universal nor methodical. They are inadequate and misleading. For better communication, the names must have the same meaning for all who use them. They are not favoured by the scientific community for obvious reasons, e.g. **(i)** they are confined in use and are useful for an ethnic or a linguistic group, **(ii)** they are available for a limited number of plant species, **(iii)** same taxonomic group is refemed by different names in different linguistic areas and even different parts of the same area and **(iv)** the same name is also applicable for different taxa.

(III) Scientific Names

The above objections generally raised by the scientific community are set aside for all botanists have agreed to have uniform scientific names for each taxon. Certain rules are passed by the international scientific community for nomenclatural purpose. These are available in the form of ICN. The scientific names are methodical and help communicate uniformly throughtout the world. They are thus international in character and scope. The scientific names, moreover, are based on the Latin language. It is known universally by scholars and students. It is specific and exact in meaning. It is particularly pertinent to the needs of descriptive phases of the natural sciences. It employs the Roman alphabets and thus obviates the confusion that would result if written in other alphabets. It is relatively unbiased being language of no country presently.

(IV) *Binomial Nomenclature*

Before the middle of the 18th century, the plant names were commonly polynomials. This system of naming gradually evolved into binomials. We are aware that binomial nomenclature was initially devised by Caspar. Bauhin (1560 - 1624). However, it was not used by him consistently. The trail was later on picked up by Carolus Linnaeus, a Swedich botanist. He used binomials consistently for all plants described in his '*Species Plantarum*' (1753).

This system postulates that the name of every species of plant consists only of two words, the first word designates the genus to which the plant belongs and the second word a particular species of that genus. This second word of a plant's name is a specific epithet; the two words in combination comprise the species name and form a binomial. For example, in *Abrus precatorius 'Abrus'* is a generic name and '*precatorius*' is the specific epithet. The binomial is a binary epithet. In the binomial the first name (generic name) is a substantive (noun), while the second (specific epithet) is an adjective or noun.

A specific manner for writing botanical names is insisted by the Rule. All generic names must begin with a capital letter, while the specific epithet should invariably begin with a small letter and it should agree with the gender (masculine, feminine or neuter) of the generic name. It is to be noted that where the context is understood the generic name can be abbreviated e.g. *Abrus precatorious* is written as *A. precatorius.*

Bionomial system of nomenclature provides a summary of affinities for us which acts as an aid to memory and enables us to make inferences about plants unknown to us. It also facilitates talking about groups of species that have certain features in common.

(V) Early Attempts For Organised Nomenclature

As stated earlier, plant names were polynomials before the middle of 18th century. They were composed of many descriptive terms or adjectives in a series. More or less such names were constituting terse description of a plant e.g. (i) *Plantago foliis ovatis glabris nudo scapo tereti spica flosculis imbricatis.* These polynomials served as labels for the taxa and also diagnosis of the species. These two functions were separated by Linnaeus (1753) in his *'Species Plantarum'*. He consistently used the binomial system of nomenclature. For example, he used the name *Plantago major* and its diagnoses were given separately. The binomials were used by him as additional trivial name for the sake of convenience. This practice was found so convenient that it continued further. Thus polynomial system was superseded by the binomial one. The binomial system postulated that the name of every species of plant consists only of two words, the first word being genus to which the plant belongs, and the second a particular species of the genus.

Explorations increased numbers of new plants with time which caused concern over methods of naming these novelties. Some elementary rules were proposed to serve as guides. The beginning of rules of nomenclature was made by Linnaeus in his *'Critica Botanica'* (1737). This work comprised an elaboration of the Aphorisms of his *'Fundamentica Botanica'* (1736). The Aphorisms constituted his principle of nomenclature which were closely parallel to contemporary rules of nomenclature. He made real beginning of a sound nomenclature for plants in his *'Phillosophica Botanica'* (1751) in which his views were amplified. After the time of Linnaeus, most authors avoided assigning different names to identical species or identical names to different species.

After Linnaeus, the first significant work *'Theorie elementaire de botanique'* (1813) by Augustin de Candolle appeared. It explained organography and nomenclature. He gave explicit instructions on nomenclatural procedures. He also pointed the fallacies of proposals and practices of other botanists of his time. His set of rules were the forerunner of 'Lois de la nomenclature botanique' Proposed by his son Alphonse de Candolle (1867). He made valuable efforts to bring together crystallized knowledge of botanists world over. He took efforts to organize an Internationa Botanical Congress at Paris the same year. He explained his ideas about nomenclature and laid emphasis on author citation 'Kew Rule' was also proposal by scientists working at Kew. This also received acceptance. Subsequently, international cogresses are held and different codes are on record. Various rules have been adopted at different cogresses held periodically and are being held. These efforts have resulted in 'The International Code of Botanical Nomenclature' (ICBN, present ICN). This is dilated later.

(VI) International Code of Nomenclature (ICN)

A group of botanists in the United States organized a meeting at Rochester in 1892 and formulated a set of rules referred to as 'Rochester Code'. This code is based principally on earlier Paris Code (1867). The fundamental differences between these two codes were in the establishment of **(i)** the concept of types as a scientific and fundamental basis by which to determine the precise application of names (each species to be based on a specified herbarium specimen to be known as the nomenclatural type of the binomial, and larger units of classification to be based on a species genus or family), **(ii)** the strict application of the rules of priority which admitted no legislation comparable to a 'Kew Rule', **(iii)** the acceptance of all binomials resulting from application of the rule of priority, even though the specific name repeated the generic name (tautonymy), **(iv)** the intepretation of rule of priority to apply not only to the date of publication but also to the precedence of a name within a publication, ruling that if two names on a given page were later treated as belonging to the same plant, the first name on the page had precedence over the second name. Two opposing schools of thought were established. One was headed by Dr. Britton and his students, the other by students of Asa Gray and those concurring with the more conservative philosophies prevalent among the botanists influenced by Kew. The fifth international Botanical congress was held in 1930 at Cambridge. Efforts were made to settle down the differences between the Vienna Code and the American Code (1907). A general agreement regarding the internationally acceptable rules of plant nomenclature was reached at this meeting of IBC. A code of nomenclature came into being for the first time. This code is called the International Code of Botanical Nomenclature (ICN). On a regular basis, the amendements or modifications as proposed by IBC at subsequent meeting are being incorporated.

Plants are named according to the principles laid down in ICN. The ICN cover all plants including fungi, slime – moulds and blue-green algae. In case of Euglenophyta, there is possibiliy of some overlap with the other codes since they are variously referred. The current ICN contains rules loosely laid down at the IBC held at Cambridge, UK in 1930. However, these rules in turn are derived from earlier ones starting from Paris Congress (1867). In recent times, the ICN is ratified and amended during the meetings of the Nomenclature Section of each IBC. The current version of the ICN (1988) proceeds from the 14th IBC held at Berlin in 1987. It provides an internationally acceptable code of practice on nomenclature so as to produce strong elements of stability. It is set out in the form of 76 main Articles (Rules), with numerous Recommendations, Notes and Examples subordinate to them. The code is divided into three divisions. Divison I provides a set of six principles. Division II deals with articles and recommendation and Division III includes provisions for modifications of the code alongwith three other appendices.

(VII) Codes of Nomenclature

Linnaeus proposed some rules for naming plants in his 'Fundamentica Botanica 'and 'Critica Botanica'. It is also a knowledge that first proper set of rules for nomenclature of plants was, however, drafted by Alphonse de Candolle and passed by the International Botanical Congress at Paris in 1867. Since then at regular intervals meetings of IBC were held. Additions of rules and improvements were made at each time. Different codes to date, have appeared on record. Their gradual developments over the past should be acknowledged to every one those interested especially in the history of nomenclature. The developments as contained in the various codes are highlighted briefly underneath.

1. **Paris Code (1867):** Alphonse de Candolle made significant efforts to organize an international congress at Paris in August 1867. In the same year, he published

his 'Los de la nomenclature botanique' (Laws of Botanical Nomenclature). Many European and American botanists participated in the said event. The laws adopted by the Assembly were recommended as the best Guide for Nomenclauture in the Vegetable Kingdom These laws were referred as the 'Paris Code' after the name of the French capital where they were adopted. These were also called 'de Candolle Rules' as they were originially prepared and proposed by Alphonse de Candolle. The starting point for all nomenclature fixed with Linnaeus, the rule of priority as basic with no provisons for exceptions, valid publication, author citation and terms as to be applied to categories of plants were some of the important subjects of resolutions passed. However, later American and British botanists followed the unwritten laws, called 'Kew Rule'. There was a dispute especially when a species was to be transferred to another genus. According to the Kew Rules, if no epithet previously had been given to the species in the new genus, the author was free to give it a new epithet in the new combination. This was contrary to the directives of the present Rules. It is to be noted that Kew Rule was followed while compiling the first volume of 'Index Kewensis'

2. **Rochester Code (1892):** Dr. N.L. Britton led a group of American botanists. They developed a set of rules to govern nomenclature. These rules were discused at the meetings of Botanical Club hold in Rochester, New York. They based these rules on modifications of the Paris Code. There were some basic differences between these two codes in respect of : **(i)** the concept of types as a scientific and fundamental basis by which to determine the precise application of names, **(ii)** the strict application of the rules of priority which admitted no legislaition comparable to a 'Kew Rule', **(iii)** the acceptance of all binomials resulting from application of the rule of priority even though it is a tautonym, **(iv)** the interpretation of the rule of priority to apply not only to the date of publication but also to the precedence of a name within a publication, ruling that if two names on a given page were later treated as belonging to the same plant, the first name on the page had precedence over the second name.
3. **Vienna Code (1905):** The Third International Botanical Congress met in Vienna in June 1905. Prior to it, in 1900 Paris Convention serious thought was given to the nomenclatural problems. The advocates of the Rochester Code again tried to incorporate their basic tenets in the revised rules. The revised rules were based on those of Paris Code (1867) but they were in modified and amplified form at the time of Vienna Congress. Some new changes were incorporated such as : **(i)** Species Plantarun by Linnaeus established as the starting point for naming vascular plants, **(ii)** Nomina generica conservanda by which generic names having a wide use would be conserved over earlier but less well-known names, **(iii)** banning tautonyms, **(iv)** requirement that names of new taxa be accompanied by a Latin diagnosis. Vienna Code, however, did not gain international acceptance because the most adherents

of the Rochester Code refused to abandon their cause and would not subscribe to the Vienna Rules.

4. **American Code (1507):** During the Vienna Cogress (1505) followers of Rochester Code emphasized on its merit still because of some drawbacks especially tautonyms. However, the idea of following principles of Rochester Code was abandoned. The Congress meanwhile suggested that wide usage of any name should be considered valid rather than the name of the same individual published prior to it. This was termed as the principle of 'nomina generica conservanda'. This resolution dissatisfied more the followers of Rochester Code and they refused these international rules. This matter of controversy between the codes was unsolved until 1930 when the Cambridge Congress rules agreed unanimously to follow Vienna Congress and accepted the rejection of American Code.
5. **Brussels Code (1912):** The fourth IBC was held at Brussels in 1916. It resolved the establishement of different starting points for prority of names of non-vascular plants, the recongnition of value of type concept and clarification of phraseolgy of the Vienna Rules.
6. **Cambridge Code (1935):** During the fifth IBC held at Cambridge in 1930, the basic difference between the Vienna Code and American Code were reconciled. They finally approved provisions such as: **(i)** the type concept should be pursued, **(ii)** there should be a list of 'nomina generica conservanda', **(iii)** tautonyms are not admissible and, **(iv)** Latin diagnosis is required after Jan 1, 1932.
7. **Amsterdam Code (1947):** The sixth IBC was held at Amsterdam in 1935. A few significant changes were made. First, it was resolved that from Janaury 1, 1935, names of new groups of recent plants (except Bacteria) are considered validly published only when they are accompanied by Latin diagnosis. Secondly, the proposal of the list of 'nomina specifica conservanda' was rejected.
8. **Stockholm Code (1932):** The seventh IBC met at Stockholm in 1950. A certain number of definitions on types was introduced. For the first time, the term 'taxon' (pl.taxa) was introduced to designate any taxonomic entity or group.
9. **Paris Code (1956):** At the eighth IBC meeting in Paris in July 1954, 'types' (Appendix IV) were greatly emphasized. However, it was resolved that the code should be published in English, French and German Languages. The Preamble and principles were separate from the Rules and Reconmednations of the Code. Appendix II (nomina generica conservanda *et* rejicienda) was amended and supplemented.
10. **Montreal Code (1961):** During the meating of the ninth IBC held at Montreal in August 1959, a committee was appointed to study the question of family names (nomina familiarum conservanda for Angiosperm as Appendix II was introduced). It was also specified that the naming of fossil plants should follow the same lines as that of recent ones.

11. **Edinburgh Code (1966):** In the earlier meeting of IBC held at Montreal, a committee was constituted. Its report was submitted at the meeting at the tenth IBC held at Edinburgh in August 1964. The subject matters approved were : **(i)** the starting point for family names was decided A.L. de Jussueu's Genera Plantarum (1789). Inclusion of family names in the list formed a ready reference providing correct names of families together with a type genus for each family, **(ii)** in the list of 'nomina familiarum conservanda 'a few names are found with a spelling different from the long-established names (e.g. Capparaceae for Capparidaceae, Haloragaceae for Haloragidaceae, Cannabaceae for Cannabinaceae etc.), **(iii)** the rule of Latin for new diagnoses was finally settled and no amendments were proposed.

    A committee was constituted to prepare a glossary of the main technical terms occurring in the code which resulted in 'An Annotated Glossary of Botanical Nomenclature'.

12. **Seattle Code (1972):** The eleventh IBC was held at Seatle in August 1969. Some refinements and precision were concerned with the issue such as : (a) the tautonymus designations of taxa between genus and species and below species, (b) the perennial question of superfluous names and a reorgansiation of the rules for hybrids, (c) introduction of the word autonym in the Code.

13. **Leningrad Code (1978):** During the meeting of the twelfth IBC held at Leningard in July 1975 some recommendation were made. These differed on some minor points from the Seattle Code. These are: **(i)** the concept of organ genera is eliminated for fossil plants, **(ii)** the Code does not apply to names of organisms treated as bacterial and does apply to all other organisms treated as plants, **(iii)** the principle of automatic typification is extended to those names of taxa above the family rank that are ultimately based on generic names and the application of the priority principle is recommended while selecting among names thus typified, **(iv)** a name or combination published before 1953 without indicating a rank is considered validly published but imperative in questions of priority except for homonymy and certain names to be accepted at the varietal rank, **(v)** Article 69 of the previous Code is modified on the basis of type method and Articles 70-71 dealing with discordant elements and monstrosities were deleted but the Article numbers are retained to facilitate the use of the Code, **(vi)** the section on orthography is thoroughly rewritten, **(vii)** individual paragraphs on all Articles and Recommendation are numbered in a decimal –like system, some being rearranged.

14. **Sydney Code (1983):** The thirteenth IBC was held at Sydney in August 1981. No significant decisions were made.

15. **Berlin Code (1988):** The fourteenth IBC was held at Berlin in 1986. For the first time, 'nomina specifica conservada' was introduced in the Congress. Two Articles 66 and 67 were removed from the Code. Two names *viz.,Triticum*

aestivum Linn. and *Lycopersicon esculentum* P. Miller were decided to conserve against the rule of priority.

16. **Tokyo Code (1994):** The fifteenth IBC met at Tokyo in 1993. The Code was translated into French, German, Italian, Chinese, Japanese, Slovak and Russian languages. Apart from this, extensive renumbering took place and hence its preface included a tabulation comparing the placement of its provisions with those of the Berlin Code.
17. **St. Louis Code (1999):** The sixteenth IBC was held at St. Louis Missouri in July – August 1999. This Code is made available in several languages. The Code is divided into Rules, Articles and Recommandeations. Rules are set up to put the nomenclature of the past into order and to provide space for the future. The recommendations deal with subsidiary points. Their objective is to bring about uniformity and clearness. Names contrary to a recommendation are not examples to be followed. The Rules and Recommendations apply to all organisms whether fossil or non-fossil including fungi and excluding bacteria.
18. **Vienna Code (2005):** This code does not differ substantially in overall presentation and arrangement from the St.Louis code. There have been few additions and modifications. Deliberations and decisions of the Nomenclature Section of the XVII International Botanical Congress was held in Vienna, Austria, July 2005. The Vienna Congress was fairly conservative in nomenclatural matters in comparison with some earlier Congresses. Relatively few changes were accepted, but a small number of significant ones and many useful clarifications are noteworthy: (i) Changes with respect to names of fossil plants, pleomorphic fungi and fungi. (ii) Glossary of terms was the first time included in the Vienna Code and appeared as Appendix-VII. (iii) The recommendation by the Committee for Spermatophyta that the name *Acacia* be conserved with a type from Australia. (iv) To avoid ambiguity, the word "type" is replaced by "lectotype, neotype, or epitype". (v) Italicised plant names were debated but ultimately maintained. This code pens further for author citations of scientific names, valid publication, new combinations, rules of determining a rank, etc.
19. **Melbourne Code (2011):** The Nomenclature Section of the XVIII of International Botanical Congress was hold in the University of Melbourne, July 2011. The significant change in the Code are: (i) Change in name 'International Code of Botanical Nomenclature' (ICBN) to 'International Code of Nomenclature' (ICN) for algae fungi and plants. (ii) Approval to electronic publication of all nomenclatural acts from 1st January, 2012. (iii) Agreed to provide a description and/or diagnosis either in Latin or English published on or after 1st January, 2012. (iv) Approval to one fungus, one name. (v) Approval to one fossil, one name. (vi) Agreed to conserve the name *Accia*. (vii) Approval new rule for registration of fungal names published on or after 1 January 2013.

(viii) Approval to electronic form of Appendices of the Code and few other rules.

20. Shenzen Code (2017): The Nomenclature Section of the XIX International Botanical Congress was held at University Town Shenzen, July, 2017. Total 397 proposals were submitted for consideration to amend. This is the largest number presented since the Stockholm Congress of 1950. Some of the major decisions accepted are: (i) Agreed on a fragment for the registration of new names and nomenclatural acts of plants and algae. (ii) An expanded Division III on governance of the Code i.e. how the rules for naming algae, fungi and plants can be changed. (iii) The Nomenclature Section voted to put all a dedicated chapter of the Code, and that chapter will in the future be changed and improved by the Nomenclature Session of an International Mycological Congress, which takes place every four years. (iv) Inclusion of the Appendix-I (Articles dealing with names of hybrid organisms) in the body of the Code.

(VIII) Preamble of ICN

Prior to the Paris Code (1956), the ICN included the Preamble, Principles, Rules and Recommendations. However, during the deliberations of the eighth IBC held at Paris in July 1954, the Preamble and Principles were separated from the Rules and Recommadations. The ICN has now the Preamble which gives broad guidelines for the use of the Code. It also emphasises reasons and purpose. The Preamble states as:

1. Botany requires a precise and simple system of nomenclature used by Botanists in all countries dealing with on the one hand, with the terms which denote the ranks of taxonomic groups or units, and on the other hand, with the scientific names which are applied to the individual taxonomic groups of plants. The purpose of giving a name to a taxonomic groups is not to indicate its character or history, but to apply a means of referring it and to indicate its taxonomic rank. The code aims at the provision of a stable method of naming taxonomic groups, avoiding and rejecting the use of names which may cause error or ambiguity or throw science into confusion. It avoids the useless creation of names.

1. The Principle form the basis of the system of Botanical Nomenclature.
2. The detailed provisions are divided into *Rules and Recommendations*. Examples are added to the rules and the recommendations to illustrate them.
3. The object of the Rules is to put the nomenclature of the part into order and to provide for that of the future, names contrary to a rule can not be maintained.
4. The Recommendations deal with subsidiary points, their object being to bring about greater uniformity and clearness, especially in future nomenclature, names contrary to a recommendation can not, on that account, be rejected, but they are not examples to be followed.
5. The provisions regulate the modification of this code from its last decisions.

6. The *Rules and Recommendations* apply to all organisms treated as plants (except Bacteria) whether fossil or non-fossil. Nomenclature of Bacteria is governed by the ICN. Special provisons are needed for certain groups of plants. The International Code of Nomenclature of cultivated plants (1980) was adopted by the International Commission for the Nomenclature of Cultivated Plants. Provisons for the names of hybrids appear in Appendix I.
7. The only proper reasons for changing a name are either a more profound knowledge of the facts resulting from adequate taxonomic study or the necessity of giving up nomenclature that is contrary to the rules.
8. In the absence of a relevant rule or where the consequence of rules are doubtful, established custom is followed,
9. This (i.e. present) edition of the code supersede all previous editions

(IX) Principles of ICN

The code is divided into three divisons. Division I provides a set of six principles which form the very basis of the system of nomenclature. They are pertinently discussed in the following:

(i) **Principle I :** Botanical nomenclature is independent of zoological nomenclature:

Species Plantarum by Carolus Linnaeus (1753) includes names of plants as well as of animals. There is another separate code for zoological nomenclature (ICZN). Certain organisms, originially considered under the plant kingdom, have been transferred to the animal kingdom and *vice - versa*, The ICN, in such cases is equally applicable to the names of taxonomic groups treated as plants whether or not these groups were originally considered under the plant kingdom. Special provisons are made for such taxa. Article 45 accepts the original dates of publication of taxa transferred from animals to plants, e.g. *Amphiprora* once published as a genus of animals in 1843 is transferred to the plant kingdom in 1844. It got priority from 1843 and not 1844, in botanical nomenclature. According to the Article 65 of ICN, if a taxon is transferred from the plant kingdom to the animal kingdom, its name retains its status in botanical nomenclature for the sake of homonymy. This name is not applied for any other taxon. Also, the name of plant can not be rejected simply beause it is the name of an animal.

(ii) **Principle II:** The application of name of a taxonomic group is determined by means of nomenclatural types.

The naming of taxonomic groups is determined by means of nomenclatural types. A nomeclatural type is that element to which the name of a taxon is permanently attached either as a correct name or as a synonym. However, the 'type or typus' is not necessarily the most typical or representative element of a taxon. It is simply that element with which the name is permanently associated. This principle was applicable only to names of taxonomic groups of the rank of family and below. This concept has been now extended to names of taxa of higher ranks also when the names are ultimately based on generic names. The genera that typify the respective alternative names should be taken as types of the family name which are not based on generic names, for example, **(i)** genus *Areca* is the type genus of the family Palmae, alternative name is now therefore Arecaceae, **(ii)***Poa* of Gramineae, alternative is now Poaceae, etc, The 'type' can be defined as an 'element' on which the original validating description of a taxon is based. The Code has thus greatly emphasized on 'nomenclature type' in order to bring about stabilization of names.

(iii) Principle III : The nomenclature of a taxonomic group is based upon priority of publication.

Many taxa have more than a single name. Only one name is correct and others are synonyms. These many names arise because different authors have named plants in ignorance of each other's activities. Sometimes, taxa once thought distinct are not so later. New names coined are also the results of different taxonomic judgement. These situations have been greatly rectified by this principle known as 'principle of priority', sometimes also called 'Rule of Priority' Each taxon can have only one correct name the earliest one that is in accordance with the rules of the Code. For example, Bentham and Hooker recognized '*Ruppia maritima* L. and *Ruppia rostellata* Koch (1824). On examination of the Linnaean type of *R. maritima* has established that Koch's *R. rostellata* is identical with the true *R. maritima* L. (1753). The name *R. rostellata* must therefore be replaced by *R. maritima*, which is earlier name.

(iv) Principle IV : Each taxonomic group with a particular circumscription, position and rank can bear only one correct name, the earliest that is in accordance with the Rules, except in specified cases.

A properly formulated name can have any status in botanical nomenclature only when it satisfies two basic conditions. First the name must be published in a medium appropriate to the requirements of the Code. Secondly, the name should be published in accordance with the provisions of the Code. The name is said to be 'effectively published' if the first condtion is fulfilled. The name is said to be 'validly published' if the second condition is fulfilled. These two condtions differ from each other. Effective publication deals with standards applicatble to the works in which the names are published and not to the names themselves. Valid publication takes into account the standards applicable to names themselves. Articles 29-45 of the ICN lay down conditions for effective and valid publication. These are explained separately later in this chapter.

(v) Principle V : Scientific names of taxonomic groups are treated as Latin regardless of their derivation.

A very fundamental provision of the ICN is that scientific plant names must be Latin ones, or treated as such even though derived from other languages. This provision at once resolved the problem of multiplicity of different languages which makes common names so unsatisfactory as a means of international communication. Latin has been chosen as the language for many reasons. It is no longer spoken as a native tongue by any humanity and hence it avoids any element of national bias. In past, for a long time, it was the common Language of scholars and students in Europe and lot of botanical literature has been written in Latin. Moreover, many European Languages originated from it. More importantly, Latin is specific and exact in meaning. Also, it is pertinent to the needs of descriptive phases of the natural sciences beause of its preciseness and conciseness. It is written in Roman alphabet and therefore obviates the confusion that would result if these names were written in other alphabets.

(vi) Principle VI : The rules of nomenclature are retroactive unless expressly limited.

To introduce element of stability, the ICN provided an worldwide acceptable code of practice on nomenclature. The Code presently includes 76 main Articles (rules), with numerous Recommendations, Notes and Examples subordinate to them. The Articles (rules) are retroactive unless otherwise stated. For example, **(i)** the requirement for description of new taxa to be in Latin dates only from1935 in the case of bryophytes and vascular plants, and 1959 in the case of algae, **(ii)** there is a limitation of 'Principle of Priority' for Spermatophyta and Pteridophyta. 1st May 1953 is taken to be the beginning. It means that one can unearth the earliest name only as far back as 1753. Thus, in certain cases, the rules are expressely limited, otherwise they are retroactive.

(X) Rule of Priority

The process of identifying and naming of plants was independent especially prior to organized nomenclature. As a consequence, the plant species distributed widely in different parts of the world received different treatements. They were identified independently in ignorance about the activities of the other botanists or due to lack of international communication. The same species was named differently, e.g. Indian coral tree (*Erythrina variegata* L.) has nearly 200 botanical names. Usually useful plants have numerous names. All these caused confusion amongst the botanists. The ICN therefore serves to provide a stable method of naming taxonomic groups. Article 11 of the ICN states that each family or taxon of lower rank with a particular circumscription, position and rank can bear only one correct name. In such circumstances, the rule of priority, which is also one of the principle of ICN, is applicable.

For example, the name *Lawsonia inermis* was given to a taxon by Linnaeus in 1753. The same taxon was named by Lamarck as *Lawsonia alba* in 1789. When the Rule of Priority is applied to this circumstance, the name given by Linnaeus gets priority over the name given by Lamarck. Similarly, *Trianthema portulacastrum* Linn, (1751) has priority over *Trianthema monogyna* Linn. (1767). Thus priority of publication is important part of the Rules of nomenclature.

(XI) Limitations of Priority

The Rule of Priority greatly rectified the confusion caused due to many names given to a single taxon. It gave opportunity to many taxonomists to reveal the earliest legitimate names. The recent nomeclatural history indicates that taxonomists showed much enthusiasm and interest in revealing the earliest names. In the course of time, the Rule of Priority has had a rather unexpected result. Instead of stabilising the process of nomenclature, it became a most serious cause of instatbility in plant nomenclature. The rule ultimately had to be modified. The Rule of Priority has now certain limitations. It can not be applied to all names published.

The Code (Article13) provides valid publication of names of different groups of plants as beginning on different dates. For Spermatophyta and Pteridophytes it is May 1, 1773, the date publication of Species Plantarum by Linnaeus. While applying the Rule of Priority names published before this date need not be considered. It means that one can reveal the earliest name only as far back as 1773, and the names used prior to this date have to be disregarded. The priority for mosses except Sphagnaceae starts with Hedwig's Species Muscorun, 1st January 1801. There are also other circumstances for limiting the Rule. For example, more frequently a very well established name for a taxon is found by later reaserch to be incorrect. The change of a generic name necessitate the coining of a great many new combination for the species included under it. To avoid this situation, generic names found to be incorrect according to ICN, but whose substitution would cause much inconvenience, are able to be conserved, and the names which would otherwise be correct rejected. Such names are known as 'nomina consevanda' and 'nomina rejicienda' respectively. For example, the genus *Combretum* Loefl. (1768), with hundred of species described under it is conserved against *Grislea* L. (1753), which has hardly been used since the time of Linnaeus. Thus the name *Butea* Willd. 1802 is conserved against its earlier name *Plaso* Adans. (1763), *Cajanus* DC. (1813) against *Cajan* Adans. (1763), etc.

(XII) Typification

It is the process of designating or indicating a type. The ICN insists that the application of botanical names be determined by means of nomenclatural types. The method of determining the application of names by means of types is known as the type method. A nomenclatural type (*typus*) is that constituent element of a taxon to which the name of the taxon is permanently attached whether as a correct name or as a synonym. It is the type method which decides, for example, the application of names when one taxon is divided up into two or more, or when the circumscription of a taxon is widened so as to include two or more already named taxa. If a taxon is so circumscribed as to include the type of a name, then that name must apply to that taxon. It is later decided that the type of the name is outside the range of variation of the taxon, then that name can no logner be applied to that taxon. It must then be applied to the taxon in the range of variation of which its type is considered to fall.

The type of a name is defined as the element on which the description validating the publication of that name was based. The type of the name of a species is usually simply the herbarium specimen from which the original description validaing the name was drawn up. The type of the name of a genus is the species on which the original description validating the name was based. The type of the name of a family or taxon of higher rank is the genus on which the original description validating the name was based. It is important to remember that only names have types and not the taxa. A type is thus merely a nomenclatural device.

(XIII) Working of Type Method

What has been said earlier about type, type method or typification will become more readily understood when some example of how the type methods works. We will consider the species *Helitorpium indicum* Linn. as we generally cosnsider this taxon noted consisting of two quite different elements. For discriminating these two elements, it is necessary to decide which one of them *H. indicum* proper. The specimen which matches with the type material based on which Linnaeus described the species should be attributed the Linnaean binomial. That is type material in this case goes properly with the name *H. indicum* Linn. Consequently, the other element was described under a new name *H. keralense* Manilal and Sivarajan (1972). Whenever taxa are subdived, the type materials are of particular significance in nomenclature. The part with which original type goes, inherits the original name or epithet.

(XIV) Kinds of Nomenclatural Types

This subject was acted on at The Seventh International Botanical Congress met at Stockholm in 1950. Certain definitions of types were introduced and officially designated:

1. ***Holotype*:** A holotype is the one specimen or other element used by the author of the name, or designated by him, as the nomenclatural type.
2. ***Lectotype*:** A lectotype is a specimen or other element selected from the original material to serve as the nomenclatural type, when the holotype was not designated at the time of publication or when the holotype is missing.
3. ***Neotype*:** A neotype is a specimen selected to serve as the nomenclatural type of a taxon in a situation when all material on which the taxon was based is missing.
4. ***Paratype*:** A paratype is a specimen cited with the original description, other than the holotype.
5. ***Isotype*:** An isotype is a specimen believed to be a duplicate of the holotype.
6. ***Syntype*:** A syntype is one of two or more specimens or elements used by an author when no holotype was designated or in lieu of a holotype or when one of two or more specimens were designated simultaneously as the type.
7. ***Typotype*:** A typotype is a specimen collected at the same station as was the type i.e, from the type locality.

8. ***Epitype*:** An epitype is a specimen selected to serve as an interpretive type when the all other types (holotype, lectotype, neotype, etc) could not effectively be indentified to name the taxon, it is called 'epitype'

(XV) Effective and Valid Publication

The ICN requires that names must be published in order to have any standing in the scientific naming of plants. Publication *sensu* ICN needs the fulfillment of two conditions, effective publication (Article 29-31) and valid publication (Article 32-45).

The name must be published in a medium appropriate to the requirements of the Code. If this requirement is fulfilled, it is said that the name is 'effectively published'. Publication of name is effected only by distribution of printed matter to the general public or at least to botanical institutions with libraries accessible to botanists generally. Printed matter means proper books or recognized scientific periodicals. Manuscripts, duplicated sheets, newspapers or magazines, trade catalogues, microfilm and other such semi-permenant or not easily accessible media are not allowed as means of effective publication. This requirement ensures that if we want to check back to the original publication of a name it will be relatively easy to find the publication in which it appeared.

The name should be published in accordance with the provisions of the Code. It this requirement is fulfilled, it is said that the name is 'validly published'. Valid publication deals with the standards applicable to names themselves. The ICN lays down certain conditions as below:

(i) Publication of a name must be accomparied by a description of the taxon or by a reference previously and effectively published description.

(ii) A new name published of 1-1-1953 must have clear indication of the rank of the taxon concerned.

(iii) Names published on or after 1-1-1935 must be accomparied by a Latin diagnosis or description or a reference to an earlier one.

(iv) Publication on or after 1-1-1958 of a name of a new taxon of the rank of a family or below must indicate nomenclatural type.

(v) In case of new combinations the full reference of the *basionym* must be given.

(XVI) Type Concept and Population Concept

Circumscription of a taxon on the basis of its nomenclatural type is called the type cocept. This nomenclatural type is a holotype. It may not always represent the average characters of species, subsepecies or variety. It represents either of the extremes in the range of variation in respect of quantitative characters and hence can not aid in satisfactory identification of those plants which represent the other extreme of this taxon. Moreover, after a lapse of several decades and centuries the types may not faithful to the description of the taxon.

The true concept of a species be formed on the basis of a population study of plant, a suggestion which is acceptable to many. Such study could provide for natural variations, range of dimensions and even for some abnormal individuals. This study is called the population concept.

Although there is a deficiency in type method, the types have served the very purpose of concept for identification of specimens. It represents better any taxonomic group than a description or illustration. No one can opt indefinitely for maintaining living populations. A satisfactory suggestion is to provide descriptions of the taxa covering larger number of individual representative of a population. Average description and range of variations may be also provided. Thus a taxonomist is expected to draw the description on the basis of available variation in the description on the basis of available variation in the specimens studied by him. He should also establish new taxa only when the distinguishing characters are distinct, stable and significant. It can be briefly stated that the type concept and the population concept are not contradictory but complementary.

(XVII) Rejection of Names

A name or epithet in accordance with the Articles of the ICN is said to be legitmate. A name which is contrary to the Ariticles is said to be illegitimate. The latter one is obviously rejected. Thereare different circumstances in which the plant names are rejected. However, it must be remembered that a name or epithet must not be rejected, changed or modified merely beause it is disagreeable or inappropriate. The circumstances which suggest rejection of names are:

(i) A superfluous name is to be rejected. A name published for a taxon which includes the type material of another taxon of the same rank for which there is already a legitimate name or epithet is said to be nomeclaturally superfluous. For example, (i) *Eclipta erecta* Linn. (1771) is a superfluous name, while *Eclipta prostrata* (Linn.) Linn. (1771) is already a legitimeate name.

(ii) A later homonym is to be rejected. A name is called *homonym* when the same name is attributed to different taxa based on different type materials e.g. *Capsicum frutescens* Roxb. is a later homonym of *C. frutescens* Linn. and has to be rejected.

(iii) A tautonym is to be rejected. When the specific epithet repeats exactly the generic name then, it is called tautonym e.g. *Nasturtium nasturtium* is a tautonym and is illegitimate. It is hence rejected.

(iv) A name '*nomen nudum*' is to be rejected. A name is said to be *nomen nudum* when it is published without a proper and complete description e.g. (i) *Odina wodier* Roxb., (ii) *Centranthera humifusa* Wall, etc.

(v) A name of a taxon must be rejected if it becames a permanent source of confusion or error on account of its different meaning or senses e.g *Quercus rubra* L. is rejected from use for any plant.

(vi) A name of a taxon is rejected if it is based on a monstrosity or a type consisting of two or more discordant elements.

(vii) When there is an older valid name based on another member of the same group, called 'metanym', it is to be rejected.

(viii) When there is an older valid name based on the same type, called 'typonym', it is to be rejected.

(XVIII) Retention and Choice of Names

Different kinds of realignments of taxa sometimes became necessary because of in- depth taxonomic studies. In some circumstances, the taxa need to be remodelled, divided or united. For instance, **(i)** the original name is to be retained for the one which includes the type species in case a genus is divided into two or more genera, e.g the genus *Ludwigia* is divided into two genera *viz.,Ludwigia* and *Fissendocarpa.* The name *Ludwigia* is retained with the one which included type species, **(ii)** the rule also applies similarly in case of species or infraspecific taxa are divided e.g. **(a)***Helitropium indicum* Linn. is divided into two, the one which included type retained the name *H. indicum* L., while another is named as *H. keralense* Manilal and Sivarajan **(b)***Portulaca pilosa* consisted two different subspecies, the name which included the type of the species retained the name *P. pilosa* ssp *pilosa* (Geesink, 1969), whereas another is known as ssp *grandiflora.* The rule is similarly applied at varietal level. **(c)***Portulaca pilosa* ssp *pilosa* is composed of two varieties, the one which included type of the species is P. pilosa ssp. *pilosa* var. *pilosa* and the remaining one is var. *tuberosa* (Sivarajan, 1981b), **(iii)** Legitimate epithets are retained when a subdivision of a genus or a species is transferred to another genus or another generic name for the same species and when an infraspecific taxon is transferred to another species, unless such epithets are occupied already in their respective positions or earlier legitimate epithets are available for them, for example, **(a)***Saponaia* sect *Vaccaria* DC. when transferred to *Gypsophila* is *Gypsohila* (DC). Godr.**(b)***Melochia cordata* Burm. *f.* under the genus *Sida* became *S. cordata* (Burm.*f.*) Borssum. **(c)***Ecbollium linneanum* var. *latervirens* Clarke under the new name for the species *E. viride* became *E. viride* var *latervirens* (Clarke) Raizada, **(iv)** the oldest legitimate name or the earliest legitimate epithet is retained in case of two or more taxa of the same rank are united, unless it contravenes other provisions of the ICN. For example, **(a)***Derris* Lour. (1790) and *Pongamia* (1803) are united under the correct name for the genus *Derris* Lour. **(b)***Sida rhombifolia* Linn. (1753) and *S. rhomboidea* Roxb. ex Flem. (1810) when united is *S. rhombifolia* Linn.

(XIX) Change of Names

One of the most annoying things to the non-taxonomist is a change of name. Whenever a user of plant name is acknowledged that the name by which he has long known a plant has been replaced by another name, he naturally feels bitter. But this must be understood that once it is detected that an illegitimate name is in use, the sooner it is rectified, the better it is. Changes become necessary after careful indepth study of the circumstances and materials. Stability and clarity can be reached out by a vigorous application of the rules of the ICN. Name changes result from three main causes : (I) nomenclatural reasons (II) taxonomic reasons (III) reasons of misidentification.

1. **Nomenclatural Reasons:** Most nomenclatural changes originate in one of three ways : (i) although the ICN is modern, most of its provisions are retroactive In past, many names have come into use which are not now in accordance with the modern Code, and have to be rejected, (ii) taxonomists may have introduced names which are not in accordance with the Code, (iii) source of name-changes is one which has perhaps caused most changes in recent years.
2. **Taxonomic Reasons:** These arise from taxonomic research itself. They are inevitable accompaniment of a natural system of classification which is ever changing or it is made ever more predictive, comprehensive and useful
3. **Reasons of Misidentification:** A plant which has become well-known under a certain name is later found misidentified. It means the plant has been wrongly named and has become known by a misapplied name which really applies to a different plant.

(XX) Conservation of Names

Changes of names annoy the users. The nomenclatural changes sometimes interfere with the free availability and flow of information. This should be avoided whenever possible. The ICN has also recognized this situation and has provided a method in the interest of nomenclatural stability. Provision is made for the conservation of names of genera, families, taxa of intermediate rank and even species. This provision is known as the conservation of names. For instance, the change of a generic name could necessitate the coming of a great many new combinations for the species included under it. To avoid this, generic names found to be incorrect according to ICN, but whose substitution would cause much inconvenience, are able to be conserved and the names which would otherwise be correct are to be rejected. The names conserved are known as '*nomina conservanda*' and those rejected are called '*nomina rejecienda*'. For example, *Chimonathus* Lindley (1819) is conserved over *Meratia* Loisel (1818). Conservation of a name is thus the last limitation of the principle of priority.

(XXI) New Names

The ICN not only governs the formation and usage of all scientific plant names, it cares for all possible circumstances that come in way of flow of information. It takes into account the old names as well as new names. The latter become necessary under the circumstances: **(i)** when a new taxonomic group e.g. family, genus or species, is founded, **(ii)** when an already existing name contravenes the Rules of the Code, **(iii)** when it is felt that it should be transferred from one genus to another, that is new combination is necessary, **(iv)** when a taxon has been referred variously at different times by different names and thus leads to ambiguity, in such circumstance, it becomes desirable to have a new name.

Apart from these, the new names should also fulfil certain requirement of effective and valid publication of a name as stated earlier.

(XXII) Legitimate Names

The legitimacy of names, whether published prior to the ICN or after it, matters much in botanical literature. It is important in taxonomic considerations as well as in nomenclatural studies. Legitimate names are those which are in accordance with all the relevant rules of the ICN and are therefore available for consideration as the acceptable name of a taxon. Illegitimate names are the ones that are not published in accordance of the rules of the Code and are not therefore available for the above said reason. A legitimate name is the accepted name or correct name according to the rules of ICN. Certain categories of names e.g. synonyms, homonyms, tautonyms, *typonyms*, etc. and superfluous names are illegitimate. These have no standing in scientific literature.

(XXIII) Names of Cultivated Plants

Names of cultivated plants are governed for the most part by the same rules as those of wild plants. The rules of ICN are applicable upto specific rank. There are, however, differences at infraspecific levels. A separate volume of the 'International Code of Nomenclature of Cultivated Plants (ICNCP) with 57 articles, is available produced by a committee of the International Union of Biological Sciences'. Most of the rules in the ICNCP are from the relevant ones in the ICN. But below the species level there is a provision for the recognition of only one rank the cultivar, instead of the normal (taxonomic) hierarchy. Cultivars are entities at infraspecific levels and are of varied nature. They may be clones, self-fertilised pure-lines or cross-fertilised assemblages characterized by one or more attributes.

Cultivated plants differ from wild ones in that they generally do not occur in natural populations. They generally exist in artificial stands controlled by man. It follows, therefore, that taxonomic infraspecific categories (e.g. subspecies, varietas, etc.) are not really applicable to cultivated plants. They are replaced by the category 'cultivar', an international term. There are some provisions of ICNCP regarding cultivar names. They are : (i) After 1-1-1959 new cultivar names must be 'Fancy' names in modern languages. They can no longer be given Latin names. Previous to the said date Latin names ware given and they are still in use. The only exceptions to this are names of botanical taxa reduced to cultivar rank, **(ii)** if the botanical name of the plant to which they belong is changed cultivar names, however, must remain unchanged, **(iii)** it is not permissible for two or more cultivars in the same cultivar class to bear the same names, **(iv)** new cultivar names must not now be the same as a botanical or common name of a genus or a species, **(v)** cultivar names must be published by the distribution of printed or duplicated matter dated at least to the year. After 1-1-1959 they must be published with their description in any language, **(vi)** It is recommended that cultivar names be registered with a recognized registration authority which undertakes to keep a list of cultivars for the plants concerned. Cultivars are subject to national and international regulations protecting the rights of plant breeders.

(XXIV) Names of Hybrids

The names of hybrids are largely governed by the same rules of ICN. However, there are also special ariticles for them. For example (i) interspecific hybrids are designated by a formula – the two species names separated by a multiplication sign. e.g. *Calystegia sepium* x *Calystegia silvatica.* In addition, a binary name may be given, if an author so wishes, consisting of the generic name and a hybrid epithet separated by a multiplication sign e.g. *Calystegia* x *lucana.* The binary name is equivalent in rank to a species. It is governed by the same rules of ICN, (ii) variants of an interspecific hybrid may be described as nothospecies, nothovarieties etc. e.g. *Salix x rubens* nothovar. b*asfordiana.* (iii) Intergeneric hybrids are also designated at the generic level by a formula e.g. (a) *Coeloglossum* x *Gymnodenia.* If required additionaly by a hybrid-genus name preceded by a multiplication sign e.g. x *Gymnoglossum* (b) *Triticum* x *Secale* = x *Triticale.* At the species level, they may be designated by a formula either as *Coeloglossum viride* x *Gymnodenia consopea* or as binary name e.g. x *Gymnoglossum jacksoni.*

(XXV) Author Citation

The names of taxa, particularly at and below the genus level, are usually followed by the name of the author (the authority) who first published the name e.g. *Annona* L. or *Annona squamosa* L. These authorities are generally abbreviated. In these cases, L. stands for Linnaeus. The names of the authors placed after the name of the taxon is known as the author citation. The name of a taxon is regarded incomplete when it is written down without the names of authors who first validly published the name. This helps verify the date of publication and impart precision in nomenclature. Citations also provide bibliographic reference and are helpful in recognizing homonyms. e.g. *Festuca incrassata* Salzm. ex Lois. and *Festuca incrassata* L. The two names refer to two different taxa but we would not have been able to recognize this. But this is possible for the citation of author's names appended to the respective plant names. There are some rules for author citation explained below :

(a) the names of authors are generally abbreviated. They are not underlined or italicized e.g.*Tribulus terrestris*, (b) when the name of the plant is jointly published by two authors, their names are linked by means of et or an ampersand **(&)** e.g. *Aristida funiculata* Trin. *et* Rupr. or *Panicum paianum* Naik & Patunkar, (c) Citation is normaly restricted to the first author and is followed by *et al.* when there are more than three authors who first published the name, (d) when an author validly publishes a name and ascribes it to another person, the name of the latter is connected to the name of the person who validly published the name by an *exe.g. Indigofera duthiei* Drum. *ex* Naik (e) when rank or position of a genus or taxon of lower rank is altered, its name or epithet is retained. The name of the author who first published the name or epithet (basionym) must be cited in parenthesis followed by the name of the author who effected the change. This is known as double citation.e.g. *Butea monosperma* (Lamk.) Taub. (1894) and basionym: *Erythrina monosperma* Lamk. (1785).

(XXVI) Phylocode

The origin of *phylocode* is very recent. It seems appropriate to know events about its development particularly in relation to botany. A taxon name could be defined by reference to a part of a phylogenetic tree. The theory of phylogenetic nomenclature was actually built on this suggestion. The development was initiated since the early 1980's. The developments are ultimately published in a series of theoretical papers in the 1990's by the papers by de Queiroz and Gautheir. Subsequently, phylogenetic nomenclarture was discussed in several other papers. Few symposia focused on it. Important ones of these are: symposium held at San Diego, California (USA) in 1995; symposium at the Claremont, California (USA), 1996 and another one at St. Louis, Missour. (USA), 1999. The preparation of the phylocode began in 1993 and a workshop was held at the Harward University Herbaria, Cambridge, Massachusetts, USA, 1998. An intial draft of the phylocode was provided to the participants, which was prepared by Philip Cantino and Kervin de Queiroz. It was revised during the deliberation of the said workshop. Some terminologies and rules were derived from the Biocode and several rules from the pre-existing Botanical and Zoological Codes. There are many rules of the Phylocode which are totally unique also. Thus draft of Recommandtions is completed.

The pre-existing code is unable to provide names which clearly refer to a particular clade or species. This is its main deficiency. The name of a species often changes whenever a species is referred to a different genus as a result of phenetic or phylogenetic considerations. *Supraspecific* names under pre-existing code are associated with clade. They are defined in terms of ranks and types. They therefore fail to retain their associations with particular clades. An entity whose hyothesised composition has remained unchanged may be given a different name under the pre-existing codes based on considerations of rank in case of a clade or genus. The Phylocode or Phylogentic Code of Biological Nomenclature has been proposed. It is designed to name species and clades in such a way so that it can be used concurrently with pre-existing codes as the sole code governing the names of taxa. The Code is not meant to replace existing names but to provide an alternate system by which both the existing and newly proposed names can be applied together. The names of clades will be redefined in terms of phylogenetic relationships rather than taxonomic rank as done under the Linnaean system of nomenclature. The starting date of the Phylocode is cited as 1-1-2000 in the draft code. The phylogentic system of nomenclature has the characteristics : (i) the system is rankless since the assignment of rank is not part of naming process, (ii) it provides presently rules for naming clades. In future, it will provide rules for naming species also. In the present circumstances, the naming of species is governed by the pre-existing codes, (iii) species and clades are not ranks. The former is a segment of a population lineage, while the latter is a *monophyletic* group of species, (iv) According to the Phylocode, a supraspecific name is given a phylogenetic definition and is applied to the clade which fits that definition, (v) taxonomists are allowed to restrict the application of names with respect to clade composition, (vi) there is, however, operational difference in determining synonymy and homonymy in the existing system and phylogenetic system.

Although so, there are few advantages of phytogenetic nomenclature e.g. (a) Phylocode permits naming the intermediate ranks such a superfamily, etc., (b) it improves nomenclatural stability, (c) the abandonment of ranks in phylocode eliminates the error caused by many taxonomists who consider taxa at the same rank.

13

Herbarium

(I) Development of Herbarium

Mankind is dependent on the plants for all his basic requirements of life since his appearance on the earth. The idea of sharing his experience and knowledge related to utility of plants with next generations must have struck to him. This must have necessitated him to preserve plants, especially of more significance to mankind, for future reference. Eventually, in later periods, the early thinkers or plant scientists attempted at hearing about morphology, growth, multiplication, evolution, various utilities, besides mode of nutrition and defense. They naturally became more attentive in developing the techniques of herbarium preparation and preservation.

The earliest preserved *herbarium specimen* is available at Rome, collected by Gherardo Cibo, a naturalist and pupil of Luca Ghini in the year 1532. The art of preparing dry herbarium specimens was initiated by Luca Ghini (1490-1556), an Italian from Bologna. He subsequently recommended this method of plant preservation. He collected and prepared herbarium specimens in Italy. This paved the way for the establishment of the first herbarium of the world in 1545 in the University of Pauda (Italy). The first botanic Garden was also established there in the same year. In the middle of 16th century, his three students, Aldrovandi and Cesalpino, and Turner from England, also made their own herbaria. Herbarium was raised by Caesalpino, currently in Firenze. It can be compared with his book 'de Plantis Libri XVI' which introduced a scientific approach to the study and classification of plants. An Englishman, Jodn Falconer, who (probably) met Ghini in Italy, also his herbarium in 1533. More than twenty herbaria have been established prior to 1600 in various cities in Europe. 'Isagoges' a treatise in Botany and first recorded written publication belonged to Adrian Spieghel, a native of Brussel. It dealt with method of drying plants under a press, selection of paper and some accurate information(*cf.* Rao and Murti, 1995; Mondal, 2005).

'Materia Medica' by Discorides (1st century A.D.) is an account of the medicinal uses of nearly 100 plant species. During period of renaissance in Italy, the Italians began teaching Botany. They developed the first over botanical garden. They also prepared a 'Book' of Gardens or 'Horti Sicci' (*cf.* Rao and Murti, 1995). Pitton de Tournefort (1956-1708) applied the word '*herbarium*' (pl. herbaria) for the first time in his book 'Elemens'. At the end of 1790,many herbaria were established for the advancement of scientific knowledge such as Rijksherbarium, Leiden University, Netherland, (1575), Museum National d' Histoire Naturalle, Paris (1635), British Museum (Natural History), (1753), Swedish Museum of Natural History, Stockholm, Sweden (1758), University of

Cambridge, England (1761), Botanical Museum, Lund, Sweden (1770), Linnean Society of London (1788), etc. (*cf.* Rao and Murti, 1995; Mondal, 2005). When Heinrich Van Rheede (1637-1692), a Dutch Governer at Malabar (India), was botanists in Malabar region, during his period, the concept of the Herbarium making developed fairly well. British botanists also prepared a ' Hortus – Siccus' nearly in the same period and sent to London. Amongst these were James Petiver, Samual Brown, Edward Bulkley and Pluterent. The herbarium specimens were accumulated in British Museum.

(II) What is Herbarium

The concept of preserving plant specimens in a dried form for future studies and reference was unknown until the middle of 16th century when the Italian botanists Luca Ghini (1490-1556) initiated and recommended this method. By the late 19th century, herbarium making had become a major feature of systematic botany. Davis and Heywood (1963) phased out the plant taxonomic researches as (i) Exploratory, (ii) Consolidatory, (iii) Biosystematic and (iv) Encyclopaedic phase. They remarked further that 'The Herbarium, essentially, has been a tool of taxonomists during the first two phases of taxonomic research, but it does not outlive its importance in the last two phases, as is exhibited by its being the source of material for anatomical, palynological and chemotaxonomical studies, where it continues as the sole testimony on the correct identity and nomenclature of the plants. Earlier Davis (1961) also stated that ' it is usually only in the herbarium that we can compare all the related species of a genus in the same place, in the same state and at the same time. According to Fosberg and Sachet (1965), modern Herbarium is a great filing system for information about plants both primary in the form of actual specimens and secondary in the form of published information, pictures and recorded notes. Modern Herbarium, in the words of Radford (1986), is a research training and service institution that serves as a reference centre, documentation that serves as a reference centre, documentation facility, and data structure. He further adds that it includes all kinds of plants, their microscopic slides, photographs, photomicrographs, wood specimens, camera lucida drawings, field notebooks, diaries, letters, manuscripts, reprints and botanical illustrations, etc. Today, we are moving toward 'Electronic Herbarium'. Herbarium has now become a multipurpose institution. Everybody, whether botanists or not, consult herbaria for one or other purpose. Several definitions are generally now available in various standard texts, research literature and publications. Some glimpses are provided to study them in the following:

(i) *Herbarium* is a collection of dried, pressed plants mounted on sheets bearing a detailed data label and stored in strong cupboards in systematic sequence (*cf.* Stace, 1989).

(ii) *Herbarium* is a place where plant material is preserved using various techniques and arranged in the sequence of an accepted classification (Naik, 1984).

(iii) *Herbarium* is a store – house of plant specimens collected from far and wide (or different places) mounted on appropriate sheets, arranged according to some known system of classification and kept in pigeon – holes of classification, and kept in usually especially prepared for the purpose (Jain and Rao, 1977, Mondal,2005).

(iv) *Herbarium* is a museum or a storehouse of pressed and dried plant material mounted on sheets and filed in cupboards in a systematic order for reference purposes (Almeida, 1995).

(v) *Herbarium* is a storehouse of plant specimens. In it dried, pressed, preserved and mounted plant specimens are arranged in a sequence of an accepted system o classification for future reference and study (Johri and Sneh Lata, 2005).

(vi) A collection of dried and preserved plants arranged according to classification system and available for study or reference is known as herbarium (Sharma, 1993).

A sum total of above definitions indicate that the various authors have used words (terms) suggestive of some techniques (drying, pressing, mounting, preserving), arrangement according to an accepted system of classification, and purpose (study, reference). One or two words/ terms have also missed in their definitions, few have even ignored the purpose for which an herbarium is generally established. The term 'systematic order' in one of the definitions above does not mean necessarily as a classificatory system of plants, (it can be also alphabetical and chronological way). Words (terms) may vary from author to author, but serious objections can be raised against when the purpose and system of classification are not clearly mentioned. A definer should take into account these essential elements of being herbarium.

(III) Kinds of Herbaria

Herbarium may be large having national or international reputation and holding a large collection of various plant groups from far and wide. It may be small holding of collection of a restricted area e.g. a state, region or a group of plants or family. Thus the importance, significance of herbarium and services rendered by it depend on the kind of herbarium and also on the locality where it is situated. We can distinguish between different kinds of herbaria according to the purpose which they serve, for example, (i) Herbaria of Organisations (as Botanical Survey of India), (ii) Regional Herbaria, (iii) Local Herbaria, (iv) Herbaria of Institutions, Universities and Colleges, etc (v) Herbaria of medicinal plants and drugs (vi) Herbaria of crop plants and weeds in cultivated fields (as in Agricultural Colleges and Universities), etc.. Davis and Heywood (1963) distinguished herbaria based on purposes they serve: (i) Major Herbaria: These are mainly National Institutes. They adequately cover the flora of world and have dual purpose of identification and research, (ii) Regional (or Local) Herbaria: These concentrate on the flora of their own and adjacent countries. University Herbaria and many museums come under this kind. They furnish important collections for the preparation of state or country floras. However, small university herbaria are rather a special case as they are primarily concerned for teaching and post – graduate research, (iii) Working Herbaria: These are designed expressly for identification. They consist of the species from a particular area arranged within their genera and represented by only one or very few 'representative' sheets so that the collection can be consulted quickly. Working herbaria often form part of a major herbarium [e.g. British Museum (Natural History), London] or small Universities or Institutions serving adequately the needs of identification. Cronquist (1968a) categorized the herbaria as *viz.*; (i) their own research programmes, (ii) service as repositories of type and other historical plant materials (iii) the loan of specimens for study of other institutions, and (iv) the training of graduate students. Brenon (1968) opined that the national herbaria were established especially to fulfill public needs as they are supported by public funds. He also remarked that this service must take a high place among other functions. The public interests embrace a wide range of people and bodies.

Important Herbaria of The World

Name & Abbreviation	Number of Specimens
1. Royal Botanic Garden, Kew London, UK (K)	(Approx.) 65,00000
2. V.L. Komrow Botanical Institute Museum National d' Histoire USSR (LE)	5000000
3. Naturelle Laboratoire de Phanerogamme, Paris, France (P)	5000000
4. Conservation Et Jardin Botaniques, Geneva, Switzerland (G)	4000000
5. U.S. National Museum Washington, USA (US)	3000000
6. New York Botanic Garden, New York, USA (NY)	3000000
7. Vienna Botanischer Gaertn Vienna, Austria (W)	2500000
8. National History Museum, Chicago, USA (F)	2350000
9. Royal Botanic Garden, Edinburgh, UK (E)	2500000
10. Missouri Botanic Garden, St. Louis, USA (MO)	2357000
11. Arnold Arboretum, Boston, USA (A)	700000
12. Botanical Research Institute of Texas, Texas (BRIT)	500000
13. University of Florida Herbarium, Florida (FLAS)	25000
14. Herbarium of the Rangoon University, Rangoon, Burma (RANG)	15,000
15. Central National Herbarium Shibpur, Howrah, India (CAL)	2500000

(IV) Functions of Herbarium

Modern herbarium is a great filling system for information about plants both primary in the form of actual specimens and secondary in the form of published information, pictures and recorded notes (*cf.* Fosberg and Sachet, 1965). The first herbarium in the world was established in 1545 in the University of Pauda, Italy. The progress made and the services rendered by herbaria in the world would not have even imagined by the founders of the first herbarium. An herbarium has now emerged as and functioned in a way similar to an educational institution. Herbarium can offer many services to the needy students, teachers, research workers, naturalists, ecologists and human society at large. The various literary sources have vividly explained the role of an herbarium, they are briefed as follows :

1. It acts as a readily approachable reference centre for learning and teaching in Botany and its various domains like taxonomy, morphology, anatomy, palynology, ecology, economic botany etc.
2. It provides data in respect of distribution of plants, their flowering and fruiting periods, utilities, vernacular names, abundance and rarity, and endemism.
3. It is a valuable tool in deciphering the identity of plant materials and even their authentifications.
4. It especially helps to maintain 'type specimens' in the respective region or from other areas as well.
5. It provides materials for research. Critical studies on various species and genera lead to the establishment of the correct identity, which may have been wrongly identified by earlier research workers.
6. The herbarium plays a key role in various fields such as ethnobotany, economic botany, aerobiology, pharmacognosy, molecular biology, ecology, environmental science, morphology and anatomy, besides floristic surveys. It helps for interdisciplinary research.
7. Botanists working in different disciplines at various herbaria, universities or institutes can request for the loan of herbarium specimens from other herbaria.
8. It is the basis for research and the preparation of 'floras and monographs'. It informs on wide range of taxa and a large sample of examples of each taxon under one roof.
9. It helps preserve voucher specimens, which have been examined for studies in cytology, phytochemistry, etc. The future research workers can check the authenticity of the material investigated.
10. Herbarium acts as a conservatory of material and data in case of natural catastrophies or other factors have totally destroyed the vegetation. The collections in the herbarium provide evidence of what once existed there.
11. The role of herbarium is quite substantial in 'alpha' taxonomy and still continues in 'omega' phase of taxonomy.
12. Ethnobotanical investigations can be conducted based on the herbarium specimens deposited in the herbaria. The labels inform local uses and vernacular names. The latter also reveal sometimes man-plant interactions of a specified region.
13. It helps in estimating biodiversity of a region, state or even globally.

14. It aids in assessing conservation status of a taxon.
15. It serves as a source for search of new genetic material for improvement of cultivated stock.
16. It helps in assessing and cataloguing of all species of economic potential such as medicinal, food, commercial species, etc.
17. It aids in development of computer data base on species and helps thereby to maintain active links to international network of systematic resources and electronic base.
18. It is useful to fulfil public needs by way of supplying materials and scientific information in respect of plants, and also through training programmes, exhibitions etc.
19. A well established herbarium may disseminate their knowledge and share their experiences or views in respect of plant collection from the field, different techniques in processing the plant material in the field and in the laboratory, herbarium sheet making, preservation, etc.
20. A well known, reputed herbarium attracts attention of tourists, naturalists, journalists, academicians, etc. Thus it can create interest in the society.

In a nutshell, herbarium thus plays a significant role in the development of academic and scientific bodies, serves human society in varied ways and helps in growth and upliftment of a country. It serves many many-fold functions. Fosberg and Sachet (1965) rightly remarked that 'the characteristic of a herbarium may be summarised by regarding it as a great organism into which is maintained a flow of specimens and unorganized data. The staffs of botanists, technicians, clerks and other workers furnish the driving force and controlling influence. With an adequate staff and with appropriate support the organism maintains an active growth. There is continuous production of results in the form of services to other human enterprises and other additions to the general fund of organized knowledge that is science.

(V) Type Specimens : Importance and Care

A nomenclatural type (typus) is that constituent element of a taxon to which the name of the taxon is permanently attached, whether as a correct name or as a synonym. International Code of Botanical Nomenclature (ICN) has made it compulsory to deposit type specimens. In case of a discrepancy between various descriptions of the element of a taxon the type specimen help clarify the situation. The type specimen of ataxon are, therefore, are very important. They are specially cared by the curators of herbaria. They are kept, in some herbaria, in separate and special cases. This avoids over unnecessary handling and allows more adequate inspection for possible insect infestation. Generally, such type specimens are placed within a protective cover of some sort. The most protective housing is desired for them. They should be placed in a reasonably fireproof structure. They should not be used or handled any more than is essential. Usually, the authorities of the herbaria do not permit their being sent out on loan to other botanists or institutional.

14

Botanic Gardens, Museum and Library

(A) Botanic Gardens

(I) What are Botanic Gardens

Botanic gardens are primarily outdoor collections of labelled, different living plants collected from various parts of the world and systematically landscaped. These include not only trees, shrubs, herbs, climbers and other plants of wild varieties, but also ornamental and cultivated ones, as well as medicinal and economically important plants. They contain glass-houses, green-houses, library, herbarium, research laboratories, ornamental flower beds, lakes and many miscellaneous resources like paintings, photographs, illustrations, reprints, etc. They also primarily serve scientific functions such as germplasm banks through collection of botanical varieties, plant introduction, education, research and training in many taxonomic, horticultural and other botanical disciplines. Moreover, they are a centre for documentation, reference, data storage, conservation and immense importance to many others, apart from botanists. Well established botanical gardens are members of the International Association of Botanic Gardens (IABC) established in1954. They are different from public parts or recreation parks being primarily meant for scientific functions.

Public parks and gardens differ from botanic gardens in that they are meant for recreation and primary centres of aesthetic beauty and public amenity. Botanical Gardens primarily serve scientific functions. They also, serve although secondarily, as centres of aesthetic attraction being well planned.

A botanic garden is a well-tended area displaying plants labelled with their botanic names. It may contain particular plant collections such as cacti and succulent plants, herb gardens, plants from particular parts of the world, etc. There may be greenhouses, shadehouses, again with special collections such as alpine plants, tropical plants or other exotic plants. Visitor services at a botanic garden might include tours, educational displays, art exhibitions, book rooms, open-air theatrical and musical performances and other entertainment.

Botanic gardens are often run by universities or other scientific research organizations and often have associated herbaria and research programmes in plant taxonomy or some other aspects of plant science. Their main role is to maintain documented collections of living plants for the purposes of scientific research, conservation, display and education. However, this will depend on the resources available and the special interests pursued at each particular garden. Botanic gardens, as cultural and scientific organisations, have responded to the interests of botany and horticulture. Nowadays most botanic gardens display a mix of the themes mentioned and more having a strong connection with the general public. They afford opportunity to provide visitors with information relating to the environmental issues being faced by society or especially those relating to plant conservation and sustainability.

The role of major botanic gardens worldwide has been considered so broadly similar as to fall within textbook definitions. The Liberty Hyde Bailey Hortoriumof Cornell University (1976) defined a botanic garden as a controlled and staffed institution for the maintenance of a living collection of plants under scientific management for purposes of education and research, together with such libraries, herbaria, laboratories, and museums as are essential to its particular undertakings. Each botanic garden naturally develops its own special fields of interests depending on its personnel, location, extent, available funds, and the terms of its charter. It may include greenhouses, test grounds, herbarium, arboretum, and other departments. It maintains a scientific as well as a plant-growing staff, and publication is one of its major modes of expression.

A botanic garden may be an independent institution, a governmental operation, or affiliated to a college or university relating to a teaching program. In any case, it exists for scientific ends and is not to be restricted or diverted by other demands. It is not merely a landscaped or ornamental garden, although it may be artistic, nor is it an experiment station or yet a park with labels on the plants. The essential element is the intention of the enterprise, which is the acquisition and dissemination of botanic knowledge. A contemporary botanic garden is a exculusively protected natural green area where a managing organization creates landscaped gardens. It holds documented collections of living plants and/or preserved plant accessions containing functional units of heredity of actual or potential value for purposes such as scientific research, education, public display, conservation, sustainable use, tourism and recreational activities, production of marketable plant-based products and services for improvement of mankind.

There are now about 1800 botanic gardens worldwide and arboreta in about 150 countries of which about 400 are in Europe, 200 in North America, 150 in Russia and an increasing number in East Asia. These gardens attract about 150 million visitors a year so it is hardly surprising that many people gained their first exciting introduction to the wonders of the plant world in a botanic garden. Botanic gardens, in past, exchanged plants through the publication of seed lists. This was a means of transferring both plants and information between botanic gardens. This trend continues today although the possibility of genetic piracy and the transmission of invasive species has received greater attention in recent times.

The International Association of Botanic Gardens (IABG) was formed in 1954 as a worldwide organisation affiliated to the International Union of Biological Sciences. Recently coordination has also been provided by Botanic Gardens Conservation International (BGCI) which has the mission «To mobilise botanic gardens and engage partners in securing plant diversity for the well-being of people and the planet". BGCI has over 700 members – mostly botanic gardens – in 118.countries and strongly supports the Global Strategy for Plant Conservation by producing resources and publications, and by organizing international conferences and conservation programs, etc.

(II) Botanic Gardens:Historical Development

The botanic gardens of the 16th and 17th centuries were medicinal gardens.The idea of a botanic garden changed to encompass displays of the beautiful, strange, new and sometimes economically important plant trophies being returned from the European colonies and other distant lands. In the 18thcentury, they became more educational in function, demonstrating the latest plant classification systems devised by botanists working in the associated herbaria as they tried to order these new treasures. In the 19th and 20th centuries, the trend was towards a combination of specialist and eclectic collections demonstrating various aspects of both horticulture and botany.

Near-eastern royal gardens set aside for economic use or display and containing at least some plants gained by special collecting trips or military campaigns abroad, are known from the second millennium BCE in ancient Egypt, Mesopotamia, Crete, Mexico and China. In about 2800 BCE the Chinese Emperor Shen Nung sent collectors to distant regions searching for plants with economic or medicinal significance. It has also been suggested that the Spanish conquest of Mesoamerica influenced the history of the botanic garden as gardens in Tenochtitlan established by king Nezahualcoyotl. Gardens in Chalco and elsewhere greatly impressed the Spanish invaders, not only with their appearance but also because the indigenous Aztecs employed many more medicinal plants than did the classical world of Europe. Early medieval gardens in Islamic Spain resembled botanic gardens of the future an example being the 11th century Huerta del Ray garden of physician and author Ibn Wafid (999–1075 CE) in Toledo. This was then taken over by garden chronicler Ibn Bassal until the Christian conquest in 1085 CE. Ibn Bassal then founded a garden in Seville, most of its plants being gathered on a botanic expedition that included Morocco, Persia, Sicily and Egypt. The medical school of Montpelier was also founded by Spanish Arab physicians and by 1250 CE it included a physic garden. However, the site was not given botanic garden status until 1593.

Physic Gardens: Modern botanic gardens, in the modern sense, developed from physic gardens, whose main purpose was to grow herbs for medicinal use. Such gardens have a long history. In Europe, for example, Aristotle (384 BCE – 322 BCE) is said to have had a physic garden in the Lyceum at Athens, which was used for educational purposes and for the study of botany and this was inherited, or possibly set up, by his pupil Theophrastus, the 'Father of Botany'. Though these ancient gardens shared some of the characteristics of modern botanic gardens, they are generally regarded as being the medieval monastic physic gardens that originated after the decline of the Roman Empire at the time of Emperor Charlemagne (742–789 CE). These contained a *hortus*, a garden used mostly for vegetables. Another section was set aside for specially labelled medicinal plants and this was called the *herbularis* or *hortus medicus*— more generally known as a physic garden, and a *viridarium* or orchard. These gardens were probably given impetus when Charlemagne issued a capitulary, the Capitulary de Villis, which listed 73 herbs to be used in the physic gardens of his dominions. Many of these were found in British gardens although they only occurred naturally in continental Europe, demonstrating earlier plant introduction.Pope Nicholas V set aside part of the Vatican grounds in 1447 for a garden of medicinal plants that were used to promote the teaching of botany. This was a forerunner to the University gardens at Padua and Pisa established in the 1540s. Certainly the founding of many early botanic gardens was instigated by members of the medical profession.

16th and 17th Centuries European Gardens: In the 17th century, botanic gardens began their contribution to a deeper scientific curiosity about plants. If a botanic garden is defined by its scientific or academic connection then the first true botanic gardens were established with the revival of learning that occurred in the European Renaissance. These were secular gardens affilated to universities and medical schools, used as resources for teaching and research. The superintendents of these gardens were often professors of botany with international reputations.This factor that probably contributed to the creation of botany as an distinct discipline rather than a descriptive adjunct to medicine.

Southern Europe: Botanic gardens of Southern Europe were attached with university faculties of medicine and were founded in Northern Italy at Orto botanico di Pisa (1544), Orto botanico di Padova (1545), Orto Botanico di Firenze (1545) and Orto Botanico dell'Università di Bologna (1568). The physicians (referred to in English as apothecaries) in these gardens delivered lectures on the Mediterranean "simples" or "officinals" that were being cultivated in the grounds. Student education was no doubt encouraged by the relatively recent advent of printing and the publication of the first herbals. All of these botanic gardens still exist, mostly in their original lands.

Northern Europe: The tradition of these Italian gardens passed into Northern Europe where similar gardens were established in the Netherlands (Hortus Botanicus Leiden, 1587; Hortus Botanicus (Amsterdam), 1638), Germany (Alter Botanischer Garten Tübingen, 1535; Leipzig Botanic Garden, 1580; Botanischer Garten Jena, 1586; Botanischer Garten Heidelberg, 1593; Herrenhäuser Gärten, Hanover, 1666; Botanischer Garten der Christian-Albrechts-Universität zu Kiel, 1669; Botanic Garden in Berlin, 1672), Switzerland (Old Botanic Garden, Zürich, 1560; Basel, 1589); England(University of Oxford Botanic Garden, 1621; Royal Botanic Garden Edinburgh, 1670; Chelsea Physic Garden, 1673) and in France (Jardin des plantes de Montpellier, 1593; Paris, 1597; Jardin des Plantes, Paris, 1635), Denmark (University of Copenhagen Botanic Garden, 1600); Sweden (Uppsala University, 1655) and also in Spain (Botanic Garden of Valencia, Spanish, 1567).

Plants were being imported to these major Western European gardens from Eastern Europe and nearby Asiaduring the 16th and 17th centuries. (http:/www.botanicgardens.eu/action/actionplaneu1htm). These provided many bulbs and found a place in the new gardens where they could be conveniently studied by the plant experts of the day. For example, Asian introductions were described by Carolus Clusius (1526–1609) who was director, in turn, of the Botanic Garden of the University of Vienna and Hortus

Botanicus Leiden. Many plants were being gathered from the Near East, particularly bulbous plants from Turkey. Clusius laid the foundations of Dutch tulip breeding and the bulb industry. He established one of the earliest formal botanic gardens of Europe at Leyden where his detailed planting lists have made it possible to recreate this garden near its original location. The '*hortus medicus*' of Leyden in 1601 was a perfect square divided into quarters for the four continents. But by 1720 however it was a rambling system of beds, struggling to contain the novelties rushing in and it became better known as the '*hortus academicus*'. His '*Exoticorum libri decem*' (1605) is an important survey of exotic plants and animals that is still consulted today. The inclusion of new plant introductions in botanic gardens meant that their scientific role was now widening as botany gradually asserted its independence from medicine.

Paris Jardin des Plantes was a centre of interest with the greatest number of new introductions to attract the public in the mid to late 17th century. In England, the Chelsea Physic Garden was founded in 1673 as the «Garden of the Society of Apothecaries». The Chelsea garden had heated greenhouses and in 1723 appointed Philip Miller (1691–1771) as Head Gardener. He had a wide influence on both botany and horticulture as plants poured into it from various parts of the world. The garden's golden age came in the 18th century, when it became the world's most richly stocked botanic garden. Its seed-exchange programme was established in 1682 and still continues in recent times.

More plants were being brought back to Europe as trophies from distant lands and these were triumphantly displayed in the private estates of the wealthy, in commercial nurseries and in the public botanic gardens with the increase in maritime trade. Heated conservatories called 'orangeries', such as the one at Kew, became a feature of many botanic gardens. Industrial expansion in Europe and North America resulted in new building skills, so plants sensitive to cold were kept over winter in progressively elaborate and expensive heated glasshouses and conservatories.

The Cape, Dutch East Indies: During the 18th century, ericas, geraniums, pelargoniums, succulents and proteaceous plants were introduced from the Cape of South Africa.The Dutch trade with the Dutch East Indies resulted in a golden era for the Leiden and Amsterdam botanic gardens and a boom in the construction of conservatories.

The Royal Gardens, Kew were founded (1760)), initially as a part of the Royal Garden set aside as a physic garden. William Aiton(1741–1793), the first Curator, was taught by garden chronicler Philip Miller of the Chelsea Physic Garden. His son Charles became first Curator of the original Cambridge Botanic Garden (1762). In 1759, the «Physick Garden» was planted and by 1767 it was claimed that «the Exotick Garden is by far the richest in Europe». Gardens such as the Royal Botanic Gardens, Kew (1759) and Orotava Acclimatization Garden (Spanish), Tenerife(1788) and the Real Jardín Botánico de Madrid (1755) were set up to cultivate new species returned from expeditions to the tropics. They also helped found new tropical botanic gardens. From the 1770s, following the example of the French and Spanish, amateur collectors were supplemented by official horticultural and botanic plant hunters. This was the great era of European, and especially British, imperialism which was "the means by which ideas, knowledge, styles of economy, politics and goods were given international reach and justified as the spreading of civilisation through development and improvement.

England was importing many woody plants from North America at this time, The popularity of horticulture had increased enormously by the horticultural and botanic collecting expeditions overseas. These were fostered by the directorship of Sir William Hooker and his keen interest in economic botany. Under the directorship of Sir Joseph Banks, at the end of the 18th century, enjoyed a golden age of plant hunting, sending out collectors to the Southern African Cape, Australia, Chile, China, Ceylon (present Sri Lanka), Brazil and elsewhere, and acting as 'the great botanic exchange house of the British Empire'.From its earliest days to the present, Kew has in many ways exemplified botanic garden ideals. It is respected worldwide for the published work of its scientists, the education of horticultural students, its public programmes, and the scientific underpinning of its horticulture.

Plant Classification: Several plants awaiting description were often enlisted in garden catalogues. Carl Linnaeus established the system of binomial nomenclature during this period.This facilitated the listing process. Names of plants were authenticated by dried plant specimens mounted on card (a *hortus siccus* or garden of dried plants) that were stored in buildings called herbaria. These taxonomic research institutions were frequently associated with the botanic gardens. Many of these by then had «order beds» to display the classification systems being developed by botanists in the gardens, museums and herbaria. Botanic gardens had by this time become scientific collections as botanists published their descriptions of the new exotic plants.These were also recorded for posterity in detail by botanic illustrations. In this century, botanic gardens effectively dropped their medicinal function in favour of scientific and aesthetic priorities and the term «botanic garden» came to be more closely associated with the herbarium, library and then laboratories.

The late 18th and early 19th centuries were marked by the establishment of tropical botanic gardens as a tool of colonial expansion (for trade and commerce and, secondarily, science) mainly by the British and Dutch, in India, South-east Asia and the Caribbean. This was also the time of SirJoseph Banks's botanic collections during Captain James Cook'scircumnavigations of the planet. His explorations of Oceania which formed the last phase of plant introduction on a large scale.

Tropical Botanic Gardens: Presently, there are nearly 230 tropical botanic gardens with a concentration in southern and south-eastern Asia. The first botanic garden established in the tropics was the Pamplemousses Botanic Garden in Mauritius, in 1735 to provide food for ships using the port. This was followed by the West Indies (Botanic Gardens St. Vincent, 1764) and in 1786 by the Acharya Jagadish Chandra Bose Botanic Garden in Calcutta, India founded during a period of prosperity when the city was a trading centre for the Dutch East India Company. Other gardens were constructed in Brazil (Jardim Botânico, Rio de Janeiro, 1808), Sri Lanka (Botanic Garden of Peradeniya, 1821 and on a site dating back to 1371), Indonesia (Bogor Botanic Gardens, Kebun Raya Cibodas, 1817), and Singapore (Singapore Botanic Gardens,

1822). These had a considerable effect on the economy of the countries, especially with respect to the foods and medicines introduced. The importation of rubber trees to the Singapore Botanic Garden initiated the important rubber industry of the Malay Peninsula. Teak and tea were introduced to India during this period and breadfruit, pepper and starfruit to the Caribbean. The investigation of the local flora for its economic potential to both the colonists and the local people was included in the charter of these gardens. Many crop plants were introduced by or through the association with European botanic gardens such as Kew or Amsterdam. These included cloves, tea, coffee, breadfruit, cinchona, sugar, cotton, palm oil and Theobroma cacao (for chocolate). The rubber plant was introduced to Singapore during these times. Especially in the tropics, the larger gardens were frequently associated with a herbarium and museum of economy. The Botanic Garden of Peradeniya had considerable bearing on the development of agriculture in Ceylon (Sri Lanka) where the Para rubber tree (*Hevea brasiliensis*L.) was introduced from Kew. Kew reported it from South America. Other examples include cotton from the Chelsea Physic Garden to the Province of Georgia (1732) and tea into India by Calcutta Botanic Garden. The transfer of germplasm between the temperate and tropical botanic gardens was undoubtedly responsible for a wide variety of agricultural crops currently used in several regions of the tropics.

Australia: Early in the 19th century, the first botanic gardens in Australia were founded e.g. The Royal Botanic Gardens, Sydney (1816) the Royal Tasmanian Botanic Gardens (1818) the Royal Botanic Gardens, Melbourne (1845); Adelaide Botanic Gardens, (1854), and Brisbane City Botanic Gardens, (1855). These were founded essentially as colonial gardens of economic botany and acclimatization and economic botany.

New Zealand: Major botanic gardens in New Zealand include Dunedin Botanic Gardens, (1863) Christchurch Botanic Gardens, (1863) and Wellington Botanic Gardens, (1868).

Hong Kong: Hong Kong Botanic Gardens, (1871) (renamed Hong Kong Zoological and Botanic Gardens in 1975), up from the Government Hill in Victoria City, Hong Kong Island

South Africa: The Kirstenbosch National Botanic Garden, (1913) has a site dating to 1848.

United States: The first botanic garden *viz,* Bartram's Garden, was founded in USA (1730) near Philadelphia and in the same year the Linnaean Botanic Garden at Philadelphia itself. In 1859, the Missouri Botanic Garden was founded at St Louis and is now one of the world's leading gardens specializing in tropical plants. This was one of several popular American gardens including Longwood Gardens (1798), Arnold Arboretum (1872), New York Botanic Garden (1891),Huntington Botanic Gardens (1906), Brooklyn Botanic Garden (1910), International Peace Garden (1932), and Fairchild Tropical Botanic Garden (1938).

Russia: Russia has more botanic gardens than any other country, with a major garden in every republic, each with many satellite gardens. Better-known gardens are Batumi Botanic Garden, Saint Petersburg Botanic Garden, (1714), M.M. Gryshko National Botanic Garden, (1936) and Moscow Botanic Garden of Academy of Sciences (1945). These gardens are notable for their structures that include sculptures, pavilions, bandstands, memorials, shadehouses, tea houses and suchlike.

Ukraine: Ukraine has a well-respected collection at Nikitsky Botanic Garden, Yalta founded in 1812.

Civic and Municipal Botanic Gardens: Many civic or municipal botanic gardens were established in the 19th and 20th centuries. These did not develop scientific facilities or programmes but the horticultural aspects were emphasized and the plants often labelled. They were botanic gardens in the sense of building up collections of plants and exchanging seeds with other gardens around the world. They tended to become, little more than beautifully maintained parks and were, indeed, often under general parks administrations.

Community Engagement: The second half of the 20th century witnessed increasingly sophisticated educational, visitor service and interpretation services. Botanic gardens started to cater for many interests. They, often include botanic exhibits on themes of evolution, ecology or taxonomy horticultural displays of attractive flowerbeds and herbaceous borders, plants from different parts of the world, special collections of plant groups like bamboos or roses, specialist glasshouse collections such as tropical plants, alpine plants, cacti and orchids, as well as the traditional herb gardens and medicinal plants. Specialised gardens e.g. the Palmengarten in Frankfurt, Germany (1869), had the largest orchid and succulent plant collections. There was a renewed interest in gardens of indigenous plants and areas dedicated to natural vegetation. There was an increase in revenue-raising public entertainment including music, art exhibitions, special botanic exhibitions, theatre and film. This was supplemented by the advent of 'Friends' organisations and the use of volunteer guides.

Plant Conservation: The heritage value of exceptional historic landscapes and plant conservation were treated with a growing sense of urgency. Some gardens were sometimes given a separate or adjoining site, to display native and indigenous plants. The Botanic Gardens Conservation in the 1970s gardens became focused on the plant conservation.Secretariat was established by the IUCN, and the World Conservation Union in 1987 with the aim of coordinating the plant conservation efforts of botanic gardens around the world. It maintains a database of rare and endangered species in botanic gardens' living collections. Many gardens hold *ex situ* conservation collections that preserve genetic variation. These may be held as: seeds dried and stored at low temperature, or in tissue culture (such as the Kew Millennium Seedbank); as living plants, including those that are of special horticultural, historical or scientific interest or by managing and preserving areas of natural vegetation. Collections are often held and cultivated with the intention of reintroduction to their original habitat. The Center for

Plant Conservation at St Louis, Missouri coordinates the conservation of native North American plant species.

(III) Functions of Botanic Gardens

The kinds of living plants from both local and foreign countries, the facilities or provisons made available in the gardens and the objectives of establishing such gardens themselves entell or reflect the various functions to be served by them. These can be summed up as :

(i) The diverse living collections of plants are the basis for on-site teaching, research and training in studies in respect of genetics, evolution, systematics and morphology. They supply living plant resources for academic studies in anatomy, embryology, palynology, phytochemistry, cytology etc.

(ii) It is possible to extend adequate protection to endangered plants and means of propogation of rare plants.

(iii) They serve to acclimatize and provide protection to economically significant species. They help expand many valuable plant species or varieties of fibre-yielding, oil-yielding plants, beverage, aromatic and medicinal plants, rubber-yielding, etc.

(iv) They supply saplings or seeds of important plants on exchange basis to other gardens to encourage the introduction of exotic, useful and economic species.

(v) They provide germplasm for hybridization to yield better varieties of fruit, vegetable and flowers in commerce

(vi) They disseminate necessary information on cultivated varieties of food plants, ornamentals and such other widely used plants.

(vii) They provide technical know-hows and training for conservation of nature, growing horticultural plants, landscaping and gardening.

(viii) Apart from exotic species, they also provide information of local flora of some special interest while preparing floras, monographs or revisions.

(ix) They creat interest in plants and nature during regular exhibitions.

(x) Some gardens maintain herbaria and botanical libraries. They aid in additional studies and exposure to botanical literature.

(xi) There is an international association of Botanical Gardens and directories of botanic gardens of the world. These help exchange information and ideas of social interests,

(xii) They also hold seminars, workshops or symposia regularly or occasionally and thereby enthuse research workers in different fields. They offer a common dias for botanists, horticulturalists, plant chemist, foresters, etc.

(xiii) They also serve although secondarily as centres for aesthetic beauty and recreation.

(IV) Botanic Gardens and Their Sections

In past, the botanical gardens were arranged in a different manner than what they are today. About a century back, botanical gardens were either gardens of medicinal plants (physic gardens) or gardens laid out to exhibit plant relationships. Some of them were designed to show economic utility of plants. Such gardens were of greater interest to botanists than to the public. Several changes have occurred in raising gardens and presently they are in a position to suffice many needs of public. They are being patterned differently depending upon the purposes or functions to be served. Instead of having randomized way of gardening, we have very well planned gardens or special sections in the gardens as in the following:

1. **Arboretum:** It is section of garden on entire garden where only arborescent species are grown e,g. Linnaeus Arboretum, Minnesota (USA), Houston Arboretum and Nature Center, Texas (USA), Queen Elizabeth Arboretum, Columbia (Canada), Svinvik Arboretum (Norway) etc.
2. **Pinetum:** In this section only coniferous species are grown e.g. Indian Botanical Garden, Kolkatta (India), Conifer walk, Royal Botanic Garden, Edinburgh (Scotland).
3. **Orchidarium:** It contains only Orchids, e.g. A world of Orchids, Kissimmee. Florida (USA), National Orchidaria, B.S.I., Shillong (India).
4. **Bambusetum:** Mainly species of bamboos are raised, e.g. Bamboo garden, Royal Botanical Garden Kew (UK), Bambusetum, Indian Botanical Garden, Kolkatta (India).
5. **Rosary:** Different species of Roses are cultivated in this section of a garden. Many gardens commonly grow roses.
6. **Sections of Medicinal or Aromatic Plants:** Different species of medicinal importance or aromatic plants are cultivated in these sections.

Likewise a Cacti – house, Palm – house, Alpine – house and sections of bulbous plants, ferns, aquatic garden, bromeliads begonias, gesnerids, cycads, orchards, like mango orchard, citrus orchard, etc. are observed depending upon the purpose and geographical location of gardens.

(V) The Experimental Garden

An important role played by some gardens obviously needs attention, we can call it 'Experimental Garden'. Functions of such gardens are reflected in the kind of plants grown :

(a) **Taxonomic Research Projects:** The garden is a means of gathering together in one place a variety of species which can be usful for anatomical, palynological, embryological, chemical or breeding work, etc. at any time. It is also essential for plants to be grown in uniform conditions to check on ecophenetic variation and for growing the progeny of artificial hybridization or of wild hybrids.

- **(b) On – site Teaching:** Collection of plants are often displayed in such a way that they are useful for demonstration or self – instruction.
- **(c) Provision of Material:** Plant species are raised particularly for use outside the garden e.g. for teaching or for research work of plant chemists, physiologists, etc.
- **(d) Conservation:** Cultivation of rare or especially plants for study in relation to conservation and the setting up of gene banks is increasing. Botanic gardens are viewed as an important facility in the conservation of generic diversity.

Well – planned gardens can perform even for other experimental studies on living plants at one site. They thus serve purpose / function of experimental station or a kind of botanic gardens.

(VI) Important Gardens of the World

- (i) Royal Botanic Garden, Kew, UK. (1760)
- (ii) Royal Botanic Garden, Edinburgh, UK. (1670)
- (iii) Botanic Gardens, Pauda, Italy. (1545)
- (iv) Botanic Garden of the Academy of Sciences, Leningrad, Russia (1712)
- (v) Missouri Botanic Garden, St. Louis, USA. (1859)
- (vi) New York Botanical Garden, New York, USA. (1891)
- (vii) Jardin Botanico Muncipal of Buenos Aires, Argentina. (1892)
- (viii) National Museum of Natural History, Paris, France. (1635)
- (ix) Botanical Garden and Museum, Berlin, Germany (1646)
- (x) Indian Botanical Garden, Howrah, India (1787)

(VII) Economic, Cultural And Historic Events In Botanic Gardens

- **(i) Botanic Garden of Leiden:** Directed by Carolo Clusius, it had nearly 1,000 species from Europe and East Asia in 1601, including oriental species such as sugarcane, ginger, castor oil plant; and American species such as potato, maize, tomato, sweet and hot peppers, prickly pear, canna lily and nasturtium.
- **(ii) Royal Botanic Garden of Madrid:** Promoted many botanical expeditions to the Americas since its founding in 1755.
- **(iii) Botanic Garden and Museum of Berlin-Dahlem:** Developed as an important taxonomic centre as a result of nineteenth-century exploration of the new German colonies in Afiica and Oceania.
- **(iv) Botanic Garden of la Orotava, Tenerife:** Founded in 1788 to acclimate new and useful plants from the New World.
- **(v) Royal Botanic Gardens, Kew:** Played a decisive role in breaking the monopoly held by Holland in the Asian spice trade by cultivating numerous Asian spices in the glasshouses and by introducing their cultivation to the Caribbean.

(vi) **Botanic Garden of Amsterdam:** The Dutch managed to break the British coffee monopoly with a single coffee plant grown in the Garden.

(vii) **Botanic Garden of Palermo:** Numerous varieties of cotton have reached the market as a result of experiments carried out in the Garden.

(viii) **Botanic Garden of Valencia:** Director, Vicente Lorente, defended the Garden during the French invasion and was jailed and sentenced to death; he was saved thanks only to the intervention of the French botanist LBon Dufourny.

(ix) **Botanic Garden and Museum of Berlin-Dahlem:** Nearly all its buildings and collections were destroyed during the Second World War; the library and herbarium were set alight, reducing the collection from almost 4 million sheets to just 500,000.

(x) **Botanic Garden of Sanlucar:** The Garden fell into the hands of a popular anti-liberal revolution and was razed to the ground; it no longer exists.

(xi) **Italian Alpine Garden 'la Chanousia':** The Garden was destroyed in the Second World War and the land passed into French hands after the international borders were redefined at the end of the War; but it re-opened in 1978 at a new site.

(VIII) Historic Buildings And Architectural Features In Botanic Gardens

(i) University of Copenhagen Botanic Garden	Palm House	1872-74
(ii) National Botanic Gardens, Glasnevin	Curvilinear Glasshouse designed by Richard Turner Cactus and Succulent House	1843-69 1890
(iii) National Botanic Garden of Belgium, Meise	Balat Greenhouse designed by Alphonse Balat. Plant Palace, larger than 1 ha	1853
(iv) University of Vienna Botanic Garden	Orangery Greenhouses	1755
(v) Palmengarten, Frankfurt	Palm House, one of the largest in Europe	1869 1840-1860s
(vi) Botanic Garden of Palermo	Greenhouse donated by Queen Maria Carolina	1860-62
(vii) Botanic Garden of Valencia	Tropical Greenhouse, Spanish cast-iron	1906
(viii) Botanic Garden and Museum of Berlin-Dahlem	Tropical Greenhouse, one of the largest in the world	1890-93
(ix) Munich Botanic Garden	Victoria House	

(x) Tapada d' Ajuda Botanic Garden	Construction commissioned by the Marquis of Pombal, and includes steps, balustrades, and a stone fountain with 40 jets	1768
(xi) Royal Botanic Garden of Madrid	King's Gate by Sabatini and wrought-iron fence	18th century
(xii) Jardin des Plantes, Paris	Labyrinth Numerous statues Tomb of the botanist Daubenton (1716-1800)	1640 1890-1910
(xiii) National Botanic Garden of Belgium, Meise	Twelfth-century castle, Flemish Farm, Hunting Pavilion, Temple of Diana, Orangery, Ice Cellars, Hermit's Hut	
(xiv) Botanic Garden of Palermo	Neoclassical Gymnasium, Tepidarum and Calidarium	1795
(xv) Tropical Agriculture Garden-Museum, Lisbon	Palacio dos Condes da Calheta	
(xvi) Royal Botanic Garden of Madrid	Villanueva Pavilion	18th century
(xvii)Botanic Garden of Malaga	Mansion built for the Marquis and Marquise of Loring	18th century
(xviii) University of Copenhagen Botanic Garden	Remains of the former bastions and fortifications of the city have been transformed into a rock garden	
(xix) Botanic Garden of Cagliari	Archaeological excavations	
(xx) Royal Botanic Gardens, Kew	Kew Palace Temples by William Chambers Mock ruined arch Octagonal pagoda Palm House	1631 1758 1759 1761 1844
(xxi) Vienna Botanic Garden	Oldest alpine garden in Europe	1802

(IX) Documented Preserved Collections In Botanic Gardens

Many botanic gardens have specialised botanical and horticultural libraries, containing a wealth of books, journals, papers, letters, maps, illustrations and engravings (see Box 10). They are of interest not only in a scientific and historic context, but also in terms of artistic merit. Their collections need to be stored and conserved properly, but be accessible for consultation by scientists, historians and others who are interested.

(X) Major Botanic And Horticultural Libraries In Botanic Gardens

(i)	Royal Botanic Gardens, Kew	Includes 120,000 monographs, 4,000 periodicals, 140,000 pamphlets, 10,000 maps and unpublished information on the exploration, discovery and investigation of plants and fungi, particularly in the nineteenth and twentieth centuries; as well as illustration collections of over 175,000 prints and drawings, including originals by G. D. Ehret, Franz and Ferdinand Bauer and special collections that include works by Linnaeus and his contemporaries
(ii)	National Botanic Garden of Belgium, Meise	More than 60,000 volumes, including numerous old and precious books, e.g. rare sixteenth-century botanical works by Dodoens with beautiful illustrations
(iii)	Botanic Museum Library, Helsinki	A large archive with many notes taken by Finnish botanists in eastern Fennoscandia
(iv)	Botanic Garden of Palermo	More than 40,000 monographs and a collection of antique books About 125,000 volumes
(v)	University of Coimbra Botanic Garden	Manuscripts and drawings, including the famous colour drawings of
(vi)	Royal Botanic Garden of Madrid	American flora by Jose Celestino Mutis, of incalculable scientific and artistic value
(vii)	Botanic Garden of Barcelona	Numerous manuscripts, correspondence and notes by famous Catalan botanists, as well as an antique book collection
(viii)	Royal Botanic Garden Edinburgh	The 75,000 volumes make up the largest collection of botanical works in northern Britain, some dating as far back as 1486

(XI) Museum Collections In Botanic Gardens

(i)	Botanic Museum of Berlin-Dahlem	Sections on palaeontology and phytogeography, Egyptian section with flower garlands and other decorative plant relics and reconstructions from Pharaohs' tombs
(ii)	Loringiano Museum, Historic- Botanic Garden of Malaga	Collection of Roman ruins
(iii)	Ethnobotany Museum, Botanic Garden of C6rdoba	Plant-related objects and palaeobotanical collection of more than 150,000 plant fossils; the most complete record in existence of Spanish flora from the Carboniferous period (see Case Study 7)
(iv)	Tropical Agriculture Garden-Museum, Lisbon	Collection of more than 3,000 types of wood, the most important collection of its kind in Portugal

(v)	University of Florence Botanic Garden	Over 6,000 wood specimens and plant fossils
(vi)	Botanic Garden of Leiden	More than 15,000 spirit specimens and 25,000 wood specimens
(vii)	Botanic Museum of Helsinki	Separate fruit and seed collections (2,635 specimens) and an extensive collection of pollen and plant anatomy slides

(XII) Few Characteristic Botanic Gardens

(i) **Botanic Garden of the University of Pisa:** The Oldest Botanic Garden In Europe: The first physic garden in Europe was founded in 1543 at Pisa University. Since then, the botanic garden has been moved twice: in 1563 and in 1593. At the beginning of the twentieth century, several of the original trees remained at their original sites in the Garden. The Garden has important collections of herbaceous plants, mainly geophytes. The living collections form the basic material for research in biosystematics and taxonomy at the University. Other sections of the Garden are used principally for the cultivation of species for geobotanic research, and for research in cellular differentiation, phytochemistry, ethnobotany and the physiology and metabolism of seeds.

(ii) **Botanic Garden of the University of Padua:** World Heritage Site: World Heritage Committee (1997) recognised Padua University Botanic Garden as a World Heritage Site, the first botanic garden to be so recognised. One of the two oldest botanic gardens in Europe, dating from 1545, the Garden remains on its original site in the centre of the city of Padua in northern Italy, and is still arranged in its original design. It was founded as an adjunct to the medical school of the University, where great medical advances in human anatomy were made in the sixteenth century by the anatomist Andreas Versalius; Galileo was a professor at Padua University later in the century.

The World Heritage Committee noted that the Garden was the starting point for botanic gardens in the world and represents the cradle of science, scientific exchange and the understanding of the relationship between nature and culture. It has contributed greatly to the development of many modem scientific disciplines, notably botany, medicine, chemistry, ecology and pharmacy.

According to the founding Convention of 1972, countries that include World Heritage Sites assume a responsibility to maintain conserve them. Thus, the Italian national government is seeking parliamentary approval for a grant of 1.5 million Euros to create a buffer zone around Padua University Botanic Garden and to cater for educational and scientific developments.

(iii) **Ethnobotany Museum of Córdoba Botanic Garden:** The Museum has a collection of more than 1,000 items, as well as 500 files of oral and bibliographical information related to ethnobotany. More than 2,500 uses of the items and plants in the collection are recorded in a database. The collection includes items from Spain, Mexico, Costa Rica, Cuba, Peru, Brazil and El Chaco. The most comprehensive parts deal with the olive tree, esparto grass and the cork oak in Andalusia; peyote, corn, tequila mescal and rubber plant in Mexico; and ethnobotanical objects in general from El Chaco. There is a grape press, over 200 years old, made of solid oak from the Alpujarras Mountains, and an elaborate 'chocolate machine' for making 'European style' chocolate.

(iv) Linnaeus's Garden in Uppsala Botanic Garden: Founded in 1655 by Olof Rudbeck the elder, the Hortus Upsaliensis is the oldest botanic garden in Sweden; about 1,800 plant species were grown there at the end of the seventeenth century. After a period of neglect, it rose to fame under Carl Linnaeus, the founder of modem systematic botany and the binomial system of nomenclature, His Species Plantarum, aiming to describe all plant species in the world, is still the basis for our naming of plants. As well as writing Systema Sexualis, he made significant contributions to ethnobotany and ecology, and the first serious attempt at a natural system of classification for angiosperms. Through his contacts and his 'apostles', he obtained plants fiom all over the world, mainly as herbarium specimens, but also for the Garden. Several descriptions in Species Plantarum are based on material grown in Hortus Upsaliensis, and studies in the Garden had a profound influence on his thinking. The original construction and layout of the Garden remain, although the Orangery is no longer used for growing plants. Though few of the original plants remain, the same species are grown as in Linnaeus's time. The Garden, run by Linnaeus in 174 1-78, is a living illustration of the Linnaean sexual system, of his utilitarian outlook and of his pedagogic principles. A permanent exhibition to be opened in the Orangery illustrates the Garden's chequered history and its scientific role, presenting the development of biological theories and ideas from the time of Linnaeus to the present and with anecdotal references to his remarkable and multifaceted career.

In his home (now the Linnémuseet), situated in a comer of the Garden, many of Linnaeus's household items can be seen together with plant and animal specimens, scientific instruments and paintings. These displays contribute to the experience offered to visitors, but the Garden itself is the heart. It is said that in early mornings in May one may still encounter the man himself, puffing his pipe, self-absorbedly counting the stamens of Linnaea borealis. The past is a key to the present and vice versa. Understanding our forebears helps us to understand ourselves. Herein lies part of the importance of historic botanic gardens, over and above their architectural history. The Linnétradgarden is a national heritage site. http:/www.wikipedia.org/wiki/Botanical-garden.

(B) Botanic Museum

(I) What is Botanic Museum

Botanical museum is an institution denoted to the acquisition, preservation, study, exhibition and educational interpretation of botanicals or significance of science, economics, culture and botanical history. It is generally a non- profit making, permanent institution in the service of botanists from different disciplines, agriculturists, naturalists, foresters and all those interested in plant sciences. It is also serviceable to the development of society at large and open to public.

(II) Functions of Museum and Importance

They have a variety of functions. Botanical museums have a significant role in preserving indigenous knowledge and are treasure houses of intellectual property. An important civilized activity is to record and convey the inventions for the future. They accumulate botanicals which indicate advancement of civilization as well. Various records including old writings, old records and botanical antiques are protected scientifically. The botanical museums help depict history of plant introduction, its effects, methods of utilization of plants or their parts /they create an awareness about plants as essential ingredients of human existence and life on earth. They also indirectly or directly popularize the science of plants. They have attained immense importance in the context of large number of vanishing species whose conservation has become important through plant introduction activity with such museums. They exhibit and communicat the tangible and intangible evidence of man – plant relationships and their environment. These connect people to knowledge and information, creativity and inspiration. They become a nucleus of pleasure for its vast majority of both literate and illiterate visitors and educate them. They aid and promote teaching / learning interdisciplinary subjects.

(III) Important Museums of The World

(i) National Museum of Natural History Paris, France

(ii) National Museum of Natural History, Beijing, China

(iii) British Museum of Natural History, London, UK

(iv) Natural History Museum Vienna, Austria.

(v) Field Museum of Natural History, Chicago, USA

(vi) U.S. National Museum, Washington, USA

(vii) Indian Museum, BSI, Kolkata, India

It is worth to note that the activity of museum is many times run by the establishments of herbaria. Museums are part of herbaria and also vice – versa. They are also instituted independently or they are part of Botanical Gardens as well. Thus museums, herbaria, botanical gardens and even libraries for taxonomic literature can be had under one or separate roof.

(C) Botanic Library

(I) What is a Botanic Library

Botanical Library is a storehouse of botanical books, journals, periodicals, magazines, photographs, movies, newspapers dissertations, keys, manuals, maps, atlases, news papers, etc. All printed materials, paintings, drawings, films and other similar articles belonging to the plant sciences are also stored using systems of classification employed in libraries. It is a centre for documentation of botanical activities, and stores the data from various disciplines of botany.

(II) Functions of the Botanic Library:

1. Taxonomists / botanists procure information to verify facts about the available literature, procedure, citation, classification and other aspects of their interaction. It is thus an information centre.
2. Since botanical library is a repository of rare books / publications, first editions, original manuscripts, illustrations, photographs, etc. and is useful as a documentation centre.
3. Library is data store – house for research since it contains all the published data related to the plant sciences.
4. It is a facility to provide the technical know – how of their valuable resources. A reference service of such libraries help procure the requisite material, photocopies, microfilms, etc. necessary for the research projects in the science. Library also have inter – library loan service. Through such exchange necessary literature can be provided to the needy in their places of work.
5. Big libraries are useful as a training centre for younger generations for introducing the science of library, for informing rules of libraries, systems of catalogues, etc.

The botanical library is thus useful to the systematists as an information source, a documentation centre, a data storehouse, and also as service facility (cf. Radford.1986).

(III) Important Botanic Libraries of The World

(i) Library of Congress, Washington, USA

(ii) Library of Academy of Sciences, Leningrad, Russia

(iii) National Library, Beijing, China

(iv) Library of the British Museum, London, UK

(v) Bibliotheque Nationale, Paris, France

It is worth to note that large botanical libraries are also associated or they are part of the botanical gardens, herbaria, museums and botanical societies e.g. New York Botanical Garden (New York, USA), Missouri Botanical Garden (St. Louis, USA), Gray Herbarium (Harvard University, USA), Academy of Natural Sciences, (Phildelphia, USA), etc.

Bibliography

Abbott, H.C.de S. (1886) Certain chemical constituents of plants considered in relation to their morphology and evolution. Bot. Gaz. 11 : 270 – 272.

Action plan for botanic gardens in the European union http:/www. botanicgardens.eu/action/ actionplaneu1htm.

Adams. R. P. (1975) Numerical-Chemosystemic Studies of infraspecific variation in Juniperus pinchotii. Biochem. Syst. Ecol. 3: 71-74.

Adams, R. P. (1979) Diurnal variation in the terpenoids of Juniperus scopulorum (Cupressaceae) - Summer versus winter. Amer. J. Bot. 66: 986-988.

Adams, R. P. and A. Hagerman (1976) A comparison of the volatile oils of mature versus young leaves of Juniperus scopulorum : Chemosystemtic Significance. Biochem Syst. Ecol. 4: 75- 79.

Adams, R. P. and A. Hagerman (1977) Diurnal variation in the volatile terpenoids of Juniperus scopulorum (Cupressaceae). Amer.J. Bot. 64: 278-285.

Adams, R. P. and R. A. Powell (1976) Seasonal variation of sexual differences in the volatile oil of Juniperus scopulorym. Phytochemistry 15: 509-510.

Adanson, M. (1973) Families des Plantes. Vol.I-II. Vincent, Paris, France.

Ahmad, K. J. (1975) Cuticular studies in some species of Lepidagathis and Barleria, Bot. Gaz. 136 : 129 – 135.

Ahmad, K. J (1976) Epidermal studies in some species of Hygrophila and Dyschoristc (Acanthaceae). J Indian Bot. Soc. 41 – 52.

Ahmad, K. J. (1979) Taxonomic significance of epidermal characters in Acanthaceae. In : Progress In Plant Research, Sliver Jublice Publication (NBRI, Lucknow) Ed. T.N. Khoshoo and P. K. K. Nair. Vol, I : 135 – 160. Today and Tomorrow Printer & Publishers, New Delhi, India.

Almeida, S. M. (1995) Role of Herbaria : An Educational Institution – Blatter's Herbarium, pp.30-40. In : Recent Advances In Taxonomy (The Herbarium) (Ed. Shiva Sharma and Kundan Sharma), Surbhi Publications, Jaipur, India.

Anonymous (1974) Trends, priorities and needs in systmatic and evolutionary biology. Brittonia 26 : 421- 439.

Anonymous (1992) Earth Summit: The United Nations Conference on 'Environment and Sustainable Development' (C3-14June). Rio de Janero, Brazil.(June 3-14). https: en.wikipedia.org/wiki/Earth-Summit.

Arber, E. A. and J. Parkin (1907) On the origin of Angiosperms. J Linn. Bot., Soc. 38 : 28 – 80.

Arekal, G. D. (1963) Embryological studies in Canadian representatives of the tribe Rhinantheae, Scrophulariaceae, Canad. Jour. Bot. 41: 267-302.

Ayensu, E.S. (1974) Leaf anatomy and systematics of New World Velloziaceae. Smithsonian Contr. Bot. No. 15: 1-125.

Babcock, E. B. (1947) The genus Crepis, Parts I and II. Univ. Calfornia Publ.Bot. 21: 1-198. 22: 199-1030.

Babcock, E.B. and D.R.Cameron (1934) Chromosome and phylogeny in Crepis- II. The relationships of one hundred eight species, Univ. Calfornia. Publ. Argic. Sci. 6: 287-324.

Babcock, E.B. and G.L. Stebbins (1942) Genetic and evolutionary processes in Crepis. Amer. Nat. 76 : 337 – 363.

Babcock, E.B. and J.A.Jenkins (1943) Chromosome and phylogeny in Crepis-III. the relationships of one hundred and thirteen species.

Babcock, E.B., Stebbins, G.L. and J.A.Jenkins (1937) Chromosome and phylogeny in some genera of the Crepidinae. Cytologia, Fujji Jubilee Vol. : 188-210.

Bailey, I. W. (1951) The use and abuse of anatomical data in the study of phylogeny and classification. Phytomorphology 1: 67-69.

Bailey, I.W. and R.A. Howard (1941) The comparative morphology of the Icacinaceae. I. Anatomy of the node and internode. I. Arnold. Arbor. 22: 125-132.

Bailey, L. H. (1939) Standard Encyelopacdia of Horticulture. Vol. I – III. The Macmillan Co. New York, U.S.A.

Bailey, L. H. (1949) Manual of Cultivated Plants The Macmillan Co. New York, U.S.A.

Baillon, H. (1867-1895) Historie des Plantes, Paris, France.

Baker, H. G. (1983) An outline of the history of anthecology or pollination biology : Pollination Biology (Ed.L.Real).PP. 7 – 28. Academic Press, Orlando, U.S.A.

Baker, R. T. and H. G. Smith (1920) A research On The Eucalyptus Especially In Regard To Their Essential Oils. 2nd. New South Wales Technologial Museum, Sydney, Australia.

Baldwin, J.T. jr.(1938) Kalanchoe : The genus and its chromosome. Amer.J. Bot. 25; 572-579.

Band, L. Heyn, C. C, and U. Plitmann (1981) Distribution of cyanogenesis in Lotus Leguminosae. Taxon 30 : 601 - 608.

Banergi, I. (1961) The endosperm in Scrophulariaceae. Jour. Indian. Bot. Soc. 40 : 1-11.

Barefoot, A.C and F.W. Hankins (1582) Identification of Modern and Tertiary Woods. Claendon Press, Oxford, U.K.

Barua, P. K. and W. Wight (1958) Leaf sclereids in the taxonomy of The Camellias. I. Wilson's and related Camellias. Phytomorphology 8 : 257 – 264.

Bate-Simth, E.C (1954) Leuco-anthocyanins - 1 : 7 Detection and identification of anthocyanidins formed from leuco-anthocyanins in plant tissues. Biochem. J. 58: 122-125.

Bate – Smith, E.C.(1958) Plant phenolics as taxonomic guides. Proc. Linn. Soc 169 : 198 – 211

Bate-Simth, E. C. (1973) Systemetic distribution of ellagi-tannins in relation to the phylogeny and classificatiion of the angiosperms. Noble 25. In : Chemistry Botanical Classification, pp. 93-102.

Bate-Smith, E. C. and C. R. Metcalfe (1957) Leuco - anthocyanins – 3 : The nature and systematic distribution of tannins in dicotyledonous plants. J. Linn. Soc (Bot,) 55 : 669 - 705.

Bate-Smith, E. C. and N. H. Lerner (1954) Leauco-anthocyanins - 2 : Systematic Distribution of leuco-anthocyanins in leaves. Biochem. J. 58:126-132.

Bate-Smith, E. C. and P. Ribberau – Gayon (1959) Leuco- anthocyanins in seeds. Qualitas plant et Mater. Veget. 5: 189 - 198.

Bate-Smith, EC. And T. Swain (1966) The asperuloides and the aucubins. In: Comparative Phytochemistry (Ed. T. Swain) pp. 159-174. Academic Press. London. U.K.

Bauhin, G. (1623) Pinax Theatri Botanici.L.Regis, Paris, France.

Bauhin, Gaspard (Bauhin Caspar) (1623) TTINA= threatri botanici ---------------- Basileae Helvet. Sumptibus et typis ludovici Regis.

Bauhin, Jean and J. H. Cheierus (1619) H. C.------------ historiae plantarum generalis ------ prodrumus---------- Ebroduni, Ex Typographia Societatis Caldorianae.

Bauhin, Jean and J. H. Cherierus (1650 – 51) Historia plantarum univesales ----------- Quam recensuit et auxit -------------- Chabraeais----------------- publici fecit, fr. Lud a Graffenried ------------- Ebrodun.

Baumann, M. G. (1946) Mydocarpus und die Phylogenie der Umbelliferen-Frucht. I. Ber. Schw. Bot.Ges.56:13-112.

Behnke, H. D. (1965) Uber das phloem der Dioscoreacean unter besonderer Beevcksichtingung ihrer phloem becken II. Mittellung: Electronenoptische untersuchunged zur Feinstruktur der phloem beckens, Z. pflanzen 53: 214-244.

Behnke, H. D. (1967) Uber den Aufuau der Siebelement Plastiden einiger Dioscoreacean, Z. Pflanzen, 57: 243-254.

Behnke, H. D. (1974 a) Sieve-element plastids of Gymnospermae : Their ultastructure in relation to systematics. Pl. Syst. Evol. 123: 1-12.

Behnke, H. D. (1974 b) Elektornenmikroskopische Untersuchungen an Siebrothren, Plastids und ibre Ausage uber die systematisch Stellung Von Lophiocarpus. Bot. Sahrb. Syst. 94 : 114-119.

Behnke, H. D. (1975) The bases of angiosperm phylogeny : Ultrastuctur. Ann. Missouri. Bot. Gard. 62 : 647-663.

Behnke, H. D. (1977) Transmission electron microscopy and systematics of flowering plants. In: K.Kubitzki (Ed.) Flowering Plants : Evolution and Classification of Higher Categories, pp. 155-178. Springer-verlag, Berlin, Germany.

Behnke, H. D. and B. L. Turner (1971) On specific sieve-tube plastids in Caryophyllales: Further investigation with special reference to the Bataceae, Taxon 20: 731-737

Behnke, H. D. and R. Dahlgren (1976) The distribution of characters within angiosperm system, 2. Sieve-element plastids. Bot. Not. 129: 287-295.

Behnke, H. D. and T. J. Mabry (1977) S-type sieve-element plastids and anthocyanins on Vivianiaceae : Evidence against its inclusion into Centropsermae, Pl. Syst. Evol, 126 : 371-375

Behnke, H. D., Pop, L, and V.V.Sivarajan (1983) Sieve-element plastids of Caryophyllales : Additional investigations with special reference to the Carphyllaceae and Molluginaceae. Pl. Syst. Evol. 142: 109-116.

Belford, H. S. and W. F. Thompson (1981) Single copy DNA homologics in Atriplex. I. Cross reactivity estimates and the role of deletions in genome evolution Heredity 46 : 91-108 II Hybrid thermal stabilities and molecualar phylogeny. Hereditsa 46 : 109 - 122.

Bell, C.R. (1959) Mineral nutrition and flower to flower pollen sexine variation, Amer.J.Bot. 46: 621-624.

Benson, L. (1957) Plant Classification, D. C. Heath & Co., Boston, U.S.A.

Benson, L. (1962) Plant Taxonomy. Ronald Press. New York, U.S.A.

Benson, L. (1979) Plant Classification (2ndEd) D. C. Heath, Lexinton, Mass, U.S.A.

Bentham, G. and J. D. Hooker (1862 – 1883) Genera Plantarum Vol. I-III, Reeve & Co. Willams & Norgate, London, U.K.

Berg, R.Y. (1954) Development and dispersal of the seed of Pedicularis silvatica Nytt. Mag. But. 2: 1-60.

Berkeley, E. (1953) Morphological studies in the Celastracae, J. Elisha. Mitchell Sci. Soc. 69 : 185 – 206.

Bessey, C. E. (1883) Evolution and classification. Bot. Gaz. 18 : 329 -332.

Bessey, C. E. (1897) Phylogeny and taxonomy of Angiosperms Bot. Gaz. 24 : 145 – 178.

Bessey, C. E. (1915) The phylogenetic taxonomy of flowering plants. Ann. Missouri Bot. Gard. 2 : 109 – 164.

Bhandari, N.N. and K.Indira (1969) Studies in the Viscceac IV Embryology of Eubrachion (Hook.*et* Arn.). Engl. Bot. Notiser 122: 183-203.

Bhandari, N.N. and Nanda, Kanan (1968 a) Studies in the Viscaceae I. Morphology and embrylogy of the Indian dwarf mistletoe – Arceuthobium minutissimum Hook. *f.*Phytomorphology 18: 435-450.

Bhandari, N.N. and Nanda, Kanan (1968 b) Studies in the Viscaceae II A reinvestigation of the female gametophyte of Arceuthobium douglasii. Am. J. Bot. 55: 1028-1031.

Bhatanagar, S. P. (1970) Santalaceae. In : Proceedings of Symposium On Comparative Embryology of Angiosperms. Indian National Secience Academy, New Delhi, India. pp. 15 – 18

Bierhorst, D. W. (1971) Morphology of Vascular Plants, Macmillan, New York, U.S.A.

Blackith, R.E. and R.A. Reyment (1971) Multivariate Morphometrics. Academic Press, New York, USA.

Blackwelder, R. T. (1967) A critque of numerical taxonomy. Syst. Zool. 16 : 64 – 72.

Blair, W. F. and B. L. Turner (1972) The integrative approach to biological classification. In: J.A. Behnke (Ed.). Challeneging Biological Problems, pp. 133-217 Oxford Univ Press, New York, U.S.A.

Blakeslee,A.F. (1922) Variations in Datura due to changes in chrosome number, Amer.Nat. 56: 16-31.

Blakeslee, A.F. (1930) Extra chrosomes, a source of variations in the Jimson weed. Smithsonian Report 1930: 431-450.

Blakeslee, A.F. (1934) New Jimson weeds from old chromosomes. J.Heredity 25: 80-108.

Bocher. T. W. (1943) Studies on variation and biology in Plantago lanceolata L. Dansk. Bot. Arkiv. 11 (3) 1 – 18.

Bocher, T. W. (1972) Evolutionary problems in the Arctic flora. In: Taxonomy, Phytogeography and Evolution. (Valentive, D.H.Ed.) 101–113. Academic Press, London, U. K.

Bock, Hieronymus (1539) New Kreutterbuch Von underscheydf wurckung und namen der kreutter------------ gedruckt zu Strassburg, durch Wendel Rihel.

Bock, W. J. (1973) Philsophical foundations of classical evolutionary classification. Syst. Zool. 22 : 375 – 392.

Bold, H. C. (1967) Morphology of Plants (2nd Ed.) Harper & Row, New York, U.S.A.

Borrill, M. (1961) The pattern of morohological variation in diploid and tetraploid Dactylis. J. Linn, Soc. Bot, 56 : 441-452.

Bosemark.N.O. (1956 a) On the accessory chromosomes in Festuca pratensis -3rd. Frequency and geographical distribution of plants with accessory chromosomes. Hereditas 42: 189-210.

Bosemark, N.O. (1956 b) Cytogenetics of accessory chromosomes in Phleum phleoides. Hereditas 42: 443-466.

Botanic garden- Wikipedia-the free encyclopedia. http:/www. wikipedia.org/ wiki/ Botanical-garden.

Bradshaw, A. (1965) Evolutionary significance of phenotypic plasticity in plants. Adv. Gen. 13 : 115 - 155.

Braun, A. (1859) Flora der Provinz Bradenburg. W. Englemann, Leipzig, East Germany.

Bremekamp, C.E.B (1944) Materials for a monograph of Strobilanthinae (Acanthaceae), Verb. Nerderl. Acad. Wet. II 40 (1) : 1- 306.

Bremekamp, C.E.B. (1953) The delimitation of the Acanthaceae. Proc. K. Ned. Akad, Wet. (Ser. C) 56: 533-546.

Bremer, K. and H. E. Wanntorp (1979) Geographic populations or biological species in phylogeny construction ? Syst. Zool. 28 : 220-224.

Brenan, J. P. M. (1968) The relevance of national herbaria to modern taxonomic research. In: modern Methods In Plant taxonomy (Ed.V.H Heywood) PP,23-32. Academec Press, London, U.K.

Brewbaker, J. L. (1967) The distribution and phylogenetic significance of binucleate and trinucleate pollen grains in the angiosperms. Amer. J. Bot. 54 : 1069 – 1083.

Briggs, D. and M. Block (1981) An investigation into the use of the deme terminology. New Phytol. 89 : 729 – 735.

Briggs, D. and S. M. Walters (1984) Plant Variation and Evolution, 2nd Ed. Cambridge University Press, Cambridge, U.K.

Britton, N. L. and A. Brown (1913) An Illustrated Flora of North United States, Canada and British Possessions IInd : 211 Charles Scribner, New York, U.S.A.

Brogniart, A. (1843) Enumeration des Generes des Plantes & C. Paris, France.

Brooks, D. R. (1981) Classifications as languages of empirical comparative biology. In: V. A. Funk & D. R. Brooks (Ed.) Advances in Ciadistics. pp. 61-70. New York Botanical Garden, NewYork, U.S.A.

Brown, W.V. (1958) Leaf anatomy in grass systematics. Bot. Gaz. 119:170-178.

Brown, W.V. (1977) The Kranz syndnome and its subtypes in grass systematics. Mem. Torrey Bot. Club. 23(3) : 1-97.

Brummitt, R. K., I. K. Ferguson and M. M. Poole (1980) A unique and extraordinary pollen type in the genus Crossandra (Acanthaceae). Pollen *et* Spores 22 : 11 – 16.

Brunfels, Otto Von (1530) Herbarum Vivae eicones ------------ Argentorati apud Soannem Schottum.

Brunswick, H (1920) Uber das Vorkommen Von Grispskristallen bei der Tamaricaceae. Sber Acad. Wiss. Wien (Math. Naturw.Kl. Abst.1.) 129:115-136.

Bureau, E. (1864) Monographie der Bignoniacees. Paris, France.

Burger, W. C. (1967) Families of Flowering Plants in Ethiopia. Oklahma Agric. Exp. Sta. Bull. 45.

Burger, W. C.(1975)The special concept in Quercus. Taxon 24: 45-50.

Burkill, I. H. and D. Prain (1936) An account of the genus Dioscorea. Ann. Roy. Bot, Gard, Calcutta, 14:1-528.

Buss, P.A (Jr.) and N. R. Lersten (1975) Survey of tapetal nuclear number as a taxonomic character in Legumiosae. Bot. Gaz. 136 : 388 – 395.

Cain, A. J. (1953) Geography, ecology and coexistence in relation to the biological definition of the species, Evolution 7 : 76 – 83.

Cain, A. J. (1956) The genus in evolutionary taxonomy.Syst. Zool. 5 : 97-109.

Cain, A. J. (1958) Logic and memory of Linnaeus system of taxonomy. Proc. Linn. Soc. (London) 169 : 144 – 163.

Cain, A. J. (1959) Deductive and inductive methods in post-Linnaean taxonomy. Proc. Linn. Soc. (London) 170: 185-217.

Cain, A. J. (1976) The use of homology and anatology in evolutionary theory (with Discussion). In : M. Von Cranach Ed. Methods of Inference from Annals To Human Behaviour. PP. 25 – 35. The Hague Mouton.

Cain, A. J. and G. A. Harrison (1958) An analysis of the taxonomist's judgement of affinity. Proc, Zool. Soc. (London) 131 : 85 – 98.

Camerarius, R. J. (1694) De sexu plantarum Epistola, Tubingen, Camper, Dissertation Physiqves, Germany.

Camp, W. H. and C. L. Gilly (1943) The structure and origin of species with a discussion of intraspecific variability and related nomenclatural problems. Brittonia. 4 : 323 – 385.

Campbell, D.H. (1930) The phylogeny of the angiosperms. Ann. Bot. 44: 311-331.

Candolle, A. L. P. P. de (1855) Geographic Botanique, 2 Vols, Victor Masson. Paris, France.

Candolle, A. P. de (1813) Theorie elemantaire de la Botanique. Paris, France.

Candolle, A. P. de (1824 -1873) Prodrumus Systematics naturalis regni Vegetabilis. Vol. I – XVII. Trenttel & Wintz, Paris, France.

Carlquist, S. (1961) Comparative Plant Anatomy. Holt. Rinchart and Winston, New York. U.S.A.

Cave, M. S. (1959) Embryological characters of ltaxonomic value. Proc. 9th. Int. Bot. Congr. Montreal 2: 62

Cave, M. S. (1962) Embryological characters of taxonomic significance. Lillea 31: 171-182.

Cerceau – Larrival, M.T. (1971) Morpholgie Pollinique ed correlations Phylogenetique chez les Ombellferes. In : The Biology and Chemistry of The Umbelliferae (V.H.Heywood Ed.) pp. 109 – 156. Academic Press, London, U.K and New York, U.S.A.

Ceasalpino, Andrea (Caesalpinus, Andreas) (1583) De Plantis libri XVI ----- Florentiae, Apud Georgium Marescottum.

Charles L' Ecluse (Clusisus Carolus) (1601) Caroli Clusii atrebatis------------- rariorum plantarum historia------------- Antverpiae Ex. Offician plantiniana Apud Joannem Morteam.

Chaung, T. I. and I. R. Heckard (1972) Seed coat morphology in Cordylanthus (Scrophulariaceae) and its taxonomic significance, Amer. J. Bot. 59: 258-265.

Cheadle, V. I. (1942) The occurrence and types of vessel in the various organs of the plants in the Monocotyedonae. Amer. J. Bot. 29: 441-450.

Cheadle, V. I. (1943) The origin and certain of trends specialisation of the vessels in Monocotyledonac. Amer. J. Bot. 30: 11-17.

Cheadle, V. I. and H.Kosaki (1971) Vessels in Liliaceae. Phytomorphology 21:320-333.

Cheadle, V. I. and H.Kosaki (1972) Vessels in Cyperaceae. Bot.Gaz. 113: 214-223.

Cheadle, V. I. and J.M.Tucker (1961) Vessels and phylogeny of Monocotyledonae.In: Recent Adv.Bot.(Ed.D.LBailey)I: 161-165.

Cheadle, V.I. and N.W.Uhl (1948) Types of vascular bundles in the Monocotyledonae and their relation to the late metaxylem conducting elements. Amer. J. Bot. 35: 486-496.

Chesnais, F. (1941) De l' anciennete du genre Actinidia et de sa parente avec les Magnoliacees. Bull. Mus. Kist. Nat.Marseille 13: 202.

Chiarugi, A. (1933) Lo sviluppo del gametofito femmineo della Weddellina squamulosa Tul. (Podostemaceae). R.C.Acad. Nat. Lincei 17: 1095-1100.

Chopra, R.N. and A.J.Mukkada (1966) Gametogenesis and Pseudo-embryo sac in Indotristichia ramosissima (Wight) Van Royen. Phytomorphology 16: 182-188.

Chowdhuri, P. K. (1957) Studies in the genus Silene. Notes Roy, Bot. Gard. Edinburgh 22: 221-278.

Christensen J.E. and H.T. Horner Jr. (1974) Pollen pore development and its spatial orientation during microsporogenesis in the grass Sorghum bicolor. Amer, Bot. 61 : 604 – 623.

Clausen J., Keck, D.D.and W.M.Hiesey (1940) Experimental studies on the nature of species I. Effect of varied environments of western North American Plants. Publication Carnegie Inst. (Washington) 520 : I – VIII. 1 – 452.

Clausen, J., Keck, D.D. and W.M.Hiesey (1945) Experimental studies on the nature of species II. Plant evolution through amphidiploidy and autoploidy with example from the Madiinac Publication Carnegie Inst (Washington) 564 : i – viii, 1– 174.

Clausten, J., Keck, D.D.and W.M. Hiesey (1948) Experimental studies on the nature of species III. Environmental responses of climatic races of Achillea. Publication Coregie Inst. (Washinton) 581 : i–iv, 1 – 129.

Cleland, R.E. (1923) Chromosome arrangement during meiosis in certain Oenothera. Amer. Naturalist. 57: 562-566.

Cleland, R.E. (1936) Some aspects of the cyto-genetics of Oenothera. Bot.Rev. 2: 316-348.

Cleland, R.F. (1951) Extra diminutive chromosomes in Oenothera. Evol. 5: 165-176.

Clements, F. E. and F. L. Long (1923) Experimental Pollintion : An outline of the ecology of flowers and insects. Publ. Carnegie Inst. (Wishington) 336 : 1 – 274.

Clifford, H. T. and L. Watson (1978) Identifying Grasses. Data Method and Illnstraions. University of Queensland Press, St. Lucia, Australia.

Cocucci, A.E. (1983) New evidence from embryology in angiosperm classification. Nord. J. Bot. 3 : 67 – 73.

Cole, G. T. and H. D. Behnke (1975) Electron microscopy and plant systematics. Taxon 24: 3-16.

Cole, M. J. (1961) Interspecific relationphips and intraspecific variation of Chenopodium album L. in Britain. I. The taxonomic delimitation of the species. Watsonia 5:47-58.

Conner, E. J. H. (1976) The Seeds of Dicotyledons. Cambridge University Press, London, U.K.

Cook, O. F. (1899) Four categories of species. Amer. Naturalist 33 : 287–297.

Cook, C. D. K. (1968) Phenotypic plasticity with pariculer reference to three amphibious plant species. In : Modern Methods In Plants Taxonomy (V.H. Heywood Ed.), PP. 97 – 111. Academic Press, London, U. K..

Cook, R.B. Ivimey (1969) The phenetic relationship between species of Ononis. In: Numerical Taxonomy (Ed. A.J.Cole) Academic Press, London, UK and New York, USA, pp-68-85.

Cooke, T. (1903) The Flora of The Presidency of Bombay. Taylor and Francis, London, U.K.

Cooke, T. (1958) The Flora of The Presidency of Bombay. Vol.I-III B.S.I. Calcutta, India (Repr.Ed.).

Copeland, H. F. (1938) The kingdom of organisms. Quart. Rev. Biol. 13 : 383 – 420.

Cordus, Valerius (1561) In hoc volumine continentur Valerli Cordii-------------, Annotationes in Pedaci Dioscoridis----------, de Medica materia-------eiusdem Val. Cordi historiae stirpium lib. IIII------------ Omnia------------ Conr. Gesneri -------- collecta, et prafationibus illustrate. Argentorati excudebat Josias Rihelius.

Core, E. L. (1955) Plant Taxonomy, Prentice-Hall, Englewood Cliffs, N. J., U.S.A.

Corner, E. J. H. (1951a) The Leguminous seed. Phytomorphology 1:117-150.

Corner, E.J.H. (1951b) Review of P. Maheshwari, an Introduction to the Embrylogy of Angiosperms. Phytomorphology 1: 243.

Corner, E.J.H. (1958) Transference of function. Bot. Jour. Linn. Soc.,56:33-40.

Corner, E.J.H. (1976) The seeds of Dicotyledons. Cambridge Univ. Press, Cambridge,UK.

Correa, J. P. (1958) Morphological and embtyological studies on the Loranthaceae Viscoideae. Ph.D. Thesis. Univ Delhi, India.

Coulter, J. M. and C. J. Chamberlain (1903) Morphology of Angiosperms, Appleton, New York, U.S.A.

Cowan, J.M (1950) The Rhododendron Leaf : A Study of The Epidermal appendages. Oliver & Boyd, London, U. K.

Crawford, D. J. and K. A. Evens (1978) The affinities of Chenopodium flabellifolium (Chenopodiaceae) : Evidence from seed coat surface and flanonoid chemistry. Brittonia 30:313-318

Crawford, D.J., Stuessy, T.F. and M. Silva (1987) Allozyme divegence and the evolution of Dendroseris (Compositae : Lactuceae) on the Juan Fernandez Islands. Syst. Bot. 12 : 435 – 443.

Crete, P. (1949 a) Embrylogie des Gentianacees: Development de 1' embryon chez le Erythraea centaurium Pers. C.R.Acad. Sci. Paris 228: 1448-1449.

Crete, P. (1949 b) Embrylogie des Gentianacees: Gentiana asclepidea L. C.R.Acad. Sci. Paris. 228: 768-770.

Crete, P. (1951) Repartition et interet phylogenetique des albumens a formations haustoriales chex les angiospermes et plus particulierment ches les gamopetales. Ann, Sci. Nat. Bot. II. Ser 12: 131-191.

Crete. P. (1955) Embryogenie des Gentianacees: Development de 1' embryo chez le Chlora perfoliata L. C.R.Acad. Sci. Paris. 242: 3110-3113.

Crete, P. (1956) Menyanthees: Development de 1, embryi chez le Limnanthemum nymphoides Horfget Link. C.R.Acad. Sci. Paris 142: 3110-3113.

Cronquist, A. (1957) Outline of a new system of families and orders of dicotyledons. Bull. Jard. Bot. Brax. 27 : 13 – 40.

Cronquist, A. (1964) The old systematics : In : C. A. Leone (Ed.) Taxonomic Biochemistry and Serology. PP. 3–11. Ronald Press, New York, U.S.A.

Cronquist, A. (1965) The status of the general system of classification of flowering plants. Ann. Missouri Bot. Gard. 52:281-303.

Cronquist, A. (1968a) Evolution and Classification of Flowering Plants, Nelson, New York, U.S.A.

Cronquist, A (1968b) Relevance of mational herbaria to modern taxonomic research in the United States of America. In : Modern Methods In Plant Taxonomy (Ed.V.H. Heywood). Academic Press, London, U.K.

Cronquist, A. (1975) Some thoughts on angiosperm phylogeny and taxonomy. Ann. Miss. Bot. Gard. 62 : 517-520.

Cronquist, A. (1978) Once again, what is a species? In : L.V. Knutson, chm. Biosystematics in Agriculture, pp. 3-20. Allenheld Osmun Montclair, N, J.

Cronquist, A. (1981) An Intergrated System of Classification of Flowering Plants. Columbia University Press, New York, U.S.A.

Cronquist, A. (1988) The Evolution and Classification of Flowering Plants (Ed. II). New York Bot. Gard., New York, U.S.A.

Crovello, T. J. (1974) Analaysis of character variation in systematics. In : A. E. Radford, W. C. Dickison, J. R. Massey and C. R. Bell (Ed.) Vascular Plant Systemetics. PP. 451 – 484. Harper and Row, New York, U.S.A.

Crow, G. E. (1979) The systematic significance of seed morphology of Sagina (Caryophyllaceae) under scanning electron microscopy. Brittonia 31: 52-63.

Crowson, R.A. (1970) Classification and Biology. Atherton Press, New York, U.S.A.

Cullen, J. (1962) A taxonomic Revision of Anthyllis sect. Vulneraria (Unpublished).

Cutler D.F. (1965) Vegetative anatomy of Thurniaceae. Kew Bull. 19:431-441.

Cuvier, G. (1835) Le Regne Animal Distribue après son Organization. 2nd Ed. Crochard, Paris, France

Dadswell, H.E. (1938) The wood structure of some Australian Rutaceae with methods for their identification. Tech, Pap. For. Prod. Res. Aust. No. 25, pp.32

Dadswell, H. E. and D. I. Ellis (1939) The wood anotomy of some Austalian Meliaceae with methods for their identification. Tech. Pap. For. Prof. Res. Aust. No. 31, pp : 20,

Dahlgren, R. M. T. (1975) A system of classification of the angiosermae to be used to demonstrate the distribution of characters. Bot. Notiser (London) 128: 119-147.

Dahlgren, R. (in cooperation with B. Hansen, K.Jakobsen and K.Larden) (1976) Angiospermernes Taxonomi. Bind 4. Akademisk Forlag, Kobenhavn.

Dahlgren, R. M. T. (1977) A system of classification of the angiosperms to be used to demonstrate the distribution of character states. IN : K. Kubitzki (Ed.) Plants Systemantics & Evolution (Suppl. 1). PP. 253 – 283. Springer – Verlag, New York, U.S.A.

Dahlgren, R. M. T. (1980) A revised system of classification of the angiosperm, Bot. J. Linn. Soc. 80 : 91 – 124.

Dahlgren, R. M. T. and H. T. Clifford (1982) The Monocotyledons : A Comparative Study. Academic Press, London. U.K.

Danser, B. H. (1929) On the taxonomy and the nomenclature of the Loranthaceac of Asia and Australia. Bull. Jard. Bot. Buitenz. 10: 291-373.

Danser, B. H. (1929) Ueber die Begriffe Komparium, Kommiskuum und Konvivium and ueber die Entstehungsweise der Konvivien. Genetica. 11 : 399 – 450.

Danser, B.H. (1933 a) A new system for the genera of Loranthaceac – Loranthaceae with a nomenclatur for the Old World species of this subfamily. Verb, Akacl, West. Amsterdam 29: 1-128.

Danser, B.H. (1933 b) Thaumasianthus cine neue Loranthaceen-gattung aus den Philippinen, Rec. Trav. Bot. Neerl. 33: 467-481.

Darlington, C. D. (1937) Recent Advances In Cytology. (2nd Ed.), Blakiston, Philadephia, U.S.A.

Darlington, C.D. (1956) Chromosome Botany. Allen & Unwin, London, U.K.

Darlington, P. J. Jr. (1971) Modern taxonomy, reality and usefulness. Syst. Zool. 20: 341-365.

Darwin, C. (1859) On The Origin of Species By Means of Natural Selection or The Preservation of Favoured Races In The Struggle For Life. J. Murray London, U. K.

Darwin, C. (1862) On The Various Contriances By Which British and Foreign Orchids Are Fertilised By Inseets. Murray, London, U.K.

Darwin, C. (1876) The Effects of Cross and Self Fertilisation In The Vegetable Kingdom, Murray, London, U.K.

Daubenmire, R. F. (1959) Plants and Envirnoments. A Texbook of Plants Autoecology, London, U. K.

Davis, G.L. (1966) Systematic Embroylogy of Angiosperms, John Wiley & Sons, New York, U.S.A.

Davis, P.H and V. H. Heywood (1963) Principles of Angiosperm Taxonomy, Oliver & Boyd, Edinburgh and London, U.K.

Davis, P. H. (1960) Materials for a Flora of Turkey : IV. Ranunculaceae, II. Notes Roy. Bot, Gard, Edinburgh 23:103-161.

Davis, P. H. (1961) Hints for hard-pressed collectors. Watsonia 4 : 283-289.

Davis, P. H. (1978) The moving staircase : A discussion on taxonomic rank and affinity. Notes Roy; Bot. Gard. (Edinburgh) 36 : 325 – 340.

Davis, P. H. and M. S. Khan (1961) Aristolochia in the Near East. Notes Roy. Bot. Gard. Edinburgh 23: 515-546.

Davis, P. H. and V. H Heywood (1963) Principles of Angiosperm Taxonomy. Van Nostrand Princeton, N. J., U.S.A.

De Candolle (Augustin P. (1825), Theorie Elementaire de la Botanique. Deterville, Paris France.

De Candolle (Augustin P. (1813), A. P. Prodrumus Systematis naturalis regni Vegetabilis. Vol. 2 Treuttel & Wurtz, Paris, France.

De Candolle (Augustin P. (1867), Lose de la nomenclature botanique. Kessinger Publishing, USA.

De Launay, L. (1926) Phyolgenetische chromosomen zung verkur. Z. Sellfarsch' Microsk, Anat. 4 : 338

De Swet, J.M.J. and J.R. Harlan (1972) Chromosome pairing and phylogenetic affinities, Taxon 21 : 67 – 70.

Dede, R. A. (1962) Foliar venation patterns in Rutaceae. Amer. J. Bot. 49 : 90 -497.

Dehay, C.L. (1935) Appareillibero- lignerux foliaire des Euphorbiacees. Ann. Sci Nat. b. Ser. IO 17 : 147-290.

Dehgan. B. (1982) Comparative anatomy of the petiole and infrageneric relationship in Jatropha (Euphorbiaceae).Amer. J. Bot. 69(8) : 1283-1295.

Diels, L. (1924) Die Methoden der Phytographic und der Systematik der Pflanzen. In : E. Abderhalden (Ed.) Handbuch der biologischen Arbeutemethdoen. Abt. 11 : Methoden Zur Evforschung der Leistungen des Pflanzenorganismuse 1 : 67 - 190. : Urban and Schwarzenburg, Berlin.

Dilchor, D. (1974) Approaches to the identification of Angiosperm Leaf remains. Bot. Rev. 49(1): 1–157.

Dillemann, G. (1960) La differentiation chimique infrospecifique. Planta Medica 8 : 263 - 274.

Dittrich, M. (1968) Morphologische under suchungen an fruchten der subtribus Carduceac – Centaureinae (Conpositae). Willdenowia 5 : 67 – 107.

Dixit, S. N. (1958 a) Morphological and embryological studies in the Loranthaceac –IV[h] Ameyema Van Tiegh. Phytomorphology 8; 346-364.

Dixit, S.N. (1958 b) Morphological and embryological studies in the Loranthaceac –v[th] Lekpeostege gemminiflorus (Bl.) Bl - Phytomorphology 8: 365-376.

Dixit, S.N. (1961) Morphological and embryological studies in the Loranthaceae – VIIIth Tolypanthus Bl. Phytomorphology 11: 335-345.

Dnyansagar, V.R. (1955) Embryological studies in the Leguminosae 12. Status of the Leguminosae. Saugar Univ. J. 1: 39-53.

Dnyansagar, V.R. and A.Dhanraj (1972) Meiosis in plants of Solanum viarum Dunal with B-chromosomes. In: Biology of The Land Plants (V. Puri Ed.) pp.214-245. Sarita Prakashan,Meerut (U.P.) India.

Donoghue, M. J. (1985) A critique of the biological species concept and recommendations for phylogenetic alternative. Bryologist 88 : 172 – 181.

Dormer, K. J. (1946) Vegetative morphology as a suide to the classification of the Papilionatae. New Phytol. 45 : 145 - 161.

Dormer, K. J. (1962) The fibrous layer in the anther of Compositae. New. Phytol. 61: 150 – 153. In Senecio (Asteraceae). Bot. J.Linn.Soc. 97 : 63 – 71 in Senecio (Asteraceae). Bot. J. Linn. Soc. 97 : 63-71 Vincent, P.L.D. and F.M.Getliffe (1988) The endothecium.

Doyle, J. A.; M. Von Campo and B. Lugardon (1975) Observation on exine structure of Eucommiidites and lower cretaceous argiosperm pollen. Pollenet Spores 17 : 429 – 486.

Doyle, J.A. and L.J. Hickey (1976) Pollen and leaves from the mid-cretaceous potomac group and their bearing on early angiosperm evolution. In : Origin and Early Evolution of Angiosperm.C.C.B. Beck Ed.) New York, pp. 139-206.

Du Riet Z, G. E. (1930) The fundamental units of biological taxonomy. Svensk Bot. Tidskr. 24 : 333 – 428.

Ducker, S. C. and R. B. Knox (1985) Pollen and pollination : A historical review. Taxon 34 : 401 – 419.

Dufou, O. (1961) Structure cytolgique de pollon de quelques Angiospermes. Rev. Gen. Bot. 68 : 73 – 110.

Duke, J. A. (1961) Preliminany revision of the genus Drymaria. Ann. Missouri Bot. Gard, 48: 173-268.

DuPraw, E. J. (1964) Non-Linnean Taxonomy. Nature 202 : 849 – 852.

DuPraw, E. J. (1965) Non-Linnean taxonomy and the systematics in honeybees. Syst. Zool. 14 : 1 – 24.

Eames, A. J. (1936) Morphology of Vascular Plants, Lower Groups McGraw – Hill, New York. U.S.A.

Eames, A. J. (1953) Floral anatomy as an aid in generic limitation. Chron.Bot.14:126-132.

Eames, A.J. (1961) Morphology of the Angiosperms, McGrawl-Hill, New York, U.S.A.

East, E.M. (1928) The genetics of the genus Nicotiana Bibliogr. Genet. 4: 243-320.

Eastabook, G. F. (1978) Some concept for the estimation of evolutionary relationships in systematic botany. Syst. Bot. 3 : 146 – 158.

Edwards, P. (1976) A classification of plants into higher taxa based on cytological and biochemical criteria. Taxon 25 : 529-542.

Eglinton, G. and R. J. Hamilton (1963) The distribution of alkanes, In: Chemical Plant Taxonomy (T. Swain Ed.) pp. 187-217. Academic Press, London, U. K.

Ehrendorfer. F. (1968) Geographical and ecological aspects of infraspecific differentiation. In : Moden Methods In Plants Taxonomy (Heywood. V. H. Ed.). PP. 261 – 296. Academic Press, London, U.K.

Ehrlich, R. R. and P. H Raven (1969) Differentiation of populations. Science 165 : 1228 – 1232.

Eichler, A. B. (1883) Syllabus der Vorlesungen uber Phanarogamen Kunde 3rd Ed. Gebruder Borntraeger, Berlin. Germany.

Eichler, A. W. (1883) Syllabus der Vorlesungen uber Specielle und medicinisch Pharmaceutische Botanik. Gebruder Borntraeger, Berlin., Germany.

Eldredge, N. and J. Cracraft (1980) Phylogenetic Patterns and The Evolutionary Process : Method and Theory. In : Comparative Biology. Columbia University Press, New York, U.S.A.

Emboden, W. A. and H. Lewis (1967) Terpenes as taxonomic characters in Salvia section Audibertia. Brittonia 19:152-160.

Engler, A, H.G. and K. Prantl (1887-1915) Die naturlicten Pflazenfamilien. W.Englemann, Leipzig, East Germany.

Engler, A. A. (1886) Fuhrer durch den koniglich botanischen Gaertn der Universitat 24 Breslu. Breslu. J. U. Kern's Verlag.

Engler, A. H. G. (1892) Syllabus der Pflanzenfamilien. Gebruder Borntraeger, Berlin, Germany.

Engler, A.H.G. and Diels (1936) Syllabus der Pfanzenfamilien (11th Ed.) Gebruder Borntraeger, Berlin. Germany.

Erdtman, G. (1952) Pollen Morphology And Taxonomy : Angiospmers (An Introduction To Palynology – I) Almqvist & Wiksell, Stockhom, Sweden.

Erdtman, G. (1969) Hanbook of Palynology : Morphology – Taxonomy – Ecology : An Intoductuin To The Study Of Pollen Grains and Spores. Hafner, New York, U.S.A.

Erdtman, G. and H. Straka (1961) Cormophyte spore classification. Geol. Foren. Forhandl. 83 (1) : 65 – 78

Estabrook, G. F., Johnson C. S. and F. R. Memorris (1975) An idealised concept of the trust cladistic character. Math. Biosci, 29 : 181-187.

Faber, F.C. and C.Von (1912) Das erbliche Zusammenleben von bakterien and tropischen pflanzen, Jahrb. Wiss. Bot. 51(3) : 285-375.

Faegri, K. (1956) Recent trends in palynology. Bot. Rev. 22 : 639 – 664.

Faegri, K. (1981) A note on viscin threads Grana 20: 143.

Fahn, A. (1953) The topography of the nectary in the flower and its phylogenetical trend. Phytomorphology 3:424-426.

Farris, J. A. (1979) The imformation content of the phylogenetic system. Syst. Zool. 28 : 483-519.

Favali, M. A. and F. M. Gerola (1968) Tubular and fibrillar components in phloem of Brassica chinensis L. Leaves. Giorn Bot. Ital. 102 : 447 – 467.

Favanger,C. (1960) Etucle cytologique du Cicendia filiformis et du Microcala pussilla (Gentianaceae). Bull.Soc.Bot.Fr. 107: 94-98.

Favarger, C. and J. Condandriopoulis (1961) Essai sur pendisme. Ber. Schweiz Bot. Ges. 71 : 384 – 406.

Fceetherly, H. I. (1954) Taxonomic Terminology of The Higher Plants. Iowa State College Press, Arnes, U.S.A.

Fernald, M. I. (1950) Gray's Manual of Botany (Ed.8.). Boston and New York, U.S.A.

Fish, R. K. (1970) Megagamentogenesis in Clematis and its taxonomic and phylogentic implications Phytomorpholgy 20 : 317 – 327.

Flory, W.S. (1977) Overview of chromosome evolution in the Amaryllidaceae. Nucleus 20: 70-84.

Fosberg, F. R. and M, Sachet (1965) Manual of tropical herbaria, Regnum Vegetabite 39 : 5- 132.

Foster, A. S. and E. M. Gifford (1974) Compartive Morphology of Vascular Plants (2nd Ed.), Freeman, San Fransico, U.S.A.

Frei, E. (1955) Die Innervierung der floralen Neetarien dikotyler Pflanzenfamilien. Ber. Schweiz.Bot.Ges. 65:60-114.

French, J. C. (1958) b Pattern of endothecial wall thickening in Araceae : Subfamilies Calloideae, Lasioideeae and Philodendroideae. Bot. Gaz. 146 : 521 – 533.

French, J. C. (1985a) Patterns of endothecial wall thickening in Araceae : Subfamilies Ptohoideac and Monsteroideae. Amer. J. Bot 72 : 472 – 486.

Friedman, H. (1966) The significance of the unimportant in studies of nature and art. Proc. Amer. Philos. 110 : 256 - 260.

Frisch. K. Von (1914) Der Farbensin und Formensinn der Biene. Zool. Jahrb. Abr. Allg. Physiol. Tiere 35: 1-182.

Fritzsche, Carl Julius (1837) Uber den Pollen, Von Julius Fritzsche.St.- Petersburg:Kais. Academic der Wissenschaften. 3 : 649 – 672.

Frost, S. (1958) The geographical distribution of accessory chromosomes in Centaurea scabiosa. Hereditas 44: 75-111.

Funchs, Leonhard (1542) De historia stirpium ---------- Baseleae, iN officina Insingriniana-------------

Gagnepain, F, and E. Boureau (1946) Une nouvelle famille de Gymnosperms Sarcopodacees. Bull. Soc. Bot. Fr. 93: 313-320.

Gagnepain, F. and E. Boureau (1947) Nouvelles considerations systemetique a propos du Sarcopus aberrans Gagenpain. Bull,.Soc. Bot. Fr. 94 : 182 – 185.

Ganapatishastri, T. (1924) Kautilyam Arthshastrum (Dvitiyah Sampushtah) Trivandrum, Tamil Nadu, India.

Garber, E.D. (1958) The genus Collinsio – III. The significance as a cytotaxonomic tool. Madrono 14 : 172 – 176.

Gasson, P (1979) The identification of eight woody genera of the Caprifoliaceae by selected fcatures of their root anatomy. Bot. J. Linn. Soc.78:267-284.

Geesink, R. (1969) An account of the genus Portulaca in Indo – Australia and the Pacific (Portulacaceae). Blumea 17 : 275 – 301.

Gemeraad, J.H. and J. Muller (1970) A computer-based numerical coding system for the description of pollen grains and spores. Rev. Palacobto. Palynol. 10: 175-202.

George, T. N. (1956) Biospecies' chronospecies and morphospecies. In : P. C. Sylverter-Bradley Ed. The Species Concept In Palaeontology. PP. 123–137 Systematics Association, London, U. K.

Gerard, John (1597) The Herball or Gernerall Historie of Plantes-------- John Norton, London, U.K.

Giannsi, D. E. (1978) Generic relationships in the Ulmaceae based on flavonoid chemistry. Taxon : 331 - 344.

Giannsi, D.E. and Tsan Lans Chaung (1976) Flovonoid systematics of the genus Perideridia (Umbelliferae). Brittonis 28 : 177 - 194.

Gibbs R. D. (1958) Biochemistry as an aid to establishing the relatioship of some families of dicotyledonous plants. Proc. Linn. Soc. Lond. 169 : 216 – 230.

Gilmour, J. S. L. and J. Heslop – Harrison (1954) The deme termintnology and the units of microevolutionary change. Genetica 27 : 147 – 161.

Gilmour, J. S. L. and J. Heslop-Harrison (1954) The deme terminology and the units of microevolutionary change. Genetica 27 : 147-161.

Gilmour, J. S. L. and J. W. Gregir (1939) Demes : A suggested new terminology. Nature 144 : 333.

Gilmore, S., Peakall, R. and J. Robertson (2007) Organelle DNA haplotypes reflect crop-use characteristics and geographic origin of Cannabis sativa. Forensic Science International 172:179-190.

Gomez-Campo and Maria, T. E. (1974) The Taxonomic and evolutionary significance of some juvenile characters in the Brassicaceae.Bot.J.Linn, Soc. 69: 105-124.

Gornal, R. J., Bohm. B. A. and R. Dahlgren (1979) The distribution of flavonoids in angisperms, Bot. Notiser. 132 : 1 - 30.

Gottlieb, L. D. (1977) Electrophoretic evidence and plant systematics. Ann. Missouri Bot, Gard. 64: 161 – 180.

Govil, C. M. (1970) Convolvulaceae. In: Proc of The Symposium on Comparative Embryology of the Angiosperms, Indian National Science Academy, New Dehli, India No. 41: 246-249.

Govil, C.M (1995) Angiosperms : Floral Anatomy. In : Botamy In India : History and Progress Vol. II. pp. 37-57. (B.M.Joshi Ed.) Oxford & IBH Publishing Co.Rvt. Ltd. New Delhi, India.

Govil, C.M. and S. Lavania (1980) Floral anatomy and embryology of some species of Cuscuta L. Proc. Indian Acad. Sci. (Plant Sci) 89(3) : 219 – 228.

Govindarajalu, E. (1962) The comparative morphology of the Alangiaceae, Proc. Natl. Inst. Sci. Ind.28 (B) : 507-517.

Govindarajalu, E.(1969) The systematic anatomy of south Indian Cyperaceae : Cyperus L. sybgenus Kyllinga (Rottb.) Surinagsar. Bot. J. Linn Soc. 62:41-58.

Grant, V. (1957) The plant species in theory and practice. In: (Mayr. E. Ed.) The species problem. Am. Assoc., Adv. Sci. Publ. 50: 39-80.

Grant, V. (1963) The Origin of Adaptaions. Columbia University Press, New York, U.S.A.

Grant, W.F. (1976) The evolution of karyotype and polyploidy in arboreal plants. Taxon 25: 75-83.

Gregory, W.C. (1941) Phylogenetic and cytolgical studies in the Ranunculaceae. Trans. Am. Phil. Soc. (N.S.) 31: 443-521.

Greilhuber J. and F. Speta (1978) Quantitative analysis of c–banded Karyotypes and systematics in the cultivated species of Scilla siberica group (Liliaceae). Pl. Syst. Evol. 129 : 63 – 109.

Greuter, W. *et al.* (Eds.) (1988) International Code Botanical Nomenclature As Adopted. By The Fourteenth International Botanical Congress. Berlin, July-August 1987. Regnum Veg. 118 : IXV, 1-328.

Grew, N. (1680) The Anatomy of Plants. 2nd Ed., W. Rawlins, London.UK.

Griffith, G. C. D. (1974) On foundations of biological systematics. Acta Biotheor. 23 : 85 -131.

Gulliver, G. (1866) On raphides as natural characters in the British flora. Quart. J. Micr. Sci. (N.S.) 6 : 1.

Gunderson, A. (1950) Families of Dicotyledons. Waltham, U.S.A.

Hallier, H. (1905) Provisional scheme of the natural (phylogenetic) system of flowering plants. New Phytol. 4 : 151 -162.

Hallier, H. (1908) On the origin of angiosperms. Bot. Gaz. 45 : 196 – 198.

Hallier, H. (1912) l Origine df la Systems phyletique des angiospermes Oxposes a laide de leur arbre genealogique. Arch. Neerl. Sci. Ser. 3rd be. 1: 146-234.

Hamann, U. (1961) Merkmalsbesland und verwandts-schaftsbezienhungen der Farinosae, Ein Beitrag Zum System der monocotyledonen. Willdenowia 2:639-768.

Handro, W. (1964) Contribuicao do estudo da Venacao e anatomnia foliar das Amarantacees dos cerrasos I. Annls. Acad. barasil cienc 36:479-499.

Handro. W. (1967) Contribuicao ao estudo da venacao e anatomia foliar das Amarantacees dos cerrados II. Annls. Acad. barasil cienc. 39: 495-506.

Hanks, S.L. and D.E.Fairbrothers (1976) Palynotaxonomic investigation Fagus L. and Nothofagus Bl.: Light Microscopy, Scanning Electron Microscopy, and Computer Analysis. In: Botanical Systematics : An Occasional Series of Monograph (V.H.Heywood Ed.) Vol. I.pp.s1-s14S1.

Hansell, R. I. C. and B. E. Wing (1972) Character weights in classification. J. Theor Biol. 39 : 297-314.

Harborne, J. B. (1963) Distribution of anthocyanins in higher plants. In : Chemical Plant Taxonomy (Swain, T.ed.) pp. 359 - 388.

Harbone, J. B. (1973) Flavonoids as systematic markers in the angiosperms. In : Chemistry In Botanical Classsification (Bendz, G. and J. Santesson Eds.) pp. 103-115. New York, USA.

Harbone J. B. and D. M. Smith (1978) Anthochlors and other flavonoids as honey guides in the Compositae. Biochem. Syst. Ecol. 6 : 287 – 291.

Harborne, J.B. (1977) Flavonoids and the evolution of the angisperms. Biochem. Syst. Ecol. 5:7-22.

Harlan, J. R. and J. M. De Wet (1963) The compilospecies concept. Evolution 17 : 497 – 501.

Harling, G. (1954) The embryo sac development of Vittardinia triloba (Gaud.) DC.Sv. Bot. Tidskr. 48 : 489 – 496.

Harrinton, H. D. (1957) How To Identify Plants, Swallow Press. Chicago, U.S.A.

Haskell, G. (1949) Variation in the number of stamens of the common Chikweed. J. Genetica 49: 291-301.

Hatch, M. D. and C. R. Slack (1970) Photosynthetic Co2 fixation Ann. Rev. Pl Phys. 21: 141-162.

Hawkes, J.G. and R.N. Lester (1968) Immunological studies on the tuber bearing Solanums III. Variability within S. bulbocastanum. Ann. Bot. 32: 165-186.

Hayden, M. V. and J. D. Dwyer (1969) Seed morphology in the tribe Morindeae (Rubiaceae). Bull. Torrey Bot. Club. 96 : 704 – 710.

Hedaytullah, S. (1935) The development of female gametophyte in Moringa pterygosperme Gaertn. Proc. 22nd Indian Sci. Congr. (Calcutta) Abstr. Pt IIIrd.

Hegnauer, R. (1954) Gedanken uber die theoretische Bedeutung der chemisch – ontogenetischen und chemisch – Systematischen Betrachtung Von Arzneipflanzen. Pharma. Acta Helv. 29 : 203 – 220.

Hegnauer, R. (1958) Chemotaxonomische Betrachung VI. Phytochemie und systematik : Eine Ruckund Vorausschau auf die Entwicklung ciner chemotexnomie. Pharma Acta Helv. 33 : 287 – 305.

Hegnauer, R. (1963) The taxomic significance of alkaloids. In: Chemical Plant Taxonomy (T. Swain, Ed.). Academic Press, London, UK.

Hegnaur, R. (1966) Comparative phytochemistry of alkaloids. In : Chemical Plant Taxonomy (Ed.T.Swain) Academic Press, London, U.K.

Hegnauer, R. (1971) Chemical patterns and relationships of the Umbelliferae. In : (Heywood, V. H. Ed.) The Biology and Chemistry of The Umbelliferae., London. pp. 267 - 277.

Hegnauer, R. (1977) Cyanogenic compounds as systematic markers in Tracheophyta. Pl. Syst. Evol. Suppl. I. pp. 191-209.

Heimisch, C.,Jun. (1942) Comparative anatomy of the ssecondary xylem of the ' Gruinales' and 'Terebinthales' of Wettstein with reference to taxonomic grouping. Lilloa 8: 83-198.

Heintzelman,C.E. Jr. and R.A. Howard. (1948) The comparative morphology of the Icacinaceae.V. The pubescence and crystals. Amer. J. Bot. 35 : 42-52.

Heneen, W. K. and H. Runemark (1972) Chromosomal polymorphism in isolated population of Elymus (Agropyron) in the Aegean. Bot. Not. 125 : 419 – 429.

Hengeveld, P. (1988) : Mayr's ecological species criterion. Syst. Zool. 37 : 47-55.

Herr, J.M. (1984) Embryology and taxonomy. In: Embryology of Angiosperms (B.M.Johri Ed.), pp. 647-696. Springer-Verkg, Berlin, Germany.

Herr, W. (1973) Pseudoguianolides in Compositae. In : Chemistry In Botanical Classification (Bendz, G and D.J. Santesson Eols) pp. 153-172. London, U.K.

Heslop – Harrison, Y. (1964) New Concepts In Flowering Plant Taxonomy. Hienemann, London, U.K.

Heslop – Harrison, Y. (1981) Stigma charateristics and agiosperm taxonomy. Nord. J. Bot. 1 : 401 – 420.

Heslop – Harrison, J., Y. Heslop – Harrison and J.Barber(1975) The sfigma surface in incompatibility responses Proc. Roy. Soc. Lond. Sci. (B) Biol. Sci. 188 : 293 – 303.

Heslop – Harrison, Y. and K. R. Shivanna (1977) The receptive surface of the angisperm stigma. Ann. Bot. 41 : 1233 – 1238.

Heslop-Harrison, J. (1962) Purpose and procedures in the taxonomic treatment of higher orgaisms. In : G. G. Ainsworth and P. H. A. Sneath (Ed.) Microbial Classification. PP. 14 – 36. Cambridge University. Press, Cambridge, U.K.

Heyn, C. C. and I. Herrnstadt (1977) Seed coat structure of Old World Lupinus species. Bot.Not. 130: 435-437.

Heywood, V. H. (1954) A revision of the Spanish species of Tanacetum subsect. Leucanthemopsis Giroux. And. Inst. Bot. Cavanilles 12(2): 313-377.

Heywood, V. H. (1958) The presentation of taxonomic information : A short guide for contributors to flora of Europaea. Leicester Univ Press, Leicester.

Heywood, V. H. (1959a) A check list of the Portuguese Compositae-Chrysantheminae. Agron. Lusit. 20:205-216.

Heywood, V. H. (1959b) The taxonomic treatment of ecotypic variation. In : A. J. Cain (Ed.) Function and Taxonomic Importance. Syst. Assn. London pp. 87 -112.

Heywood, V. H. (1963) The 'species aggregate' in theory and practice, Regnum Veg. 27: 26-37.

Heywood, V. H. (1967a) Variation in species concepts. Bull. Jard. Bot. Etat. 37 : 31–36.

Heywood, V. H. (1967b) Plant Taxonomy E. Arnold. London, U.K.

Heywood, V.H. (1968) Plant taxonomy today, In: Heywood V.H. (Ed.) Modern Methods In Plant Taxomony, Academic Press, London; pp. 3-12.

Heywood, V. H. (1969) Scanning electron microscopy in the study of plant materials, Micron 1: 1-14

Heywood, V. H. (1973) Taxonomy is crisis ? Or taxonomy is the digestive system of biology. Acta Bot. Acad. Sci. Hung. 19 : 139-146.

Heywood, V. H. and K. M. M. Dakhshini (1971) Fruit structure in Umbelliferae-Caucaliade. In: V.H. Heywood (Ed.) The Biology and Chemistry of The Umelliferae, Suppl. 1 to Biol. J. Linn, Soc. 64:215-232.

Hickey, L. J. (1973) Classification of the architecture of dicotyledonous leaves. Amer, J. Bot. 60 : 17-33.

Hickey L.J. and J.A. Wolfe (1975) The basis of angiosperm phylogeny: Vegetative morphology. Ann. Miss. Bot. Gdn. 62: 538-589.

Hideux, M. and I.K. Ferguson (1975) Stereostructune de l'exine des Saxifragales: Proposition d'une description logique dt schemtique. Soc.Bot. Fr. Call Palynologie 122: 57-67.

Hideux, M. and I.K. Ferguson (1976) The stereostructure of the exine and its evolutionary significance in Saxifragaceae *sensu lato.* In: The Evolutionary Significace of The Exine (I. K. Ferguson & J.Muller Ed,) pp. 127-377. Academic Press, New York, U.S.A.

Hildtch, T. P. (1952) The seed and fruit fats of plants. Endeavour 11 : 173 – 182.

Hill, J. (1770) The construction of Timber. R. Baldwin, London, U.K.

Hill, R. J. (1976) Taxonomic and phylogenetic significance of seed coat microsculpturing in Mentzelia (Loasaceae) in Wyoming and adjacent western states. Brittonia 28: 18-112.

Hitchcock, A. S. (1925) Methods of Descriptive Systematic Botany, New York. U.S.A.

Holttum, R. E. (1968) Comparative morphology, taxonomy and evolution, Phytomorph. 17(1-4) : 36-41.

Hooke, R. (1965) Micrographia. London, U.K.

Hooker, J. D. (1855) Introductory Essay To J. D. Hooker and T. Thomson, Flora Indica, 1 – 44. London, U.K.

Hooker. J. D. (1875– 1897) The Flora of British India. Vol. I – VII. L. Reeve, London, U.K.

Hopwood, A. T. (1959) The development of Pre- Linnean taxonomy. Proc. Linn. Soc. (London) 170:230 – 234.

Horner, H.T. Jr., and C.Pearson (1978) Pollen wall and aperture development in Helianthus annuus (Compositac : Heliantheae) Amer, J.Bot. 65 : 293 – 309.

Horner, H.T. Jr. and N.R. Lersten (1968) Development, structure and function of secretory trichomes in Psychotria bactriophylla (Rubiaeae) Amer. J. Bot 55:1089-1099.

Huber, H. (1969) Die Samenmerkmale und Verwandtschaftsverhaltnisse der Liliiflorae. Mitteil. Bot. Staatssamml. Munchen 8: 219-538.

Hugo Carl Albert Fischer (1890) Bietrage zur vergleichenden morphologie dernpollenkorner. J.U.Kem's Verlag

Hunziker, J.H. (1968) Protein electrophoresis as an aid in genome analysis. The Nucleus (Calcutta) 12 (Suppl.) : 226 – 236.

Hunziker, J.H.; H.D. Behnke, I. J. Eiferrt and T. J. Mabry (1974) Halophytum ameghinoi : A betalain containing and P-type sieve-tube plastid species, Taxon 23: 537-539.

Hutchinson, J. (1926) The Families of Flowering Plants I, Dicotyledons. Arranged According To A New System Based on Their Probable Phylogeny. Macmillan, London, U.K.

Hutchinson J. (1934) The Families of Flowering Plants II, Monocotyledons. Arranged According To A New System Based on their Probable Phylogeny, Macmillan, London, U.K.

Hutchinson, J. (1959) The Families of Flowering Plants. Clarendon Press, Oxford, U.K.

Hutchinson, J. (1969) Evolution and Phylogeny of Flowering Plants. Dicotyledons : Facts and Theory. Academic Press, London, U.K.

Hutchinson, J. (1973) The Families of Flowering Plants. 3rd Edition, Clarendon Press, Oxford, U.K.

Huxley, J.S. (1940) The New Systematics. Clarendon Press, Oxford Univ, California Publ. Bot. 18: 241-291.

Iversen, J. and J. Trels – Smith (1950) Pollen Morfologiske Definitiones or Typer. Denmarks Geol. Unders. 14 : 3 – 8.

Jackon, R.C. (1957) New low chromosome number for plants. Science 126: 1115-1116.

Jackson, B. D. (1928) A Glossary of Botanic Terms : With Their Derivation and Accent (4Th Ed.) Gerald Duckworth & London, U.K.

Jackson, R, C. and C.T. Dimas (1981) Experimental evidence for systematic placement of the Haplopappus phyllocephalus complex (Compositae) Syst. Bot. 6 : 8 – 14.

Jain, S. K. (1969) Comparative ecogenetics of two Avena species occurring central California. Evol. Biol. 3: 73 – 113.

Jain, S. K. and R. R. Rao (1977) A Hand Book of Field and Herbarium Methods Today & Tomorrow's Printers and Publishers, New Delhi, India.

Jardine, N. and R. Sibson (1971) Mathematical Taxonomy. Wiley, New York, USA.

Jensen, S. R., Nielson, B. J., and R. Dahlgren (1975) Iridoid compounds, their occurrence and systematic importance in the angiosperms. Bot. Not. (Lund) 128 : 148 – 180.

Jensen, U. (1968) Serologische Beitrage zur Systemtik der Ranunculaceae Bot. Jb. 88: 204-268.

Jensen, U. (1974) The interpretatiion of comparative serological results. In: Chemistry in Botanical Classification, Nobel Symposium 25, (Bendz. G and Santesson, J.Eds.) pp.217-227.

Jepson, G. I. (1944) Phylogenetic trees. Trans. New York Acad, Sci. Ser. 6:81-92.

Johensen, D.A. (1950) Plant Embryology-Embryology of The Spermatophyta. Chronica Botanica Co, Waltham, Mass, U.S.A.

Johnson, A. M. (1931) Taxonomy of Flowering Plants. Century, New York, U.S.A.

Johnson, C. and W.V. Brown (1973) Grass leaf ultrastuctural variations. Amer. J. Bot. 60:727-735.

Johnson, M.P. and R.W.Holm (1968) Numerical taxonomic studies in the genus Sarcostemma R. Br. (Asclepiadaceae). In: Modern Methods In Plant Taxonomy (Ed. V.H.Heywood) Academic Press, London, UK, pp.199-218.

Johnson, H. H. (1975) Plant pubescence : An ecological perspective. Bot. Rev. 41: 233-258.

Johri, B. M. (1963) Embryology and taxonomy, In Recent Advances In the Embryology of Angiosperms (P.Maheshwari Ed.), International Society of Plant Morphologists, University of Delhi, Delhi, India pp. 395-444.

Johri, B.M. (1967) Angiosperms Embroylogy and taxonomy, Bull. Natn. Inst. Sci. India. 34: 263-268.

Johri, B. M. (1970) Limannthaeae. Proc. Symposium On Comparative Embryology of Angiosperms, Indian National Science Academy, New Delhi, India.

Johri, B. M. and S.P. Bhatnagar (1960) Embryology and taxonomy of the Santalales – I. Proc. Nat. Inst. Sci, India B 26: 199-220.

Johri, B. M. and S. P. Bhatnagar (1972) Loranthaceac Botanical Monograph No. 8 CSIR, New Delhi, India.

Johri, B.M. and H. Singh (1959) The morphology, embryology and systematic position of Elytraria acaulis (Linn. *f.*) Lindau Bot. Not. (Lund) : 21 – 45.

Johri, R. M. and Sneh Lata (2005) Taxonomy-1: Systematics and Morphology. Sonali Publications, New Delhi, India.

Jones, J.D.G. and R.B. Flavell (1982) The Strueture, amount and chromosomal localisation of defined repeated DNA sequenees in species of the genus Secale. Chromosome 86 : 613 - 641.

Jones, K. (1970) Chromosome changes in plant evolution. Taxon 19 : 172 – 179.

Jones, S. B. Jr. and A. E. Luchsinger (1986) Plant Systemtics (2nd Ed.) Mc Graw – Hill, New York, U.S.A.

Jorgensen, R.A., Cuelar, R.E. and W.F.Thompson (1982) Modes and tempos in the evolution of nuclear-encoded ribosomal RNA genes in legumes. Carnegie Inst. Year Book 81: 98-101.

Jorgenson, L. B. (1981) Myrosin cells and dilated cisternae of the endolplasmic reticulum in the order Capperales Nord. J. Bot 1 : 433 – 405.

Jussien, A. L. de (1789) Genera Plantarum Herissan & Barrois, Paris, France.

Just, T. (1946) The relative value of taxonomic characters. Amer. Midl. Naturalist 36 : 291 – 297.

Just. T. (1946) The use of embryological formulas in plant taxonomy. Bull. Torrey Bot. Club 73:351-355.

Kaln, Z. (1947) Sravandvaci anatome Pladu Rostlin Okolienatych (Comparative anatomuy of the fruits of Umbelliferous plants in the area of the Czechoslovak Republic Prague.

Kannabiran, B. and K.H. Krishnamurthy (1979) Stomatal ontogeny in some dicotyledonous families with special reference to its taxonomic and phylogenetic value. In: Structure, Function And Ecology of Stomata (D. N. Sen Ed.). Bishen Singh Mahendra Pal Singh, Dehradun, India.

Kapil, R.N. (1962) Some recent examples of the value of embryology in relation to taxonomy. Bull. Bot. Surv. India. 4: 57-66.

Kapil, R.N. (1970) Podostemaceae. In: Proceedings Of Symposium On Comparative Embryology of Angiosperms, Indian National Science Academy, New Delhi, India, pp. 104-109.

Kapil, R.N. and A.K.Bhatnagar (1975) A fresh look at the process of double fertilisation in angiosperms. Phytomorpholigy 25: 334-368.

Kapil, R.N. and A.K. Bhatnagar (1980) Embryology in relation to angiosperm taxonomy, Glimp. Pl. Res. V: 272-295, Rastogi Publishers, Meerut, India.

Kare Bremer, Mark W. Chase and Peter F. Stevens (1998) An ordinal classification for the families of flowering plants. Annals of The Missouri Botanical Garden 85:531-553.

Karpechenko, G.D. (1927) Polyploid hybrid of Raphanus sativus L.X.Brassica oleracea L. Bull. Appl.Bot. Genet. Breed 17: 305-410.

Karpechenko, G.D. (1928) Polyploid hybrid of Raphanus sativus L., Brassica oleracea L. Zeitscher Ind. Abstr. U. Verer-bungsi 39: 1-7.

Keating, R. C. (1979) Palynology and systematics : The 25th Systematic Synposium. Ann. Missouri Bot. Gard, 66 – 591 – 592

Kendrick, W. B. (1964) Quantitative characters in computer taxonomy. In : V. H. Heywood and J. Mcneill (Ed.) Phenetic And Phylogentic Classification. PP. 105 – 114. Systematics Association, London, U.K.

Kerner, A. Von., M. and F. W. Oliver (1894) The Natural History of plants. Vol. 2 : 629-634.

Khosla, P.K., Sareen, T.S. and P.N.Mehra (1973) Cytological studies in Himalayan Anacardiaceae. Nucleus 16: 205-209.

Knobloch, I.W. (1972) Intergeneric hybridisation in flowering plants. Taxon 21 : 97 – 103.

Knuth, M. P. (1906-1909) Handbook of Flower Pollination: Based Upon Hermann Muller's Work 'The Fertilistion of Flowers By Inseets'; Translated By J.R. Ainsworth Davis, 3 Vols., Clarendon Press, Oxford, U.K.

Kolreuter, J. G. (1761-1766) Vorlaufige Nachricht von einigen das Geschlecht der pflanxn betreffonden Versuchen und Beobachtungen. 4 Vols, Leipzig, East Germany.

Kooimen, P (1972) The occurrence of iridoid compounds : Their occurrence and systematic importance in angiosperms. Bot. Notiser. 128 : 148 – 180.

Krause, K. (1930) Liliaceae. In : Die naturliche Pflanzenfamilien (A. Engler and K. Prantl Eds.) Z.Aufl. 15a, pp. 227-386.

Kreger, D. R. (1958) Waxes. In : Encyclopedia of Plant Physiology (W. Rubland Ed.) Vol. 10, pp. 249. Springer-Verlag, Heidelberg, Germany.

Kress, W. J., Store, D. E. and S.C.Sellar (1978) Ultastructure of exine – less Pollen : Heliconia (Heliconiaceae). Amer. J. Bot. 65 : 1064 – 1076.

Kutitzki, K. (1969) Chemotaxonomische Betrachr-ugen zur Grosgliederung der Blutenpflazen, Taxon 18 : 360-368.

Kubitzki, K. (1975) Systematics and evolution of seed plants. Progr. Bot. 37 : 269-307.

Kubler, G. (1962) The Shape of Time : Remarkes On The History Of Things, Yale University Press, New Haven, U.S.A.

Kynclova-Petrocyova, M. (1970) Comparative morphology of achenes of the tribe Anthemidac Cass. (Family Asteraceae) and its taxonomic significance. Perslia (Praha) 42: 33-53.

Lack,H, Walter (2008) An Annotated catalogue of the printed illustrations by Franz Balier (1758-1840). Archieves of Natural History 31 (1):80-101.

Lam, H. J. (1936) Phylogenetic symbols, past and present (being an apology for genealogical trees). Acta Biotheor. 2 : 153-194.

Lam, H. J. (1948) Classification and the new morphlogy Acta. Biotheoretica 8 : 108 – 154.

Lawrence G. H. M. (1951) Taxonomy of Vascular Plants, MacMillan, New York, U.S.A.

Legendre, P. and P. Vaillancourt (1969) A mathematical model for entities, species and genus. Taxon 18 : 234 -252.

Lems, K. (1960) : Botanical notes on the Canary Islands. II. The evolution of plant forms in the islands : Aeonium. Ecology 41: 1-17.

Lennox, J. G. (1980) Aristotle on genera, species and 'the more and less'. J. Hist. Biol. 13 : 321-346.

Lersten, N.L. (1974) Colleter morphology in Pavetta, Neorosea and Tricalysia (Rubiaeae) and its relationship to bacteria leaf nodule symbiosis. Bot. J. Linn. Soc. 69: 125-136.

Levitszky, G. A. (1931) The karytotype in systematics. Bull, Appl. Bot. Genet. Pl. Breed. 27 : 220 – 240.

Lewis, H. (1953) Quantitative variation in wild genotypes of Clarkia. I.U.B.S. Syposium on Genetics of Population Structure, Pavia, pp. 114-125.

Lewis, H. (1967) The taxonomic significance of autopolyploidy. Taxon 16 : 265 – 271.

Lewis, W.H. (1969) Comparative cytology in relation to taxonomy.Taxon 6: 42-46.

Lindau, G. (1895) Acanthaceae. In : Engler, A. & Prantl, K. Die naturlichen pflanzenfamilien TS. Engelmann, Leipzig, East Germany.

Lindley, J. (1830 – 1840) The Genera And Species of Orchidaceous Plants. Ridgways, London, U.K.

Lindley, J. (1836) An Introduction To The Natural System of Botany. Longman, Rees, Orme, Brown and Green. London, U.K.

Linnaeus, (1736) Fundamenta Botanica Amsterdam, Salomon. Schouten Ed. 1, Sweden.

Linnaeus, C. (1735) Systema Naturae, Leiden, Netherlands

Linnaeus, C. (1737a) Critica Botanica. C.Wishoff, Amrstedam, Netherlands.

Linnaeus, C. (1737b) Genera Plantanum, Leiden, Netherlands.

Linnaeus, C. (1751) Philsophia Botanica, Stockholm, Godofir Kiesewetter, Sweden.

Linnaeus, C. (1753) Species Plantarum Vol 1-2. Impensis Laurentii Salvi, Stockhollm (5th Ed.) (1801).

Linsey, A. A. (1938) Anatomical evidence for the Menyanthaceae. Amer. Jour, Bot. 25: 480-485.

Longley A.E. (1927) Supernumerary chromosomes in Zea mays. J. Agr.Res. 35: 769-784.

Love. A. and B.M.Kapoor (1967 b) The highest plant chromosome number in Europe. Svensk. Bot. Tidskr. 61: 29-32.

M. de L'Obel (1570) Stirpium adversaria nova. Londini.

Mabry, T. J. (1966) The betaeyanins and betaxanthins. In: Compearative Phytochemistry (T. Swain Ed.) pp. 231-244. Academic Press, London, U.K.

Mabry, T.J (1977) The order Centrospermae. Ann. Miss. Bot, Gard. 64:210-220.

Mabry T. J.; I. J. Eifart; C. Chang; H. Mabry and C. Diold (1975) Mabry, T.H, Theligonaceae: pigment and ultrastruetural evidence which excludes it from the order Centrospermae. Biochom. Syst 3: 53-55.

Mabry, T. J. and H. D. Behnke (1976) Betalains and P-type sieve-element plastids. The systematic position of Dysphania R. Br. (Centrospermae). Taxon 25: 109-111.

Mabry, T. J., H. D. Behnke and I. J. Eifert (1976) Betalains and P-type sieve-element plastids in Gisekia L. (Centrospermae). Taxon 25: 112-114.

Magnus, W. (1915) Die atypische Embryonalent-wicklung der Podstemaceen. Flora 105: 275-336.

Maheshwari, P. (1945) The place of angiosperm embryology in research and teaching. Journ. Indian Bot. Soc. 24: 25-41.

Maheshwari R (1950) An Introduction To The Embrylogy of Angiosperms. Tata McGraw-Hill co. Ltd. New Delhi, India.

Maheshwari, P. (1958) Embroylogy and taxonomy. Mem. Ind. Bot. Soc. 1: 19

Maheshwari, P. (1963) History and present status of plant embyology. In : Recent Advances In The Embryology of Angiosperms (Maheshwari, P. Ed.) International Society of Plant Morpholgists. University of Delhi, Delhi. India. pp. 1 – 13

Maheshwari, P. (1964) Embroylogy in relation to taxonomy. In: Vistas In Botany (W.B.Turril Ed.) Oxford. 4: 55-97.

Maheshwari, P., Johri, B. M. and S.N. Dixit (1957) The floral morphology and embryology of the Loranthoideae (Loranthaceae). J. Madras. University. Ser B. 27 : 121 – 136.

Maheshwari, P, and B.M.Johri (1956) The morphology and embryology of Floerkea proserpinacoides Willd. with a discussion on the systematic position of the family Limnanthaceae, Bot. Mag. Tokyo Ogura Comm. Vol, 69: 410-423.

Marguillis, L. (1971) Wittaker's five kingdoms of organism : Minor revisions suggested by consideration of the origin of mitosis. Evolution 25 : 242 -245.

Marguillis, L. (1974) Five kingdom classification and the origin and evolution of cells. Evolutionary Biology 7 : 45-78.

Marsden M.P. F. and I.W. Bailey (1955) A fourth type of nodal anotomy in dicotyledons illustrated by Clerodermdron trichotomum Thunb. J. Arnold Arbor. 36: 1-51.

Marshall, D.R. and S.K.Jain (1968) Phenetic plasticity of Avena fatua and A. barbata. Amer.J.Bot. 102 : 457 – 467.

Martin, A.C (1946) The comparative internal morhology of seeds Amer. Midl. Nat. 36:513-560.

Martinet, J.f. (1777-1778) Katechismus der nature- Amster dam. Loveringh and wed. Allart-First Edition.

Mason, H. L. (1950) Taxonomy, systematic botany and biosystematics. Madrono 10 : 193-208.

Matthew, K. M. (1991) An Exursion Flora of Central Tamilnadu, India, Oxford & IBH Publishing Co, Pvt. Ltd, New Delhi, India.

Mauritzon, J. (1933) Uber die systematische Stellung der Familien Hydrostachyaceae and Podostemaceae. Bot. Notiser 1933: 172-180.

Mauritzon, J. (1934) E in Beitrag zur Embryologic der phytolaccaceen and Cactaceen. Bot. Not. 1934 : 111-135.

Mayr, E. (1948) The beating of the new systematics on genetical problems : The nature of species. Adv. Genetics 2 : 205 – 237.

Mayr, E. (1963) Animal Species and evolution. Harward University Press, Cambridge, U. K.

Mayr, E. (1966) Principles of Systematic Zoology, McGraw-Hill, New York, USA.

Mayr, E. (1969) The biological meaning of species. Boil. J. Linn. Soc. 1: 311- 320.

Mayr, E. (1974) The challenge of diversity. Taxon 23 : 3-9.

Mayr, E. (1982) The Growgh of Biological Thought : Diversity, Evolution, and Inheritance. Harward Uni. Press, Cambridge, Mass, U. K.

Mayr, E. C. (1957) Preface. In E. Mayr, Ed. The species problem, pp iii-v.: American Association For The Advancement of Science, Washingtn, D. C., U.S.A.

Mayr, E. C. (1969) Priciples of Systematic Zoology. McGraw-Hill, New York, U.S.A.

McNair, J. B. (1934) The Evolutionary status of plant families in relation to some chemical propertes. Amer. J Bot. 21 : 427 – 452.

McNair, J. B. (1935) Angiosperm phylogeny on a chemical basis. Bull. Torrey Bot. Club 102 : 515 – 532.

McNair, J.B. (1945) Some comparisons of chemical ontogeny with chemical phylogeny in vascular plants loydia 8 : 145 – 169.

Meeuse, A. D. J. (1970) Again : The descent of flowering plants in the light of new evidence from phytochemistry and from other sources. I. General Discussion. Acta Bot. Neerl. 19 : 61-72.

Melchior, H. (Ed.) (1964) A.H.G. Engler's Syllabus der Pflanzenfamilien. 12th Ed. Vol. 2. Gebruder Borntracger, Berlin, Germany.

Melville, R. (1939) Contribution to the study of British elms. II. J. Bot. London, U.K. 77 :138.

Melville, R. (1951) On the appliction of biometrical methods in plants taxonomy. Proc. Linn. Soc. Lond. 162 : 153-159.

Melville, R. (1955) Morphological characters in the discrimination of species and hybrids. In : J. E. Lousley (Ed.) Species Studies in the British Flora: 55-64; Oxford : Bot. Soc. Brit. Is.

Melville. R. (1969) Leaf venation pattern and the origin of angiosperms. Nature 224: 121-124.

Merxmuller, H. (1977) Summary Lecture. In : K. Kubitzki (Ed.) Plant Systematics and Evolution, (Suppl.1.) PP. 397 – 405. Springer – Verlag, New York, USA.

Metcalfe, C.R. and L. Chalk (1950) Anatomy of the Dicotyledons Vols. I – II. Clerendon Press, Oxford, U.K.

Metcalfe, C. R. (1971) Anatomy of The Monocotyledons. Vol. V. Cyperaceae, Clarendon Prees, Oxford, U.K.

Meyer, N.R. (1975) Ultrastructural sporodermy XII. Int. Bot. Congr. Leningrad Abstract 1: 257.

Mez, C. (1936) Morphologie und serodiagnostik. Bot. Arch. 38 : 86 – 104.

Michener, C. D. and R. R. Sokal (1957) A quantitative approach to a problem in classification. Evolution 11 : 130 – 162.

Miers, J. (1851) Observation on the affinities of the Olacaceae. Ann. Mag. Net. Hist. 8: 161-184.

Miller, I.M. Scott, A. and I.C. Gardner (1983) Leaf nodule development in Psychotria kirkii Hiern (Rubiaceae). Annals Bot. 52:719-802.

Millner, M.E.(1934) Anatomy of Silene vulgaris and Silene maritima as related to ecological and genetical problems. I Root structure. New Phytol. 33: 77- 95.

Mirov : N.T. (1961) Composition of gum turpentines of Pines. U. S. Dept. Agric. Tech. Bull. 1239.

Mishler, B. D. (1985) Transition and flora : Phylogenetic relationships of the green algae and bryophytes. Cladistics 1 : 305–328.

Mohan Ram, H.Y. (1959) Embryolgical characters in the classification of Acanthacee. Proceedings of 9th Int. Congr. Montreal 2 : 267

Mohl, H. von (1835) Sur la structure ed les formes des grains de pollen. Ann. Sci. Nat.Bot. 3 : 148 – 180, 220 -236, 304 – 346.

Mohl, H. Von (1851) Grundzuge der anatomie und physiologie der Vegetabilischen Zelle Braunschweig, Vieweg, Brunswick, Germany.

Mondal, A.K. (2005) Advanced Plant Taxonomy. New Central Book Agency (P.) Ltd., Kolkata, India.

Mondal, A.K., Mondal, S. (nee Parul) and S.Mondal (2000) Molecular taxonomy of the genus Cassia L. based on seed protein and mitochondria DNA RFLP. Phytomorphiology 50(1):15-25.

Moore, D.M. (1968) The karyotype in taxonomy In: Modern Methods In Plant Taxonomy (V.H.Heywood Ed.) Academic Press London, U.K.

Moore, D.M. (1978) The chromosomes and plant taxonomy. In: Essays In Plant Taxonomy. pp 39-56. Academic Press, London, U.K.

Morishima, H. and H. Oka (1960) The pattern of interspecific variation in the genus Oryza, its quantitative representation by statistical methods. Evolution 14:153-165.

Morton, J. K. (972) Phytogeography of the west African Mountains. In : Taxonomy, Phytogeography and Evolution (Valentine, D.H.Ed.), 221 -236. Academic Press, London, U. K.

Moudi, M. and R. Go. (2016) The comparison between nuclear ribosomal DNA and chloroplast DNA in molecular systematic study of four sections of genus Dendrobium Sw. (Orchidaceae). International Journal of Bioassays 5(3):4944-4852.

Mosaka, K. (1986) Who is the mother of the potato? – Restriction endonuclease analysis of Chloroplast DNA of cultivated potatoes. Theor. Appl Genet. 72: 606-618.

Mukesh Kumar, Stephen Sequiera and M. Ramesh (2000) The position of Indian Bamboos in Bamboo systematics. Rheedea 10(1) : 33-48.

Mukkad, A.J. (1962) Some observations on the embryology of Dicraea stylosa Wight. In: Plant Embrylogy – A Symposium. CSIR, New Delhi, India, pp. 139-145.

Mukkad, A.J. (1964) An Addition to bisporic embryo sacs- the Dicraea type. New Phytol. 63: 289-292.

Muller, H. L. H. (1883) The fertilisatron of Flowers Translated and Edited by D'Arcy Thompsin. Macmillen, London, U.K.

Muller, H.L.H. (1879) Form and funetion in angiosperm pollen. Ann. Misouri Bot. Gard. 66:593-632.

Muller, H.L.H. (1881) Alpenblumen, ihre Befructitung durch Insekten und ihre Anpassunger an dieselben. W. Engelmann, Leipzig, East Germany.

Muntzing, A. (1959) A new category of chromosomes, Proc. 10th Congr. Genet. 1: 453-467.

Muntzing. A (1957) Frequency of accessory chromosomes in rye strains from Iran and Korea. Hereditas. 43: 682-685.

Murbeck, S. V. (1933) Monographie der Gattung Verbascum. Lunds Univ. Arsskr. N. F. Avd. 229(2).

Murgai, Prem (1959) The development of embryo in Paeonia – a reinvestigation, Phytomorphology 9 : 275-277.

Murgai, Prem (1962) Embrylogy of Paeonia together with a discussion on its systematic position. In: Plant Embryology – A Symposium, CSIR, New Delhi, India, pp, 215-223.

Murrey, J. (1934) Anatomical feature in the roots of the genus Primula. Trans. Bot. Soc. Edinb. 31(3): 323-326.

Murthy,Y.S. And V. Puri (1980)Floral anatomy in relation to taxonomy. Glimp.Pl.Res. 5:222-243.

Mussleman, L. J. and W. F. Maun Jo. (1976) A survey of surface characterstics of seeds of Scrophulariaceae and Orobanchaceae using scanning electron microscopy. Phytomorph. 26: 370-378

Naik, V. N. (1974) Cytological studies in two species of Dipcadi Medik, from India. Cytologia 39: 506-511.

Naik, V. N. (1976) Chrosome behaviour and evolutionary trends in Chlorophytum (Liliaceae). Bot. J. Linn. Soc. 72 : 45-50.

Naik, V. N. (1984) Taxonomy of Angiosperms. Tata McGraw-Hill Publishing Company Ltd. New Delhi, India.

Naik, V. N. and S. M. Nirgude (1981a) Anatomy in relation to taxonomy of Chlorophytum (Liliaceae) Indian J. Bot. 4(2) : 48 – 60.

Naik, V. N. and S. M. Nirgude (1981b) Vessels in Chlorophytum (Liliaceae) Bull. Mus. Natn. Paris 4e ser., 3, sect. B Adansonia No. 2, pp. 201 – 212.

Nair, P,K.K.(1961) Pollen grains of Indian plants – I. Bull. Natl. Bot Gardns. No. 53 : 1 – 35.

Nair, P.K.K.(1970) Pollen Morphology of Angiosperms, Vikas Publishers, New Delhi, India.,

Nair, P. K. K. (1974 a) The pattern of gametophyte evolution in the plant Kingdom. New Botanist 1: 63-71.

Nair, P.K.K. (1968) Terminology for exine stratigraphy Palynol. Bull. 4611 : 20 – 22.

Nair, P.K.K. (1972) Pollen morphology and phylogenetic classification of primitive angiosperms. In: Adv. Pl. Morph. pp, 255-157. Professor V.Puri Comm. Vol. Sarita Prakashan, Meerut, India.

Nair, P.K.K. (1974 b) Palynology at the National Botanic Gardens, Lucknow – A Self Assessment. Adv. Poll. SP. Res. 1: 129-146.

Nair, P.K.K. (1974 b) The phylogeny of angiosperms : A palynological analysis. Evol. Biol. 7: 127-136.

Nair, N.C. and V. S.Sharma (1961) Organography and floral anatomy of some members of the Rhamnaceae J. Ind. Bot. Soc. 40 : 47-55.

Narayana, H.S. (1962) Post-fertlisation study on Moringa oleifera Lamk. – A reinvestigation. Phytomorphology 12: 65-69.

Narayana, H.S. (1970 a) Capparidaceae. In: Proceedings of Symposium on Comparative Embryology of Angiosperms. Indian National Science Academy, New Delhi, India. pp 78-80.

Narayana, H.S. (1970 b) Moringaceae, In: Proceedings of Symposium on Comparative Embryology of Angiosperms. Indian National Science Academy, New Delhi, India. pp 81-83.

Natho, G. (1959) Variation breite und Bastardbildung bei mitteleurropa ischen Birkensippen. Fedder Rep. 61: 211-273.

Nei, M. (1972) Genetic distance between populations. Amer – Naturalist 106 : 283–292.

Nelson, A.P. (1965) Taxonomic and evolutionary implications of lawn races in Prunella vulgaris (Labiatae) Brittonia 17 : 160 – 174.

Nelson, G. J. (1973) Classification as an expression of phylogenetic relationship. Syst. Zool. 22 : 344-359.

Newell, C. A. and Hymowitz (1978) Seed coat variation in Glycine Willd. subgenus Glycine (Leguminosae) by SEM. Brittonia 30 : 76-88.

Newmaster S. G., Subramnyam, R., Ivanoff, F. and N. C. Balasubramnyam (2006) Mechanisms of ethnobiological classifications. Ethnobotany 18: 4- 26.

Nowicke J. W. and J. J. Skvarla (1977) Pollen morphology and relationship of the Plumbaginaceae Polygonaccae and Primulaceae to the order Centrospermae. Smithsonian Contr. Bot. 37: 1 – 64.

Nowicke, J. W. and E. Ridgeway (1973) Pollen studies in the genus Cordia (Boraginaceae) Amer. J. Bot. 60 : 584 – 591.

Nowicke, J. W. and J. J. Skvarla (1979) Pollen morphology : The potential influnce in higher order sysetmatics. Ann. Miss Bot. Gard. 66 : 633 -700.

Obermeyer, A. A. (1962) A revision of the South African speeies of Anthericum, Chlorophytum and Trachyandra. Bothalia 7:669-767.

Ozenda, P. (1949) Recherches sur les Dicotyledones apocorpiques. Pub. Lab. del, Ecole Normale Sup. Serie Biol. Fasc.2nd, Paris, France.

Pai.R.M. (1992) Floral morphology and phylogeny of some monocotyledonous taxa. In: Sectional Presidential Address, Botany, The 79th Session, Indian Science Congress Association, M.S.University, held at Baroda (Gujarat), India.

Paliwal, G.S. (1966) Structure and onotgeny of stomata in some Acanthaceae. Phytomorphology 16: 527-532.

Paliwal, G.S and Lalita Kakkar (1970) Leaf anatomy of some Garrya species. Bot. J. Linn. Soc. 63: 81-90.

Palmer, J. D. and D. Zamir (1982) Chloroplast DNA evolution and phylogenetic relationships in Lycopersicon. Proc. Natl. Acad. U.S.A. 79 : 5006 – 5010.

Palser, B. F. (1961) Studies of floral morphology in the Ericales. V. Organography and vascular anatomy of several United States species of the Vacciniaceae. Bot. Gaz. 123:79 111.

Pant, D. D. (1965) On the ontogeny of stomata and other homologous structures. Plant Sci. Ser. (Allahabad) 1: 1-24.

Paris, R. (1963) The distribution of plant glycosides. In: Chemical Plant Taxonomy (T.Swain Ed.) pp. 337-358. Academic Press, London. U.K.

Patil, D. A. (1985) The nectaries in the Agavaceae. Acta. Botanica Indica 13 (2) : 289 – 291.

Patil, D. A. (2003) Flora of Dhule and Nandurbar Disticts (Maharashtra), Bishen Singh Mahendra Pal Singh, Dehradun, India.

Patil, A.M and D.A. Patil (2009) Studies on foliar trichomes in Acanthaceac. Nat. J. Life Sci. 6(1) : 37 – 52.

Patil, C. R. and D.A. Patil (2009) Obeservations on foliar trichomes in some Rubiaceae. Indian J. Applied & Pure Bio. 24(1):233-246.

Patil, D. A. and R. M. Pai (1981 a) The floral anatomy of Agave vera – cruz Mill. (Agavaceae). Marathwada University. J. Sci. 20 (13) : 14 – 16.

Patil, D.A. and R. M. Pai (1981b) The floral anatomy of Doryanthes excelsa Corr. (Agavaceae). Indian J. Bot. 4(2) : 5 – 9.Rec. Adv. Pl. Sci.

Patil.D.A. and R.M.Pai (1984) The floral anatomy of Olax scandens Roxb. (Olacaceae). Indian Bot. Reptr. 3 (1): 10-14.

Patil, D. A. and R. M. Pai (1985) Floral anatomy of Furcraea gigantea Vent. and its relationship. Indian J. Plant Sci. 3(2) : 51 – 57.

Patil, D.A. and R.M.Pai (1986) The floral anatomy of Polianthes tuberosa L, - A reinvertigation. Geobios New Reports 5 : 146 – 151.

Patil D. A. and S. B. Shisode, (2006) Nodal anatomical studies in some Malvaceae. Nat. J. Life Sci. 3(2) : 115-120.

Patil, V.P. and G. B. Deodikar (1972) Interspecific variation in chiasma frequencies and terminalization in emmer wheats. Cytologia 37 : 225 – 234

Payne, W. W (1978) A glossary of plant hair terminology. Brittonia 30 : 239-255.

Payne, W.W. (1979) Stomatal patterns in Embryophytes: Their evolution, ontogeny and interpratation. Taxon 28 (1,2/3) : 117-134.

Payne, W.W. and J.J. Skvarla (1970) Electron microscopic study of Ambrosia pollen (Compositae : Ambrosineae) Grana 10 : 87 – 100.

Perrot, E. (1897) Anatomie comparee des Gentianacees aquatiques (Menyanthees Griseb.) Bull. Soc. Bot. Fr. 40: 340-354.

Philipson. W.R. (1977) Ovular morphology and the classification of dicotyledons. Pl. Syst. Evol. Suppl. 1 : 123 – 140.

Piji, L. Van der (1960) Ecological aspect of flower evolution-I. Phyletic Evolution. 14 : 403 – 416.

Pirie, N. W. (1955) The principles of microbial classification : Summing up. J. Gen. Microbiol. 12 : 382 – 386.

Pollard, C.J (1964) The specificity of ribosomal ribonucleic acids of plants, Biochem. Biophys. Res. Commun. 17: 171-176.

Porter, C. L. (1967) Taxonomy of Flowering Plants, (2nd Ed.), Freeman, San Franciseco, U.S.A.

Potonie, R. (1967) New phylogenetic facts of fossil spores. Rev. Paleobotan Paly nol. 1 : 75 – 82

Prat, H. (1932) L'epiderme des Graminees. Ann. Sci. Nat. Bot. 14: 117-134.

Price, J.R. (1963) The distribution of alkaloids in the Rutaceae. In: Chemical Plant Taxonomy (T.Swain Ed.) pp. 429-452 Academic Press, London, U.K.

Prichard, E. C. (1955) Morphological studies in Rhamnaceae J. Elisha Mitcell Sci. Soc. 71 : 82 – 106.

Prosperi C.H and A.E. Cocucci (1979) Importancia taxonomica de la calosa de los tubos polinicos en Tubiflorae (ensayo prelimianar). Kartziana 12 -13 : 75 – 81.

Pulle, A. A. (1938) Compendium van de Terminologie, Nomenclatur en Systematick der Zaadplanten, Oosthoek Utrecht, 2nd Ed, (1950).

Puri V. (1934) A note on the embryo sac and embryo of Moringa oleifera Lamk. Proc. IndianAcademy of Sciences. B 65:249-256.

Puri V. (1951) The role of floal anatomy in the solution of morphological problems. Bot. Rev.17:471-553.

Puri V. (1952) Floral anatomy in relation to taxonomy. Agra Univ.J.Res. (Sci.)1:15-35.

Puri V. (1958) Floral anatomy and taxonomy. Mem. Ind. Bot.Soc. I. PP.15-18.

Puri V. (1962) Floral anatomy in relation totaxonomy. Bull. Bot. Surv. India. 4:161-165.

Quimby, M. W. (1939) The floral morphology of the Crassulaceae. Ph.D. Thesis, Cornell University, Ithaca, New York, U.S.A.

Radford, A. E. (1986) Fundamentals of Plant Systematics, Harper and Row New York, U.S.A.

Radford, A. E., W. C. Dickison, J. R. Massey and C. R. Bell (1974) Vascular Plant Systematics, Harper and Row, New York, U.S.A.

Radlkofer, L. (1895) Sapindaceae. In : A. Engler and K. Prantl (Ed.) Die naturlichen Pflanzenfamilien, Teil 3. Abt.5 PP. 277 – 366. W. Englemann. Leipzig, East Germany.

Raj, B (1970) Loranthaceae. In, Proceedings of Symposium on Comparative Embryology of Angiosperms.Indian National Science Academy, New Delhi, India. pp. 22-28.

Raji, A. D. (1971) Evolution of the genera of the tribe Vicieae Adanson. Bot. Journ. (Moscow) 56:978-781.

Ram (nees Glosh), Manasi (1956) Floral morphology and embrylogy of Trapa bispinosa Roxb. With a discussion on the systematic position of the genus. Phytomorphology 6: 312-323.

Ram (nee Ghosh) Manasi (1959) Morphological and embryological studies in the family Santalaceae – II. Excorpus with a discussion on its systematic position, Phytomorphology 9 : 4 – 19.

Ramadevi, D. and L. L. Narayana (1989) Floral anatomy of Balsaminaceae, In : Plant Science Reserch In India (Ed. M.L. Trivedi, B. S.Gill & S.S. Saini). Today &Tomorrows Printers & Publishers, New Delhi, India.

Ramayya, N. (1969) The development of trichomes in the Compositae. In: Recent Advances In The Anatomy of Tropical Seed Plants (K.A. Chowdhury Ed.) Hindustan Publ. Corp., New Delhi. pp. 85-113.

Ramayya, N. (1972) Classifiction and phylogeny of trichomes of angiosperms. In: Research Trends In Plant, Anatomy (Ed.A.K.m.Ghouse & Mohd. Yunus), Delhi, 91-102.

Ramayya N. and T. Rajagopal (1968) Foliar epidermis as a taxonomic aid in the flora of Hyderabad. Rart I. Portulacaceae and Aizoaceae. J. Osmania Univ. 4: 147-168.

Ramayya. N and T. Rajagopal (1971) Foliar dermotypes of Indian Aizoaceac and their use in identification. J.Indian Bot. Soc. 50 : 355-362.

Randloph, L.F. (1928) Types of supernumerany chrosomes in maize (Abstract) Anat. Rec. 41: 102.

Randloph, L.F. (1941) Genetic characterstics of the B-chromosomes in maize. Genetics 26: 608-631.

Rao, R. R. and S. K. Murti (1995) Concept and developments of herbarium with special referance to India. In : Recent Advanees In : Taxonomy (The Herbarium Ed. Shiva Sharma and Kundan Sharma). PP. 10-29 Surbhi Publictions, Jaipur, India.

Rao, R. R., T. Hussain, B. Dutt and A. Garg (1999) Palynology of Berberls (Berberidaceae). Rhccdea 9 (2) : 115 -146.

Rao, T.A. (1957) Comparative morphology and ontogeny of foliar sclereids in seed plants. I. Memecylan Phytomorphology 7 : 306 – 330.

Rao, T. A. (1980) Aspect and prospects of foliar sclereids in angiosperms. In : Current Trends In Botanical Research (M. Nagaraj and C.P. Malik Ed.) Kalyani Publishers, New Delhi, India.

Rao, T. A. (1981) Comparative morphology and taxonomic value of foliar sclereids in Limonium Tour. (Limonaceae). Proc. Ind. Acad. Sci. B 90 : 153 – 162.

Rau, M.A. (1970) Saxifragaceae. In Proceedings Of Symposium On Comparative Embryology of Angiosperms, Indian National Science Academy, New Delhi, India. pp. 90-92.

Raven, P. H. (1977) Perspectives in tropical botany: Concluding remarks. Ann. Miss. Bot Gard. 64:746.

Ray, J. C. (1686-1704) Historia Plantarum. London, U. K.

Ray, John (1704) Historia Plantarum, Tomus Primus, London, U.K.

Razi, B.A. (1949) Embryological studies of two members of the Podostemaceae. Bot. Gaz. 3: 211-218.

Razi, B.A. (1955) Some aspects of the embryology of Zelanidium olivaceum (Tul) Engl. and Lawia zeylanica Tul. Bull. Bot. Soc. Bengal 9: 36-41.

Record, S.J. (1921) Ligum- vitae : A study of the woods of the Zyogphyllaceae with reference to the true. Ligum-vitae of commerce. Bull.Yale Sch. For. 6:48.

Reeder, J.R. (1957) The embryo in grass systematics Amer. J. Bot. 44 : 756 – 768.

Reeder, J. R. (1962) The bambusoid embryo: A reappraisal. Amer. J. Bot. 49 : 639 – 641.

Reese, G. (1954) Euploidie, Aneuploidie and B-chromosomen bei Caltha vulgaris L. Planta 44: 203-268.

Regan, C. T. (1925) Organic evolution: facts and theories, Br. Assoc. Adv. Sci. Rep. 93rd Mecting, 75 – 86.

Regan, C. T. (1926) Organic evolution. Rep. Brit. Assoc. Advancem. Sci. 1925 : 75 – 86.

Reichert, E. T. (1913) Differentiation and specificity of starches in relation to genera, species, etc. Carnegie Inst., Washinton. Publication No.173.

Reichert, E. T. (1916) The specificity of proteins and carbohydrates in relation to genera, species and varieties. Amer. J.Bot. 3 : 91 – 98.

Reichert, E. T. (1919) A biochemic basis for the study of problems of taxonomy, heredity, evolution, etc. with sepecial reference to the starches and tissues of parent – stocks and hybrid stocks and the starches and and hemoglobins of varieties, species, and genera. 270 : i – xii, 1 – 834. Publication, Carnegie Inst. Washington D.C., U.S.A.

Reitasma, T, (1969) Size modification of recent pollen grains under different treatments. Rev Paleobot Palynol. 9: 175-202.

Rendle, A. B. (1904&1925) The Classification of Folwering Plants. Vol I-II. Combridge Univ. Press, (2nd Ed. 1930 & 1938), Cambridge, U.K.

Rendle, A. B. (1952) The Classification of Flowering Plants, Vol. 2 Dicotyledons.Cambridge University Press, Cambridge, U.K.

Rendle, A. B. (1959) The Classification of Flowering Plants Vol. – II. Cambridge University Press, Cambridge, U.K.

Rendle, A. B. (1979) The Classification of Flowering Plants Vol. 2nd Dicotyledons, Vikas Publishing House (P) Ltd. Sahidabad (U.P.) India,

Rhoades, M. M. and B.M. Clintock (1935) The cytogenetics of maize. Bot.Rev. 1: 202-325.

Richardson I. B. K. (1978) Endemic taxa and the taxonomist. In : Eassays In Plant Taxonomy (Street, H.E.Ed.) 245 – 262. Academic Press, London, U. K.

Rick, C.M. and D.W.Barton (1954) Cytological and genetical identification of the primary trisomics of the tomato. Genetics. 39: 640-666.

Riley, R. and V. Chapman (1958) Genetic control of the cytologically diploid behaviour of hexaploid wheat. Nature, Lond. : 182 : 713 – 715.

Robbrecht, E. (1988) Tropical woody Rubiaceae. Opera Bot. Belg.J. 1-271.

Rodriguez, R. L. (1950) A graphic representation of Bessey's taxonomic system. Madrono 10 : 214-218.

Rogers, D.J. (1963) Taximetrics-New name, old concept. Brittonia 15:285-290.

Rogers, G. K. (1985) A note on the unusual stipules in Geasonia Prance (Rubiaceae) Brittonia 37(3):320.

Rollins, R. C. (1941) Monographic study of Arabis in Western North America. Rhodora 43 : 289 – 325, 348 – 411, 425 – 481.

Rollins, R. C. (1953) Cytogenetical approaches to the study of genera. Chronica Botanica 14 : 133-139.

Rollins, R.C. (1958) The genetic evaluation of a taxonomic character in Dithyrea (Cruciferae) Rhodora 60 : 145 – 152.

Rollins, R. C. (1965) On the bases of biological clasification. Taxon 14: 1-6.

Rosanoff, S. (1866) Zur Kenntiss des Baues under Entwickelungeschichte des pollen der Mimoseae. Jahrb. Wiss. Bot. 4 : 441 – 450.

Rose, F. (1972) Floristic connections between sowtheast England and North France. In : Taxonomy, Phytogeography and Evolution. (Valentine, D.H.Ed.) 263 – 379. Academic Press, London, U. K.

Ross, H. H. (1974) Biological Systematics. Reading. Addison – Wesley Mass, U.S.A.

Rout,R.C. and D. B. Deb (1999) Taxonomic revision of the genus Pavetta (Rubiaceae) in Indian subcontiment, Bull. Bot. Surv. India 41 (1 – 4) : 1 – 182.

Rowely J. R. and J. J. Skvarla (1986) Development of pollen grain wall in Canna. Nord, J. Bot. 6 : 39 – 65.

Rudall, P. J. (1994) Laticifers in Crotonoideae (Euphorbiaceae) and evolution. Ann, Missouri Bot. Gard. 81(2) : 270 – 282.

Rudberg, P. A. (1905) Penthoraceae : 75. In North American Flora. New York Botanic Garden, U.S.A.

Rudolf-lauritzen, M. (1958) The trisomics of Antirrhinum majus L. Proc. 10th Int. Congr. Genet. 2 (Abstracts) : 243-244.

Rutgers, F.L. (1923) Embryo sac and embryo of Moringa oleifera Lamk. Ann. Jard. Bot. Buitenz. 33: 1-66.

Saklani, A.; R. R. Rao and L. B. Chaudhury (2000) SEM- characterisation of achene morphology towards the taxonomy of Indian species of Saussurea DC. (Asteraceae) Rheedea 10 (1) : 1 – 18.

Sampathkumar (1982) Studies on the cotyledonary leaves of some Convolvulaceae. Taxon 31: 53-56

Sampson, F. B. (1981)Synchronous versus esynchronous mitosis within permanent pollen tetrads of the Winteraccae. Grana 20 : 19 – 23.

Schaeppi, H. and F. Steindl (1942) Bluetenmorphologische und embrylogische Untersuchungen an Loranthoideen Vjschr. Neturf. Ges. Zurich. 87: 301-372.

Schaeppi, H. and F. Steindl (1945) Bluetenmorphologische and embrylogische Untersuchungen an einigen Viscoideen, Vjschr. Naturf. Ges. Zurich. 90: 1-46.

Schnarf, K. (1929) Embryologie der Angiospermen, In; Handbuch der Pflanzenanatomie (K. Linsbauer Ed.) Vol. 10, pt. 2. Borntraeger, Berlin, Germany.

Schnarf, K. (1931) Verg ecinde Embroylogie der Angiospermen, Borntraeger, Berlin, Germany.

Schultze-Motel, W. (1964) Santalales. In: Syllabus der Plflanzenfamilien 2nd. (by A. Engler, revised by H.Melchior). G. Borntraeger, Berlin, Germany.

Schulz, O. E. (1936) Cruciferae. In : A. Engler.Die naturlichon pflanzenfamilien (Ed.2) 17 (B): 227-658.

Schurhoff, P.N. (1924) Zytologische ungen intersuch in der Reihe Geraniales. Jahrb. Wiss. Bot. 63 : 707 – 758.

Schurhoff, P. N. (1926) Die Zytologie der Bluten pflanzen. Stuttgort, Germany.

Schwartz, R.M. and M.O. Dayhoff (1978) Origins of prokaryotes, eukaryotes, mitochondria and chloroplasts. Science 139(4327); 395 - 403.

Schwarzacher, T. and D.W. Schweizen (1982) Karyotype analysis and hetero chromatin differentiation with Geisma C-banding and flourescent counterstaining in Cephalanthera (Orchidaceae). Pl. Syst. Evol. 141 : 91 – 113.

Schweizer D. and F. Ehrendorfer (1976) Geisma banded karyotypes, systematics and evolution in Anacyclus (Asteraceae – Anthemideae). Pl. Syst. Evol. 126 : 107 – 148.

Seavey, S. R., R. E. Magill and P. H. Raven (1977) Evolution of seed size, shape and surface, architecture in the tribe Epilobieae (Onagraceae). Ann. Miss. Bot. Gard. 64: 18-47.

Selling, O. (1947) Hawaiian Pollen Statistics Vol, II The Pollen Of The Hawaiian Phanerogams. Spec. Publ. Bishop Mus. 38. Honolulu, Hawaii

Selvaraj,R. and D. Subramanian (1983) Epidermal studies in some species of Acanthaceae, J. Indian Bot. Soc. 62 : 253 – 258.

Serturner, F.W. (1817) Uber das Morphium, eine neue salzfahahige Grundlage, und die Mekonsaure, als Hauptbe standteile des Opiums. Annalen der Physik 25 : 56–90.

Seshavataram, V. (1970) Onagraceae. In: Proceeding of Symposium On Comparative Embrylogy of Angiosperms, pp. 220-225. Indian National Science Academy, New Delhi, India.

Setyskal, G. C. (1972) The meaning of the term sibling species. Syst. Zool : 21 : 446.

Shah.G.L., Danaiah, V. and M.H.Parabia (1975) The structure of pericarp of Tephrosia, In: Form, Structure & Function In Plants (Ed. H.Y.Mohan Ram, J.J.Shah & C.K.Shah) pp.187-392. Sarita Prakashan, Meerut, India.

Shamashastry, R. (1915) Kautilya's Arthashastra (Translated into English). The Government Press, Bangalore, Karnatak, India.

Sharma, A.K. (1976) A new look at chromosome and its evolution. Proc. Ind. Natl. Sci. Acad. 42(B) : 12-24.

Sharma, O. P. (1993) Plant Taxonomy. Tata McGraw-Hill Publishing Company Ltd. New Delhi, India.

Sharma, P.V. (2005) Carak Samhita (Text with English Translation) vol. 1–4. Chaukhamba Orientalia, Varansi, India.

Sharam, P.V. (2001) Samhita (with English Translation of Text and Dalhana's Commentary Alongwith Critical Notes). Vedic Book, Delhi, India.

Sharsmith, H. (1961) The genus Hesperolinon (Linaceae). Univ. Calif. Publ. Bot. 32(4): 235-314.

Shashi Kumar and G.S.Paliwal (1975) Foliar anatomy of the family Acanthaceae II. The tribe Thunbergieae and Nelsonieae. Acta Botanica Indica 3 : 121 – 131.

Sherland, F. B. (1963) The distribution of fatty acids in plant lipids. In : Chemical Plant Taxonomy (T. Swain Ed.) pp. 253-312. Academic Press, London, U.K.

Sherman, M. (1946) Karyotype evolution : A cytogenetic study of seven species and Six intespecific hybrids of Crepis, Univ. Calif. Pub.Bot. 18: 369-408.

Shisode, S.B. and D.A. Patil (2000) The foliar epidermal studies in some Celastraceae. Journal of Advances in Sciences Technology 3(1–2) : 17-22.

Shisode, S. B. and D. A. Patil (2004a) Nodal anatomy of seedlings in some Rubiaceae. J. Phytol. Res. 17(1): 29-32.

Shisode. S.B. and D.A. Patil (2004b) Foliar epidermal study in some species of Boerhavia Linn. (Nyctaginaceae) J. Swamy Bot,. Cl. 21:75-78.

Shisode, S. B. and D.A. Patil (2008a) The foliar epidermal study in some Hippocrateacee with a discussion on its taxonomic position. Flora and Fauna 14(1) : 33-40.

Shisode, S.B. and D.A. Patil (2008b) Petiolar anatomy of some Celastrales. Indian J. Applied & Pure Bio. 23(1): 185-192.

Shisode, S.B. and D. A. Patil (2010) Foliar anatomical studies in some species of Ziziphus Miller (Rhamnaceae) Deccan Current Science 3(1) : 44-49.

Shisode, S.B. and D.A. Patil (2011) Petiolar anatomy in some Rhamnaceae. Current Botany 2(9): 22 -25.

Shull, G. H. (1923) The species concept from the point of view of a Gemiticist. Amer. J. Bot. 10 : 221 – 228.

Silajak –Yakovolev, S. and D. Cartier (1982) Comparative analysis of c–banded karyotypes in Crepis praemorsa ssp. dinarica. Pl. Syst. Evol. 141 : 85–90.

Simpson, G. G. (1951) The species concept. Evolution 5 : 285-298.

Simpson, G. G. (1961) Principles of Animal Taxonomy. Columbia University Press, New York, U.S.A.

Simth P. M. (1983) Proteins, mimicry and microevolution in grasses. I : Proteins and Nucleic Acids in Plant Systematics (Jensen, U. and Fairbrothers, D.E.Eds.) pp. 311 - 323. Springer-Verlag, Berlin, Germany.

Simth, P. M. (1972) Serology and species relationships in annual bromes (Bromus L. sect. Bromus) Ann. Bot. 36:1-30.

Sinclair, W. A. (1951) The Traditional Formal Logic : A Short Account for Students (5th Ed.). Methuen, London, U.K.

Singer, C. (1953) The earliest figures of microscopic objects, Endeavour 12 : 197 – 201.

Singh Ajay (2008) Plants in Ancient Indian Civilizations. Agam kala Prakashan, New Delhi, India.

Singh, Karam and R. M.Pai (1977) Floral morphology, evolutionary trends and interrelationship of the family Convolvulaeae In : Recent Trends and Contacts Between Cytogenetics, Embryology and Morphology. Today and Tomorrows. Printers and Publications, New Delhi, India.

Singh, S.P. (1960) Morphological studies in some members of the family Pedaliaceae – 1st Sesamun indicum DC. Phytomorphology 10: 65-82.

Singh, S. P. (1963) Morphological studies in some members of the family Pedaliaceae – IInd Pedalium murex Linn. Agra Univ. J.Res. (Sci) 13: 143-162.

Singh, S.P. (1964) Morphological studies in some members of the family Pedaliaceae – IIIrd. Martynia diandra Glox. Agra Univ. J.Res. (Sci) 13: 107-128.

Singh, S. P. (1970 a) Pedaliaceae. In : Proceedings of Symposium Comparative Embryology of Angiosperms. Indian Nation Science Academy, New Delhi, India pp. 273 – 277.

Singh, S. P. (1970 b) Martyniaceae. In: Proceeding of Symposium on Comparative Embryology of Angiosperms, Indian National Science Academy, New Delhi, India.pp.278-281.

Singh, V.,Jain, D.K. and Meena Sharma (1974) Epidermal studies in Berberidaceac and their taxonomic significance. J. Indian Bot. Soc. 53: 271-276.

Singh, V.Jain, D.K.and Meena Sharma (1978) Leaf architceture in Berberidaceae and its bearing on the circumscription of the family. J.Indian bot. Soc. 57: 272-281.

Sinnott, E.W. (1914) Investigations on the phylogeny of the angiosperms. I. The anatomy of the node as an aid in the classification of angiosperms. Amer. J. Bot. 1: 303-322.

Sivcar, N.N. and Roma Sarkar (1996) Viksayur veda of Parasara : A Treatise on Plant Science (Indian Medial Science series No. 38). Sri Satguru Publications, Delhi, India.

Sivarajan, V. V. (1981a) Scanning electron microscopic studies on seed coat surface patterns of South Indian Species of Lindernia All. (Scrophulariaceae). Proc.Indian. Acad. Sci. (Pl. Sci,) 90: 499-508

Sivarajan, V. V. (1981b) Taxonomic notes on the genus Portulaca Linn. in India. J. Bomb. Nat. Hist. Soc. 78 : 256 - 260.

Sivarajan, V. V. and T. Usha (1983) On reinstating Mollugo stricta L. (Molluginaceae). Taxon 32: 123-126

Skovsted, A. (1941) Chrosome numbers in the Malvaceae- II. C.R.Trav.Lab. Carlsberg, ser physio 23: 195-242.

Skvarla, J. J. and J. W. Nowicke (1976) Ultastructure of pollen exine in Centrospermae families. Pl. Syst. Evol. 126 : 35 – 78.

Skvarla, J.J., P.H.Raven; W.F. Chissoe and M.Sharp (1978) An ultrastructural study of viscin threads in Onagraceae pollen. Pollen *et* Spores 20: 5-143.

Smirnov, E. S. (1924) Sur I' analyse de la distribution et de la correleation des caracterres dansles unites systematiques C.R. Acad. Sci. USSR, Ser. A : 81-84.

Smirnow, E. S. (1938) Species construction from a taxonomic point of view (in Russian). Zool. Zhurn. 17 : 387 – 418.

Smith P.M. (1976) The Chemotaxonomy of Plants Arnold, London, U.K.

Smith, B.N. and W.V. Brown (1973) The Kranz syndrome in the Gramineae as indicted by cabon isotope ratios. Amer. J.Bot. 60: 505-513.

Smith, F. H. and E.C. Smith (1942) Floral anatomy of the Santalaceae and some related forms. Oregon State Coll. Monogr. Steud. Bot. pp. 1 – 93.

Smith, F.H. (1942) Floral anatomy of the Santalaceae and some related forms. Oregon State Monographs. Studies In Bot. 5:1.

Smith, H.H. (1943) Effects of genome balance, polyplody and single extra chromosome on size in Nicotiana. Genetics 28: 227-236.

Smith, W. W. and G. Forrest, (1928) The sections of the genus Primula. Notes Roy. Bot. Gard. Edinburgh, 16 : 1-50.

Smith. G. M. (1938) Cryptogamic Botany. Vol. I McGraw – Hill, New York, U.S.A.

Smith-White, S. (1968) Brachycome lineariloba : A species for experimental cytogenetics. Chromosome (Berlin) 23: 359-364.

Sneath, P. H. A. (1957) The application of computers to taxonormy. J. Gen. Microbiol. 17 : 201- 226.

Sneath, P. H. A. (1976) Phenetic taxonomy of the species level and above. Taxon 25 : 437 – 450.

Sneath, P. H. A. and R. R. Sokal (1973) Numerical Taxonomy : The Principles and Practice of Numerical Classification. Freeman, San Francico, U.S.A.

Sokal, R. R. (1973) The species problem reconsidered. Syst. Zool. 2001. 22 : 360-374.

Sokal, R. R. and P. H. A. Sneath (1963) Principles of Numerical Taxonomy. : Freemen, San Francisco, U.S.A.

Sokal, R. R. and T. J. Crovello (1970) The biological species concept : A critical evaluation. Amer. Naturalist 104 : 127–153.

Sokovskaya, A.P. and N.S.Probotva (1977) On the least chromosome number (2n = 4) of Colpodium versicolor (Stev) Woronow (Poaceae) Nor. Zh. 62: 241-245.

Solbrig, O. T. (1966) Evolution and Systematics. The Macmillan Co., New York, U.S.A.

Solbrig, O.T. (1968) Fertility, sterility and species problem. In : Modern Methods In Plants Taxonomy (V.H. Heywood Ed.) pp. 77 – 96. Academic Press, London, U.K.

Solereder, H (1908) Systematic Anatomy of The Dicotyledons, Vol.I – II., Clarendon Press, Oxford, U.K.

Solereder, H. (1885) Ueber den systemaischen Werth Der Holzstruktur bei den Dicotyledonen. Munchen, Germany.

Solereder, H. and F.J. Meyer (1930) Systematische Anatomie der Monocotyledonen. Heft. VI. Scitamineae, Microspermae.Borntrager, Berlin, Germany.

Solereder,S. and F. Meyr (1933) Systematische Anatomie der Monokotyledonen. Pandanales – Helobieac-Triuridales. 1: 1- 155. Gebruder Borntraeger, Berlin, Germany.

Sneath, P.H.A. and R.R.Sokal (1973) Numerical Taxonomy. W.H. Freeman And Company, San Francisco, USA.

Sokal, R.P. and P.H.A. Sneath (1963) Principles of Numerical Taxonomy. W.H. Freeman And Company, San Francisco, USA and London, U.K.

Song, Y., Chen, Y., Lv., J., Xu, J., Jhu, S. and M.Li (2019) Comparative chloroplast genomes of Sorghum species: Sequence divergence and phylogenetic relationships. BioMed Research International Vol. 2019, Article ID 5046958, 11 pages. https://doi.org/10.1155/2019/5046958.

Soria, V.J. and C.B. Heiser Jr. (1961) A statistical study of relationships of certain species of the Solanum nigrum complex. Econ.Bot. 15:24-55.

Souges, R. (1943 b) Embrylogie des Campanulacees Developpment de 1' embrylon chez le Campanula patula. C.R.Acad. Sci. Paris 2002: 2009-2011.

Soyer, M.O. and O.K.Huppala (1974) Division and function of dinoflagellate chromosome, J.Microscopia 19: 137-146.

Sperlich. A. (1939) Das trophische parenchym, B. Exkretionsgewebe. In : K. Linsbauer (Ed.) Handbuch der Pflanzenanatomie. Gebruder Borntraeger, Berlin, Germany. pp. 18.

Spooner, D.M., Ruess, H., Simon, P. and D.Senalik (2020) Mitochondrial DNA sequence phylogny of Daucus. Systematic Botany 45(2):403-408.

Sporne, K. R. (1956) The phylogenetic classification of the angiosperms. Biol. Rev. 31: 1-29.

Sporne, K. R. (1975a) A note on ellagitannins as indicators of evolutionary status in dicotyledons. New Phytol. 75 : 613 - 618.

Sporne, K. R. (1975b) The Morphology of Angiosperms : Structure And Evolution of Flowering Plants. St. Martin's Press, New York, U.S.A.

Sprengel, C. K. (1793) Das entdeckte Geheimniss der Natur im Bau und in der Befruchtung, Berlin, Germany.

Srinivasacher, D, (1942) A contribution to the embryology of the genus Polygala. Proc Natn. Acad Sci. India 12: 1-6.

Stace, C.A. (1965) Cuticular studies as an aid to plant taxonomy. Bull. Br. Mus. Nat. Hist. (Bot) 4:1-78.

Stace, C. A. (1975) Hybridisation and the Flora of the British Isles. Academic Press. London, U.K.

Stace, C. A. (1980) Plant Taxonomy and Biosystematics. Edward Arnold, London, U.K.

Stace, C. A. (1989) Plant Taxonomy and Biosystematics (2nd Ed.) Edward Arnold, London. U.K.

Stace. C.A. (1970) Anatomy and taxonomy in Juncus suhgenus Genuini. In: New Researoh In Plant Anatomy (N.K.B. Robson *et al.* Ed.) pp. 75-82. Linnean Society, London, U.K.

Stearn, W. T. (1956) Keys; shapes of leaves, & c. In : F. J. Chittenden (Ed) Royal Horticultural Dictionary of Gardening, Supplement: 251-253; 318-322.

Stearn, W. T. (1959) An Intoduction To The Species Plantraum And Cognte Botanical Works of Carl Linnaeus In Species Plantarum : A Facsimile of the First Edition (1753), Ray Society, London, U.K.

Stearn, W. T. (1964) Problems of character selection and weighting : Introduction. In : V. H. Heywood and J. MCNeill. (Ed.) Phenetic and Phylogenetic Classification. PP. 83 – 86. Systematics Association, London, U.K.

Stearn, W.T. (1968) Observation on a computer aided survey of the Jamaican species of Columnea and Allopectus. In: Modern Methods In Plant Taxonomy (Ed. V.H.Heywood). Academic Press, London, UK. pp.219-224.

Stearn, W. T. (1983) Botonical Latin : History, Grammar, Terminology and Vocabulary (3rd Ed.), David and Chlarles, Newton Abbot, England.

Stebbins, G.L. (1950) Variation and Evolution in Plants, Columbia University Press, London, U.K.

Stebbins, G.L. (1956) Cytogenetics and evolution of the grass family. Amer. J. Bot. 43 : 890 – 905.

Stebbins, G.L. (1971) Chromosomal Evolution In Higher Plants, Arnold, London. UK.

Stebbins, G. L. (1974) Flowering Plants; : Evolution Above The Speeies Level. Arnold Press, London, U. K.

Stebbins, G. L. (1977) Process of Orgnic Evolution. 3rd Ed, Englewood Cliffs, N. J. Prentice – Hall, U.S.A.

Stebbins, G.L. and G.S. Khush (1961). Variation in the organization of the stomatal complex in the epidermis of Monocotyledons and its bearing on the phylogeny. Amer. J. Bot. 48 : 51-59.

Stenar, H. (1949) Zur Kenntnis der Embryologie und der Raphiden-Zellen bei Liliazeen. Acta Hart. Berg. 15 (3) : 45-53.

Stewart, K. D. and K. R. Mattoz (1975) Comparative cytology, evolution and classification of the green algae with some consideration of the origin of other organisms with chlorophylls A and B. Bot. Rev. 41: 104-135.

Straw, R. W. (1956) Floral Isolation in Penistemon. Amer. Nat 90 : 47-53.

Stuessy, T. F. (1972) Revision of the genus Melampodium (Compositae : Heliantheae) Rhodora 74 : 1-70.

Stuessy, T. F. (1973) Revision of the genus Baltimora (Compositae : Heliantheae) Fieldiana Bot. 36 : 31 – 50.

Stuessy, T. F. (1979) Ultrastructural data for the practcing plant systematist. Amer. Zoologist 19 : 621 – 636.

Stuessy, T. F. (1990) Plant Taxonomy: The Systematic Evaluation of Comparative Data. Columbia University Press, New York, USA.

Subodh, K. Sharma; C. R. Babu, B. M. Johro and A. Hepwordh (1977) SEM studies on seed coat patterns in Phaseolus mungo-radiatus-subloatus complex. Phytomorph. 27 : 106-111.

Subramanyam, K. (1962) Embryology in relation to systematic botany with particular referemce to Crassulaceae.In: Plant Embryology – A Symposium, CSIR, New Delhi, India, pp, 94-112.

Subramanyam, K. and N. C.Nair (1975) Trends of specialiastion in the Crassulaceae as revealed by floral anatomy. In : Form, Structure & Function In Plants, Prof. B. M. Johri Comm. Volume (Ed. H. Y. Mohan Rom, J. J. Shah & C. K. Shah) pp. 401 – 408, Sarita Prakashan, Meerut, India.

Sussenguth, K. (1954) Systematik der Spermatophyta. Fortschr. Bot. Berlin 15 : 13 – 84.

Swain, T. (Ed.) (1963) Chemical Plant Taxonomy. Academic Press, London, U.K.

Swain. T. (1975) Evolution of flaronoid compounds. In: The Flavonoids (Harborne, J. B. and T. J. Mabry Eds.), Lodon. pp. 1096 - 1129.

Takhtajan, A. L. (1942) The structural types of gynoecium and placentation. Bull. Armen. Branch. Acad. Sci. USSR. 3 – 4 (17 – 18) : 91 – 112 (In Russian).

Takhtajan, A. L. (1953) Phylogenetic principles of the system of higher plants. Bot. Rev. 19 : 1 – 45.

Takhtajan, A. L. (1964) The taxa of higher plants above the rank of order. Taxon 13:160 – 164.

Takhtajan, A. L. (1969) Flowering Plants : Origin and Dispersal. Oliver & Boyd, Edinburgh (English translation by C.Jeffrey), U.K.

Takhtajan, A. J. (1973) Evloution und Ausbreitung der Blutenpflanzen, Jena, Germany.

Takhtajan, A. L. (1980) Outline of classification of flowering plants (Magnoliphyta) Bot. Rev. 46(3) : 225-359.

Tateoka, T. (1962) Starch grains of endosperm in grass systematics. Bot. Mag. Tokyo 75 : 377 – 383.

Tayade, S.K and D. A. Patil (2003) Foliar epidermal features and their taxonomic significance in the genus Argyreia Lour. (Convolvulaeae).J. Swamy Bot. CI. 20: 15-18.

Tayade.S.K and D.A.Patil (2008) Petiolar anatomy of some species of Convolvulaceae. Indian J. Applied & Pure Bio. 23(2): 395-404.

Terentjev, P. V. (1931) Biometrische Untersuchungen uber. die morphologischen markmale von Rana ridibunda Pall. Biometrika 23 : 23 – 51.

Thakur, H. A. and D.A.Patil (2002) Nodal organization in some Euphorbiaceae, J.Swamy Bot. Cl.19:59-62.

Thakur, H. A. and D. A. Patil (2006) Nodal anatomy of some hitherto unstudied Euphorbiaccae. J. Phytol, Res 19(1) : 77-82.

Thakur, H.A and D.A. Patil (2010) Anatomy and comspectus of the genus Simmondsia (Link) R R B 54 (1) : 28-34.

Thakur. H.A. and D.A Patil (2011a) Foliar anatomical observations in some Euphorbiaceae. BioNano Frontier 4(2) : 325-332.

Thakur. H.A. and D.A. Patil (2011b) Petiolar anatomy of some unstudied Euphorbiaceae. Jour. Phytology 2(12) : 54-59.

Theophrastus (1916) Enquiry Into Plants, and Minor Works on Odours and Weather Signs, Translated By, A. Hort. 2 Vols, Haward Univ Press, Mass, U.S.A.

Thomas, P.T. (1936) Genotypic control of chromosome size. Nature 138 : 402.

Thorne, R. F. (1902) Classification and geography of the flowering plants, Bot.Rev. 53 (3) : 225-299.

Thorne, R. F. (1958) Some guiding principles of angiosperm phylogeny. Brittonia 10 : 72 – 77.

Thorne, R. F. (1963) Some problems and guiding principles of angiosperm phylogeny. Amer. Nat. 97 (896) : 287 – 305.

Thorne, R. F. (1968) Synopsis of a putative phylogenetic classification of flowering plants. Aliso 6 : 57-66.

Thorne, R. F. (1976) A phylogenetic classification of the Angiospermac. In : M. K. Hecht, W. C. Steer and B. Wallace (Ed.) Evolutionary Biology Vol. 9, PP. 35 – 106 Plenum Press, New York (U.S.A.) and London (U.K.)

Thorne, R. F. (1983) Proposed new realignments in the angiosperms. Nord. J. Bot. 3 : 85 – 117.

Thorne, R. F. (1992) Classification and geogarphy of flowering plants. Bot. Rev. 58 : 225 – 348.

Tiagi, B. (1951) A contribution to the morphology and embryology of Cuscuta hyalina Roth and C.planifolia Tenore. Phytomorphology 1: 9-21.

Tiagi, B. (1965) Development of embryo sac and the embryology in Cuscuta lupuliformis, Curr. Sci. 34: 671.

Tiagi, B. and P. K. Gupta. (1963) Floral morphology and embryology of Evolvulus alsinoides Linn. Proc. Rajasthan Acad. Sci. 10: 51-58.

Tiagi, Y. D. (1970) Paeoniaceae. In : Proceedings, Symposium on Compararive Embryology of Angiosperms. pp. 53 – 58 Indian National Science Academy, New Delhi, India.

Tiagi, Y. D. and S.Kshetrapal (1975) Studies on the vasculer anatomy of the flower of certain species of the Orobanchaceae and their bearing on the structure and evolution of the gynoecium. In: Form, Structure & Function In plants. Prof. B. M. Johri Comm, Volume pp. 319-329. (Ed. H.Y.Mohan Ram, J.J.Shah & C.K.Shah), Sarita Prakashan, Meerut, India.

Tippo, O.(1938) Comparative anatomy of the Moraceae and their presumed allies. Bot. Gaz. 100:1-99.

Tippo, O. (1942) A modern classification of plants kingdom. Chorn. Bot. 7 : 203 – 206.

Togby, H.A. (1943) A cytological study of Crepis fuliginosa, C.neglecta and their F_1 hybrid and its bearing on the mechanism of phylogenetic reduction in chromosome number.J.Genet. 45: 67-111.

Tomlinson, P.B. (1959) Structure and distribution of sclereids in the leaves of Palms. New Phytol. 58 : 253 – 266.

Tomlinson, P. B. (1960) Seedling leaves in palms and their morphological significance. J. Arn. Arb, 41: 414-428.

Tomlinson, P. B. (1961) Anatomy of the Monocotyledons. II. Palmae (C.R.Metcalfe Ed.) Clarendon Press. Oxford, U.K.

Tomlinson, P. B. (1962a) Anatomy of The Monocotyledons - III Scitamineae Clarendon Press. Oxford, U.K.

Tomlinson, P. B. (1962b) The leaf base in Palms. Its morphology and mechanical biology. J. Arn.Arb. 43:23-50.

Tomlinson, P. B. (1969) Anatomy of The Monocotyledons – III. Commelinales – Zingiberales (C. R. Metcalfe Ed.) Clarendon Press, Oxford, U.K.

Tournefort, J. P. de (1700-1716) Institutiones Rei Herbariae Vols, I-III. Typogr. Regia., Paris, France.

Tournfort, J. P. de. (1700) Institutiones Rei Herbariae, editio altera. Paris, France.

Tregunna, E.B., Smith. J. A. and W.J.S.Downston (1970) Some methods for studying the photosynthetic taxonomy of the angiosperms. Canad. J. Bot. 48: 1209-1214.

Treub, H. P. (1975) Class Liliida of superclass Monocotyidra. Taxon 24 : 453-460.

Troll, W. (1937-1939) Vergleichende. Morphologie der hoheren Pflanzenvol.I. Vegetations organe, Part 1-2, Borntraeger, Berlin, Germany.

Tschirch, A. (1889) Anegewandte Pflanzenanatomie. Urban & Schwarzenberg Wien.

Tucker, S.C. (1964) The terminal idioblast in magnoliaceous leaves. Am. J.Bot. 51: 1051-1062.

Turreson, G. (1922) The genotypical response of the plant species to the habitat. Hereditas 3 : 211-350.

Turrill, W. B. (1938) The expansion of taxonomy. Biol. Rev. 13 : 342-373.

Uphof, J. C. T. (1962) Plant Hairs. I: W, Zimmermnn & P. G. Ozenda (Ed.) Handbuch der pflanxnanatomie (Ed.2), 4(5):1-206.

Valentine, D. H. and A. Love (1958) Taxonomic and biosystematic categories, Brittonia 10 : 153-166.

Van cotthem, W.R. J. (1970) A classification of stomatal types. Bot. J. Linn. Soc. 63: 35-246.

Van Leeuwen (1971) A preliminary revision of genus Rotala (Lythraceae) in Malesia. Blumea. 19:53–56.

Van Tiegham, P, (1859) Sur les generes Actinidia ef Sauravie consideres comme types d' une famille nouvelle, les Actinidiacees. J. Bot. (Paris) 13: 170-173.

Van Tiegham, Ph. (1896) Classification des Loranthinees Bull. Soc. Bot. Fr.3: 246-286.

Van Valen, L. (1976) Ecological species, multispecies and oaks. Taxon 25 : 233 – 239.

Vaten Van. F. (1978a) Contribution to the knowledge of cyanogenesis, in Angiosperms - 3. Communication. Cyanogenesis in Liliaceae. Proc. Koninkl. Nederl. Akad. Wetensch. 81: 132-140.

Vaten Van. F. (1978b) Contribution to the knowledge of cyanogenesis in Angiosperms - 6. Communication. Cyanogenesis in some Magnolildae. Proc. Koninkl. Nederl. Akad. Wetensch. 81: 224-253.

Vavilov, N. I. (1951) The Origin, Variation Immunity and Breeding of Cultivated Plants, Chronica Botanica Co. Waltham, U.S.A.

Venkata Rao, C. (1971) Proteaceae : Botanical Monograph No.6, CSIR, New Delhi, India.

Verdcourt. B. (1958) Remarks on the classification of Rubiaceae. Bull. Jard. Bot. Elat. Brux. 28 : 209-281.

Vesque, I. (1889) Del's emploi des caracteres anatomiques dens la classification des vegetaux. Bull. Soc. Bot. France 36 : 41-77.

Vijaraghavan, M.R. (1963) Embryology of Actinidia polygama Franch, & Sav. and the systematic position of the family Actinidiaceae Proc. 50th India Sci. Congr. (Delhi) pt 3rd. Abstr. pp. 393.

Vijaraghavan. M. R. (1970) Actinidiaceae. In, Procedings of Syspotium on Comperative Embryology of Angiosperms, Indian National Science Academy, New Delhi, India. pp. 69-74.

Voss, E. (1952) The history of keys and phylogentic trees in systematic biology. J. Sci, Lab, Denison Univ. 43 (1) : 1-25.

Waddington, C. H. (1953) Genetic assimilation of an acquired character. Evolution 7 : 118 – 126.

Wagner, W.H. (1961) Problems in the classification of ferns. In: Recent Advances In Botany. University of Toronto Press, Toronto, Canada.

Wagner, W.H. (1968) Hybridisation, taxonomy and evolution, In: Modern methods In Plant Taxonomy (V.H.heywood Ed.) pp. 113-138 Academic Press, London, U.K.

Walker, J. W. (1971) Unique type of angiosperm pollen from the family Annonaceae. Science 172 : 565 – 567.

Walker, J.W. and J.A. Doyle (1975) The basis of angiosperm phylogeny : Ann. Miss. Bot. Gard. 62 : 664 – 723.

Walters, S. M. (1949) Alchemilla vulgaris L. In Britain. Watsonia 1:6-18.

Walters, S. M. (1961) The sharing of angiosperm taxonomy. New Phytol. 60 : 74-84.

Wangerin, W. (1910) Garryaceae. In: Engler's Das Pflanzenreich, iv, 56 a. w. Engelmann, Leipzig, Germany.

Warburg, E.F. (1949) Cytology and critical groups. In: British Flowering Plants and Modern Systematic Methods (A.J.Wilmott Ed.) pp. 11-16. Bot. Soc. Brit. Is.

Warburton, F. E. (1967) The purpose of classification. Syst. Zool. 16 : 241 – 245.

Wardle, P (1967) Biological flora of New Zealand 2: Nothofagus menziessii (Hook.*f*) Oerst. (Fagaceae) Silver beech. N.Z. Jl. Bot. 5: 276-302.

Warming, E. (1891) Podostemaceae. In Engler, A. & Prantl, K. Die Pflanzenfamilien Naturlichen III 2a. Engelmann, Leipzing, Germany.

Warmke, H. E. (1941) Chromosome continuity and individuality. Gold Spring Harb. Symp. Quant. Biol. 9 : 1- 6.

Watanabe, K. and S, Smith-White (1987) Phyletic and evolutionary relationships of Brachycome lineariloba (Compositae). Pl. Syst. Evol. 157: 121-141.

Watson, L. (1971) Basic - taxonomic data : The need for organisation over presentation and accumulation. Taxon 20 : 131-136.

Webster, G. L. (1967) The genera of Euphorbiaeeae in the southwestern United State. J. Arnold Arbor. 48:303-361.

Webster, G.L and E.A. Rupert (1973) Phylogenetic significance of pollen nuclear number in the Euphorbiaceae.Evolution 27 : 524 – 531.

Webster, G. L., Brown, W. V. and B. N. Smith, (1975) Systematics of photosynthetic carbon fixation pathways in Euphorbia. Taxon 24 : 27 - 33.

Wendelbo, P. (1961) Studies in the Primulaceae - I. A Monograph on The Genus Dionysia. Arnok. Univ. Bergen, Mat. - Naturv. Serie No.3.

Went, F.A.F.C (1908) The development of the ovule, embryo sac and egg in Podostemaceae. Rec. Trav. Bot. Neerl. 5; 1-16.

Went, F.A.F.C (1910) Untersuchungen uber Podostemeceen I[st]. Verh. Akad. Wet. Amst. Sec. II. 16: 1-88.

Went, F.A.F.C (1912) Untersuchungen uber Podostemeceen IInd. Verh. Akad. Wet. Amst. Sec. II. 17: 1-19.

Went, F.A.F.C (1926) Untersuchungen uber Podostemeceen IIIrd. Verh. Akad. Wet. Amst. Sec. II. 25: 1-58.

Wernham, H. F. (1912) Floral Evolution : With particular reference to the sympetalous dicotyledons: XI-Summary and Conclusion. Evolutionary Genealogy and some principles of classification. New Phytol. 11 : 373-397.

Wettstein, R. Von (1901) Handbuch der Systematischen, Botanik, Gerbuuder Borntraeger, Berlin : 4th Ed. (1930-1935) Vol. I-II. Franz Deuticke, Leipzig, East Germany and Vienna, Austria.

Wettstein, R.Von (1935) Handbuch der Systematischen Botanik. 4. Aufl. Franz Deuticke, Leipzig, East Germany.

Wiley, E. O. (1978) The evolutionary species concept reconsidered. Syst. Zool. 27 : 17–26.

Wiley, E. O. (1981) Phylogenetics : The Theory And Practice of Phylogenetic Systematics. Willey, New York, U.S.A.

Willis, J. C. (1960) A Dictionary of The Flowenns Plants and Ferns. 6th Ed. Cambridge University Press, Cambridge, U.K.

Winterstein, E. and G. Trier (1910) Die Alkaloids, eine Monographie der naturlichen Basen, Borntraeger, Berlin, Germany

Wittaker, R. H. (1969) New concepts of kingdoms of organisms. Science 163 : 150-160.

Wodehouse, R. P. (1935) Pollen Grains : Their Structure, Identification And Significance In Science And Medicine. McGraw – Hill Book Co., New York, U.S.A.

Wujek, D. E.; P. Cheema; S. S. Bir and F. J. Menapace (1977) Achene micromorphology as a systematic aid to the taxonomic recognition of Cyperus cytotypes. Rheedea 27: 93-99.

Yakovlev, M. S. (1951) On the similarity of embryogenesis of angiosperms and gymnosperms. Trud. Bot. Inst. Acad. Nauk. 2:356-363. (In Russian).

Yokovlev, M.S. and M.D. Yoffe (1957 a) Speciality of embryogenesis of Paeonia. Bot. J. 42 : 1491 – 1502 (In Russian).

Yokovlev, M.S. and M.D. Yoffe (1957 b) On some peculiar features in the embryology of Paeonia L. Phytomorphology 7: 47-87.

Yokovlev, M.S. and M.D. Yoffe (1959) On a new type of embryogenesis In Paeonia. Congr. Internatl. Bot. 9th 2: 437-438.

Yokovlev, M.S. and M.D. Yoffe (1961) Further studies of the new type of embryogenesis in angiosperms. Bot. J. 46: 1402-1421.

Yokovlev, M.S. and M.D. Yoffe (1965) The embryology of the genus Paeonia of Angiosperms In: Flower Morphology, and Reproductive Processes of Angiosperms (M.S. Yakovlev, M.S. Ed.) Moscow and Lemngard, Russia.

Zavada, M.S.(1984) Pollen wall development of Austrobaileya maculata. Bot. Gaz. 145 : 11 – 21.

Zeven, A. C. and P.M.Zhukosky (1975) Dictionary of Cultivated Plants and Their Centres of Diversity. Cenre for Agricultural Publishing and Documentation. Wageningen, The Netherlands.

Zohary, M. (1956) The genus Tamarix in Israel. Trop. Woods, No. 104: 24-60.

e-Bibliography

(i) Western European Gardens

htpp:/www.botanicgardens.eu/action/actionplaneu1htm

(ii) Linnekad garden

htpp:/www.wikipidia.org/wiki/Botanical-garden

Index-I: Latin Plant Names

C

D

E

F

G

Q

R

S

T

U

V

W

Y

Z

Index-II: Taxonomic Terms

Index-III: Botanic Terms